AF732679

Governing the Climate Change Regime

This volume, the second in a series of three, examines the institutional architecture underpinning the global climate integrity system. This system comprises an interrelated set of institutions, governance arrangements, regulations, norms and practices that aim to implement the United Nations Framework Convention on Climate Change (UNFCCC). Arguing that governance is a neutral term to describe the structures and processes that coordinate climate action, this book presents a continuum of governance values from 'thick' to 'thin' to determine the regime's legitimacy and integrity. The collection contains four parts, with part one exploring the links between governance and integrity, part two containing chapters which evaluate climate governance arrangements, part three exploring avenues for improving climate governance and part four reflecting on the road to the UNFCCC's Paris Agreement. This book provides new insights into understanding how systemic institutional and governance failures have occurred, how they could occur again in the same or different form and how these failures impact the integrity of the UNFCCC. This work extends contemporary governance scholarship to explore the extent to which selected institutional case studies, thematic areas and policy approaches contribute to the overall integrity of the regime.

Dr Tim Cadman is a Research Fellow at the Institute for Ethics, Governance and Law at Griffith University, Australia.

Dr Rowena Maguire is a Senior Lecturer in the School of Law and Co-Chair of the International Law and Global Governance Research Program at the Queensland University of Technology, Australia.

Professor Charles Sampford is Foundation Dean of Law and Foundation Director of the Institute for Ethics, Governance and Law (IEGL), Griffith University, Australia.

Law, Ethics and Governance Series

Series Editor: Charles Sampford

Director, Institute for Ethics, Governance and Law,
Griffith University, Australia

Recent history has emphasised the potentially devastating effects of governance failures in governments, government agencies, corporations and the institutions of civil society. 'Good governance' is seen as necessary, if not crucial, for economic success and human development. Although the disciplines of law, ethics, politics, economics and management theory can provide insights into the governance of organisations, governance issues can only be dealt with by interdisciplinary studies, combining several (and sometimes all) of those disciplines. This series aims to provide such interdisciplinary studies for students, researchers and relevant practitioners.

Recent titles in this series:

International Rule of Law and Professional Ethics
Edited by Vesselin Popovski
ISBN 978-1-4724-2803-5

Responsibility to Protect and Sovereignty
Edited by Charles Sampford and Ramesh Thakur
ISBN 978-1-4094-3782-6

Protection of Refugees and Displaced Persons in the Asia Pacific Region
Edited by Angus Francis and Rowena Maguire
ISBN 978-1-4094-5540-0

Intellectual Liberty
Natural Rights and Intellectual Property
Hugh Breakey
ISBN 978-1-4094-4711-5

Health Workforce Governance
Improved Access, Good Regulatory Practice, Safer Patients
Edited by Stephanie D. Short and Fiona McDonald
ISBN 978-1-4094-2921-0

For a full list of titles in this series, please visit www.routledge.com

Governing the Climate Change Regime

Institutional Integrity and Integrity Systems

Edited by Tim Cadman, Rowena Maguire and Charles Sampford

Routledge
Taylor & Francis Group
LONDON AND NEW YORK

First published 2017
by Routledge
2 Park Square, Milton Park, Abingdon, Oxon OX14 4RN

and by Routledge
711 Third Avenue, New York, NY 10017

Routledge is an imprint of the Taylor & Francis Group, an informa business

© 2017 selection and editorial matter, Tim Cadman, Rowena Maguire and Charles Sampford; individual chapters, the contributors

The right of Tim Cadman, Rowena Maguire and Charles Sampford to be identified as the authors of the editorial material, and of the authors for their individual chapters, has been asserted in accordance with sections 77 and 78 of the Copyright, Designs and Patents Act 1988.

All rights reserved. No part of this book may be reprinted or reproduced or utilised in any form or by any electronic, mechanical, or other means, now known or hereafter invented, including photocopying and recording, or in any information storage or retrieval system, without permission in writing from the publishers.

Trademark notice: Product or corporate names may be trademarks or registered trademarks, and are used only for identification and explanation without intent to infringe.

British Library Cataloguing in Publication Data
A catalogue record for this book is available from the British Library

Library of Congress Cataloging in Publication Data
Names: Sampford, C. J. G. (Charles J. G.) editor. | Cadman, Timothy, 1966– editor. | Maguire, Rowena.
Title: Governing the climate change regime : institutional integrity and integrity systems / edited by Charles Sampford, Tim Cadman, Rowena Maguire.
Description: New York : Routledge, 2016. | Series: Law, ethics and governance | Includes index.
Identifiers: LCCN 2016028687 | ISBN 9781138216440 (hardback) | ISBN 9781315442365 (e-book)
Subjects: LCSH: Climatic changes–Law and legislation
Classification: LCC K3585.5 .G68 2016 | DDC 363.738/7456–dc23
LC record available at https://lccn.loc.gov/2016028687

ISBN: 978-1-138-21644-0 (hbk)
ISBN: 978-1-315-44236-5 (ebk)

Typeset in Galliard
by Apex CoVantage, LLC

Contents

List of illustrations vii
List of tables viii
List of contributors ix
Acknowledgements xv
List of abbreviations xvi

PART I
Exploring the links between governance and the integrity of climate change 1

1 **Introduction: Governing the climate regime** 3
TIM CADMAN, CHARLES SAMPFORD, ROWENA MAGUIRE AND HUGH BREAKEY

2 **Governance values and institutional integrity** 16
HUGH BREAKEY, TIM CADMAN AND CHARLES SAMPFORD

PART II
Evaluating governance arrangement within the UNFCCC 45

3 **The climate governance regime complex: Institutions, actors and discourses** 47
PHILIPP PATTBERG, LISA SANDERINK AND OSCAR WIDERBERG

4 **Institutions, institutional practices, and global climate change governance: The case of REDD+** 70
EERO PALMUJOKI

5 The global governance system for climate finance: Towards greater institutional integrity? 93
EMILIE BÉCAULT AND AXEL MARX

6 Governance of climate displacement within the UN climate regime 118
ROWENA MAGUIRE

7 Human rights and equity: Governing values for the international climate regime 136
KATE DOOLEY AND GITA PARIHAR

8 Tradable permits and emissions reduction under the Paris Agreement: Governance values and integrity challenges 155
TANVI MANI

PART III
Improving governance within the UNFCCC 175

9 Enhancing good governance within the international climate regime through human rights principles 177
BRIDGET LEWIS

10 Climate governance accountability challenges: Lessons from multilateral climate finance 197
LISA ELGES

11 Co-producing climate-smart agriculture knowledge through social networks: Future directions for climate governance 212
JOSEPH MACHARIA AND HOPE JOHNSON

12 International climate change policy and the contribution of civil society organizations 235
PETER J. GLYNN

Afterword: The long road to Paris: Insider and outsider perspectives 250
KLAUS RADUNSKY AND TIM CADMAN

Index 267

Illustrations

Figures

1.1 Theoretical model for the evaluation of contemporary global governance (Cadman 2011, adapted) 5
3.1 Global climate change governance triangle (2015) 51
3.2 Global climate change governance decagon (2015) 52
3.3 One-mode network of global climate change governance institutions (2015) 57
3.4 Discursive fragmentation in global climate governance architecture 65

Graph

3.1 Centrality of global climate change governance institutions (2015) 58

Tables

2.1 Principles, criteria and indicators in the Cadman framework 30
2.2 This table expresses in an indicative way the expansion of governance values, from Thin, to Thickish, to fully Thick 30
3.1 Total of global climate change governance institutions per zone and per role (2015) 54
3.2 Distribution of institutions across seven zones: results for sample and ideal types (low and high fragmentation) 55
3.3 Indicator 1 – Overall discursive structure of global climate change governance (dominance of meta-discourses ≥ 25,0%) (2015) 61
3.4 Indicator 2 – Discursive coherence (homogeneity/heterogeneity) within each zone of the climate governance triangle (2015) 61
3.5 Key narratives of global climate change governance based on high frequency and case-occurrence (2015) 62
3.6 Key narratives derived from zone 1 of the global climate change governance triangle based on high frequency and case-occurrence (2015) 63
3.7 Key narratives derived from zone 2 of the global climate change governance triangle based on high frequency and case-occurrence (2015) 64
3.8 Results of measurements for all three dimensions of institutional fragmentation/integration 67
4.1 Approaches by some countries and country groups towards REDD 81
6.1 Options for loss and damage mechanism components 128

Contributors

Emilie Bécault is a research project manager at the Leuven Centre for Global Governance Studies, University of Leuven. She studied at the University of Chicago, the Ohio State University and the University of Leuven. Her research mainly focuses on global climate change governance, global democracy and international climate finance. She recently co-edited a book titled *Global Governance and Democracy* (Edward Elgar with J. Wouters, A. Braeckman and M. Lievens).

Hugh Breakey is a Research Fellow in moral philosophy at Griffith University's Institute for Ethics, Governance and Law. His work stretches across the sub-disciplines of political philosophy, normative ethics, moral psychology, governance studies and applied philosophy. He is the author of *Intellectual Liberty: Natural Rights and Intellectual Property* (Ashgate) and the co-author (alongside Charles Sampford and Ramesh Thakur) of *Enhancing Protection Capacity: Policy Guide to the Responsibility to Protect and the Protection of Civilians in Armed Conflicts.* Hugh's works explore the ethical challenges arising in such diverse fields as peacekeeping, institutional governance, climate change, sustainable tourism, private property, medicine and international law; and he has been published in journals including *The Philosophical Quarterly, The Modern Law Review* and *Political Studies.* He has taught philosophy and ethics at several universities and is currently the President of the Australian Association of Professional and Applied Ethics.

Tim Cadman is a Research Fellow with the Institute for Ethics, Governance and Law at Griffith University in Queensland, Australia. He specialises in the governance of sustainable development, including forest management, climate change and finance. His first monograph, *Quality and Legitimacy of Global Governance: Case Lessons from Forestry*, was published in 2011, followed by an edited collection, *Climate Change and Global Policy Regimes: Towards Institutional Legitimacy*, in 2013. His active research includes developing quality of governance standards for REDD+ and forest management in Nepal and Papua New Guinea and tracking the governance and integrity of the UN climate change regime. He established the Forest Stewardship Council in Australia in

2002 and also assists multi-stakeholders in environmental conflict mediation and negotiation.

Kate Dooley holds a MSc from Imperial College London. She is currently undertaking a PhD at the University of Melbourne, where her research focuses on the environmental integrity of terrestrial carbon accounting and the justice and equity implications of land-based climate mitigation. Kate also works as an advisor and consultant to government and non-governmental organisations, where she provides expertise on UN climate negotiations regarding land use, carbon trading, equity and climate justice.

Lisa Elges gained her PhD from Berlin's Freie Universität in 2015 and currently serves as Head of Global Climate Policy at Transparency International. She is an expert in environmental and climate finance governance and accountability, specialising in tort liability and anti–money laundering. She has worked for the Japanese Parliament, the Asia Crime Prevention Foundation, UN Asia and the Far East Institute for the Prevention of Crime and Treatment of Offenders. She graduated from the International Christian University (Tokyo, Japan) in 1993 and completed her LL.M in international human rights law at the University of Essex (UK) in 1997. She has authored numerous publications, including her forthcoming book, *Stratospheric Ozone Damage and Legal Liability.*

Peter J. Glynn is a consultant advising international institutions on climate change policy and related organisational development. His particular interest is the labour market impacts of climate change policy. Peter completed his PhD at Bond University in 2014; for his thesis he researched the employment and workplace impacts of climate change policy, with particular attention to the role of employers' organisations and trade unions. His professional background is in labour relations, labour market planning and association management. Prior to returning to academic research he was engaged by the International Organisation of Employers to advise on its policy in respect to climate change. He then worked with the International Labour Organisation to develop a programme to introduce these concepts and issues to employers' organisations, with a focus on developing countries. His pursuit of a PhD added the theoretical dimension to the practice, facilitating better engagement with the policy process.

Hope Johnson is currently a research associate examining emission units and property law for an Australian Research Council–funded project with the Commercial and Property Law Research Centre at the Queensland University of Technology (QUT). She completed her PhD with the support of QUT's Faculty of Law and the Institute for Future Environments. The focus of her thesis was international agricultural law and food policy. Her research interests include sustainable agriculture, world food security and land use policy. She is a member of the International Law and Global Governance Research Program at QUT and the Intellectual Property and Innovation Law Program at QUT.

Bridget Lewis is a Senior Lecturer at Queensland University of Technology's School of Law, where she is a member of the International Law and Global Governance Research Program. Bridget's principal research interests relate to international human rights law, particularly the intersection between human rights and the environment. She has also published in the areas of environmental displacement, climate justice and indigenous rights.

Joseph Macharia is a PhD candidate within the Faculty of Business at the Queensland University of Technology and an Ashoka Fellow. His research interests concern information and communication technologies used in agribusiness. He is the founder of Mkulima Young, an online platform in agribusiness which has support from the Rockefeller and Ashoka Foundations.

Rowena Maguire is a Senior Lecturer in the School of Law at the Queensland University of Technology and Co-Chair of the International Law and Global Governance Research Program at QUT. Rowena's principal research interests and publications concern international climate and forest regulation, with a focus on equitable design and implementation. Currently, Rowena is working on an Australian Research Council–funded project exploring global governance and integrity systems within the international climate regime.

Tanvi Mani is a student enrolled in the final year of the B.A., LL.B (Hons) degree. She has worked on issues of post-conflict development, disaster risk reduction, privacy legislation and the effective use of big data for comprehensive governance solutions. She is currently pursuing an internship at the United Nations University – Centre for Environment and Human Security, Bonn. She is a contributing author of a Pro Poor Study conducted by the Munich Climate Insurance Initiative, analysing the efficacy of Index Insurance schemes in enhancing the resilience of poor and vulnerable communities to weather-related risks. During her time at the UNU she also worked on a project to analyse the legal systems of the Pacific Island countries and to recommend suitable interventions for facilitating effective disaster risk reduction responses and enhancing resilience. She would like to pursue a career in Public International Law, with a focus on climate policy development.

Axel Marx is deputy director of the Leuven Centre for Global Governance Studies, University of Leuven. He studied in Leuven, Hull and Cambridge and holds a PhD from the University of Leuven. His research mainly focuses on global governance, certification, sustainability standards, non-state market regulation, human rights, trade governance, good governance, international cooperation, international development, research methodology, comparative case methods and qualitative comparative analysis. His international academic publications have appeared, inter alia, in *European Political Science Review, Regulation and Governance, Political Research Quarterly, Journal of World Trade, Research in Sociology of Organizations, Journal of Socio-Economics, International Journal of Human Rights, Journal of Human Rights Practice, International Labor*

Review, World Bank Legal Review, Asia-Europe Journal, Journal of Business Research and *Globalizations and Sociological Methodology.* He recently co-edited the books *Global Governance and Private Standards* (Edward Elgar with Miet Maertens, Jo Swinnen and Jan Wouters, 2012), *Global Governance of Labor Rights* (Edward Elgar with J. Wouters, G. Rayp and L. Beke, 2015) and *Global Governance through Trade* (Edward Elgar with J. Wouters, B. Natens and D. Geraets, 2015). As an expert to policymakers he has contributed to over fifteen reports for, inter alia, the United Nations Industrial Development Organization, the International Labor Organization, the European Commission, the European Parliament, the Committee of the Regions, the Belgian Federal Government and the Belgian Flemish Government.

Eero Palmujoki is a lecturer in International Relations at the School of Management, University of Tampere, Finland. He has acted as a professor of IR at the University of Tampere and the University of Turku. He has been a project director in several research projects concerning environmental change and development funded by the Academy of Finland and the Ministry for Foreign Affairs of Finland (MFA). He has written two books and several articles on international relations and development in Southeast Asia. His current research focuses on international society and environmental governance, on which he has published several articles, including, for example, (together with Pekka Virtanen) 'Global, National, or Market? Emerging REDD+ Governance Practices in Mozambique and Tanzania', *Global Environmental Politics* 16(1), 2016, 'Fragmentation and Diversification of Climate Change Governance in International Society', *International Relations* 27(2), 2013, 'International Organizations and the Rhetoric of Climate Change', *Journal of Indian Ocean Region* 6(2), 2010, and 'Global Principles for Sustainable Biofuel Production and Trade', *International Environmental Agreements: Politics, Law and Economics* 9(2), 2009.

Gita Parihar was Head of Legal at Friends of the Earth England, Wales & Northern Ireland until July 2016 and now works as an independent legal consultant. In her role at Friends of the Earth she provided legal policy advice to its campaigns and has a long track record of taking public-interest environmental cases in the UK. She provided legal advice to Friends of the Earth international during the UN climate negotiations on matters such as human rights, equity and loss and damage, up to the Paris agreement and beyond. She specialises in Aarhus rights at both the domestic and international level. Gita is a qualified lawyer in the UK and holds a Master's degree in international environmental law and human rights.

Philipp Pattberg is professor of transnational environmental governance and policy at VU University Amsterdam, The Netherlands. He specialises in the study of global environmental politics, with a focus on climate change governance, biodiversity, forest and marine governance, transnational relations, public-private partnerships, network theory and institutional analysis. Pattberg's current research scrutinises institutional complexity, functional overlaps and fragmentation across environmental domains (http://fragmentation.eu/). At VU University Amsterdam, Pattberg heads the Department of Environmental Policy Analysis, a team of more than twenty-five researchers. Pattberg is Chair of the Board of the Global Environmental Change Section of the German Political Science Association (http://www.ak-umwelt.de/) and a Senior Research Fellow of the international Earth System Governance Project (http://www.earthsystemgovernance.org/). Pattberg is also the General Director of the Netherlands Research School for Socio-Economic and Natural Sciences of the Environment (SENSE). From 2006 to 2011 Pattberg was the deputy-director of the Global Governance Project, a joint research programme of twelve European institutes with about forty affiliated researchers (www.glogov.org).

Klaus Radunsky received his PhD in analytical chemistry and physics at the University of Vienna in 1979. He started working for the Austrian Federal Government in 1975, helping to establish the country's Background Air Quality Monitoring Network. He moved to the Federal Environment Agency in 1985, where he worked as the technical expert for air quality, and later as Head of Unit of the Emission Trading Registry, where he remained until 2014. He is now the Austrian representative in a range of UNFCCC-related bodies, including COP, CMP, SBSTA and SBI, as well as the IPCC, and is also a member of the Adaptation Committee. He now heads the Austrian Federal Environment Agency's body responsible for overseeing the country's GHG Inventory and is the reviewer of the country's National Emission Inventories under the UNFCCC. He leads the Austrian Standardization Institute's working group on GHG monitoring and management (the national 'mirror' committee of ISO's TC207/SC7).

Charles Sampford gained his DPhil from Oxford in 1985 and was invited to come to Queensland as Foundation Dean of Law at Griffith University starting in March 1991. He led the 1998 bid for the Key Centre for Ethics, Law, Justice and Governance (the only Australian centre in law or governance to receive centre funding from the Australian Research Council). In September 2004, he became the Convenor of the ARC-funded Governance Research Network and Director of the Institute for Ethics, Governance and Law (IEGL), established on the initiative with the UN University that came to include QUT, ANU, OP Jindal (Delhi) and Strathmore (Nairobi). He has written over 130 articles and chapters and has completed 31 books and edited collections in law, philosophy and governance. His work on climate change and environmental governance goes back to an invited keynote at the World Council of Churches colloquium on carbon trading in 2000 and a keynote on governance for biodiversity at a UNESCO conference hosted by Columbia's Earth Institute. In 2008, Transparency International invited him to provide an overview paper for the International Anti-Corruption Conference in which he applied his integrity systems approach to the global financial crisis and what he termed 'the global carbon crisis'.

Lisa Sanderink is a PhD candidate in the department of Environmental Policy Analysis at the Institute for Environmental Studies (IVM), Vrije Universiteit, in Amsterdam. Her interests lie with global environmental governance, particularly in the fields of climate change and energy. She is a member of the CLIMENGO project (www.climengo.eu), which is funded by the Swedish Energy Agency. This project aims to map the institutional complexity of global climate and energy governance, evaluate its effectiveness and legitimacy and develop a knowledge base for decision makers. Currently, Lisa Sanderink is studying the so-called climate-energy nexus and the differences and synergies within its institutional constellations by means of innovative methods of automated content analysis (ACA). Lisa Sanderink holds a B.A. in European studies, with a specialisation in International and European law, from the University of Amsterdam, and a M.Sc. in Environment and Resource Management from the Vrije Universiteit Amsterdam.

Oscar Widerberg is a researcher and PhD candidate at the Institute for Environmental Studies (IVM) at the Vrije Universiteit Amsterdam. He works on global environmental governance, primarily studying interactions between international and transnational climate governance. He is a research fellow with the Earth System Governance Project and a member of several projects and networks working on climate governance, including the FORES Reference Group on international climate policy, comprising representatives from industry, government, academia and NGOs; the Galvanizing the Groundswell of Climate Action network's Data and Analysis stream; INOGOV's Early Career Investigators' Network (ECIN) and the CLIMENGO project (www.climengo.eu). He also participated in several high-level climate dialogues, such as the OECD's Climate Change Expert Group Global Forum and UNFCCC side events. Oscar has published articles on global climate governance in peer-reviewed journals such as *WIREs Climate Change, Millennium* and *Global Policy*; and he is a reviewer for several academic journals, including *Climate Policy, Global Policy, Global Environmental Politics* and *Environmental Politics*. Oscar also works as an associate consultant at Trinomics, carrying out assignments on climate change, environment and energy policy for international public institutions such as the European Commission, World Bank and the European Parliament. He holds a M.Sc. in Environmental Science from Utrecht University, the Netherlands, and a BSSc in International Relations from Malmö University, Sweden.

Acknowledgements

The research was conducted under the Australian Research Council (ARC) Discovery Project 2014–2017 grant. Chief investigators of this project are Charles Sampford, Hugh Breakey, Tim Cadman and Rowena Maguire. The views expressed herein are those of the authors and are not necessarily those of the Australian Research Council.

Abbreviations

°C	degrees centigrade
AAUs	Assigned Amount Units
ACROPOLIS	Academic Research Group for Policy Support
AOSIS	Alliance of Small Islands States
APA	Ad Hoc Working Group on the Paris Agreement
ARES-CCD	L'Académie de Recherche et d'Enseignement supérieur – La Commission Coopération au Développement de l'ARES
AWG-LCA	Ad Hoc Working Group on Long-Term Cooperative Action
BeFinD	A consortium of four research centres at Belgian Universities
BFIs	bilateral financial institutions
BINGO	Business and Industry Major Group
C&I	criteria and indicator
CBD	Conventions on Biological Diversity
CBDR	common but differentiated responsibilities
CCD	Combatting Desertification
CDM	Clean Development Mechanism
CDP	Carbon Disclosure Project
CDP	climate displaced person
CER	Carbon Emission Reduction
CESs	certified emissions reductions
CIEL	Center for International Environmental Law
CIF	Climate Investment Funds
CMA	Conference of Parties Meeting as Parties to the Agreement
CMP	Conference of the Parties serving as the meeting of the parties to the Kyoto Protocol
CO_2	carbon dioxide
COP	Conference of the Parties
CPI	Climate Policy Initiative
CSA	Climate-smart agriculture
CSO	Civil Society Organization
DBSA	Development Bank of South Africa
DC	District of Columbia
DGD	Belgian Directorate General for Development Cooperation

ECOSOC	Economic and Social Council
EEZ	Exclusive Economic Zone
EIA	Environmental Impact Assessment
ERUs	Emission Reduction Units
ES	English School
EU ETS	European Emission Trading Scheme
FAO	Food and Agriculture Organization
FAR	Fifth Assessment Report
FCPF	Forest Carbon Partnership Facility
FPIC	free, prior and informed consent
G77	Group of 77 is a loose coalition of developing nations, designed to promote its members' collective economic interests and to create an enhanced joint negotiating capacity in the United Nations.
GACSA	Global Alliance for Climate-Smart
GCF	Global Climate Finance
GDP	gross domestic product
GEF	*Global Environment Facility*
GGS	global governance system
GHG	greenhouse gas
GNI	gross national income
GNP	gross national product
Gt	Gigatons
HAB	historical adjusted baseline
ICAP	*International Carbon Action Partnership*
ICESCR	International Covenant on Economic, Social and Cultural Rights
ICJ	International Court of Justice
ICTs	information and communications technologies
IET	international emissions trading
IIED	International Institute for Environment and Development
INDCs	*Intended Nationally Determined Contributions*
IOA	Internal Organizational Arrangements
IOM	International Migration Organization
IOs	international organizations
IPCC	Intergovernmental Panel on Climate Change
IR	international relations
JI	Joint Implementation
JISC	Joint Implementation Supervisory Committee
KP	Kyoto Protocol
LDC	least developed countries
LDCF	Least Developed Countries Fund
LPAA	Lima Paris Action Agenda
LULUCF	land use, land-use change and forestry
MRV	monitoring, reporting and verifying

NAMA	*Nationally Appropriate Mitigation Actions*
NAPAs	National Adaptation Programs of Action
NDBs	national development banks
NDC	nationally determined contributions
NMAs	non-market-based approaches
NMM	new market-based mechanisms
NSA	non-state actors
OAS	Organization of American States
ODA	official development assistance
OECD	*Organisation for Economic Cooperation and Development*
OHCHR	Office of the High Commissioner for Human Rights
PB	project based
PC&I	principles, criteria, and indicator
PIJ	Public Institutional Justification
PIJ-HR	Public Institutional Justification-Human Rights
QDA	Qualitative Data Analysis
REDD	Reducing Emissions from Deforestation and Forest Degradation
REEEP	*Renewable Energy and Energy Efficiency Partnership*
SBI	Subsidiary Body for Implementation
SBSTA	Subsidiary Body for Scientific and Technological Advice
SCCF	Special Climate Change Fund
SCF	Standing Committee on Finance
SD	standard deviation
SISs	Small Island States
TCG_MP	The Climate Group (Members Principles)
UK	United Kingdom
UNCED	United Nations Conference on Environment and Development
UNDHR	United Nations Declaration of Human Rights
UNDP	United Nations Development Programme
UNEP	United Nations Environment Programme
UNFCCC	United Nations Framework Convention on Climate Change
UNGA	United Nations General Assembly
US	United States
USD	United States currency
VCS	Verified Carbon Standard
VLIR-UOS	Flemish University Council – University Development Cooperation
WB	World Bank
WIM	Warsaw International Mechanism for Loss and Damage
WMO	World Meteorological association
WTO	World Trade Organization
WWF	World Wildlife Fund

Part I

Exploring the links between governance and the integrity of climate change

1 Introduction

Governing the climate regime

Tim Cadman, Charles Sampford, Rowena Maguire and Hugh Breakey

Governing the climate regime

The Paris Conference of Parties (COP) has sought to initiate a global regime for the mitigation of greenhouse gas (GHG) emissions and, where necessary, adaptation to them. The governance challenges are immense – in the governance of the negotiations, in the governance of the complex set of mechanisms that those negotiations are generating, in the integrity mechanisms to ensure as far as possible that those mechanisms fulfil their functions (leading to a rudimentary Global Carbon Integrity System/Regime).[1] In the light of the dynamic relations between state and non-state actors within the climate negotiations, there is a growing recognition that more research is needed to explore the social dimensions of governance quality in climate mechanisms, particularly around decision-making processes. These take note of the social-political nature of climate governance and explore governance quality by looking at the participatory structures, deliberative processes, substantive products, and longer-term outcomes as the determinants of regime effectiveness.[2] *Effectiveness* is understood as the measure of institutional performance against a set of desired objectives.[3]

1 See Charles Sampford, "Global Transparency: Fighting Corruption for a Sustainable Future: From National Integrity Systems to Global Integrity Systems," in *TI Commissioned Conference Overview Discussion Paper for 13 IACC* (Athens, 2008); Charles Sampford, "Global Lessons from Fitzgerald: From State and National to Global Integrity Systems," in *The Fitzgerald Legacy: Reforming Public Life in Australia and Beyond*, ed. Colleen Lewis, Janet Ransley, and Ross Homel (Brisbane: Australian Academic Press, 2010).

2 See Timothy Cadman, *Quality and Legitimacy of Global Governance: Case Lessons from Forestry* (New York and Houndmills, Basingstoke, Hampshire: Palgrave Macmillan, 2011), p. 42. For commentary on the social dynamics of climate change, see Jon Barnett, "Adapting to Climate Change: Three Key Challenges for Research and Policy – an Editorial Essay," *Wiley Interdisciplinary Reviews: Climate Change* 1, no. 3 (2010). For an exploration of governance issues through a specific mechanism, see Esteve Corbera and Heike Schroeder, "Governing and Implementing Redd+," *Environmental Science & Policy* 14, no. 2 (2011); Mary Thompson, Manali Baruah, and Edward Carr, "Seeing Redd+ as a Project of Environmental Governance," *Environmental Science and Policy* 14, no. 2 (2011).

3 Christopher Elliott, *Forest Certification: A Policy Perspective* (Bogo: Center for International Forestry Research, 2000), p. 23.

In this case, the desired objectives concern themselves with 'good' governance and constitute a 'thick' set of governance values, as outlined late in this chapter and more extensively in Chapter 2. Following the approach delineated there, the governance quality of the climate regime is not attributed to any single institutional arrangement, such as transparency, even though this is, of course, important. Evaluating the performance of an institution on the basis of a restricted set of values would constitute a somewhat 'thin' assessment; consequently this volume explores some of the broader parameters affecting quality of governance and institutional integrity and their possible impacts on the policy responses to climate change.

It is suggested here that the social and political interactions occurring within an institution's participatory structures and deliberative processes also contribute to institutional *legitimacy*. However, legitimacy has proven to be an analytical challenge for governance theory, and while some excellent scholars have delineated some of its normative and empirical dimensions, they also have maintained that research in this field remains incomplete.[4] Acknowledging these observations, it is suggested here that legitimacy can be seen, at least in the UNFCCC (UN Framework Convention on Climate Change), as a desired attribute. It would be difficult to justify a regime that was not sanctioned by those that ratified it. As Chapter 2 discusses, legitimacy may be investigated by exploring the structural and procedural components (inputs) of the institution's governance system and how they interact one with another (throughputs), as well as the rules, standards, etc., this interaction creates (outputs). The 'thickness' of the governance values incorporated will also contribute to the longer-term behavioural and problem-solving impacts of the institution on the ground (outcomes). This is an integrating model of institutional legitimacy discussed by previous governance theorists (Scharpf 1997, Kjaer 2004, Schmidt 2012), as depicted in Figure 1.1.[5]

Institutional integrity and its relationship to governance

In the companion volume to this book, Breakey and Cadman build on Sampford's work to analyse institutional integrity and to consider its several interrelated

4 See Frank Biermann et al., *Earth System Governance People, Places, and the Planet: Science and Implementation Plan of the Earth System Governance Project Report No. 1* (Bonn: International Human Dimensions Programme on Global Environmental Change, 2009), http://www.ihdp.unu.edu/docs/Publications/ESG/IHDP_ReportNo20_ESG_ReportNo1.pdf; Frank Biermann and Aarti Gupta, "Accountability and Legitimacy in Earth System Governance: A Research Framework," *Ecological Economics* 70, no. 11 (2011).

5 Anne Mette Kjaer, *Governance* (Cambridge and Malden, MA: Polity Press, 2004), pp. 7–9; Fritz W. Scharpf, *Games Real Actors Play: Actor-Centered Institutionalism in Policy Research* (Boulder: Westview Press, 1997), p. 153; Vivien A. Schmidt, "Democracy and Legitimacy in the European Union Revisited: Input, Output and 'Throughput'," *Political Studies* 61 (2012).

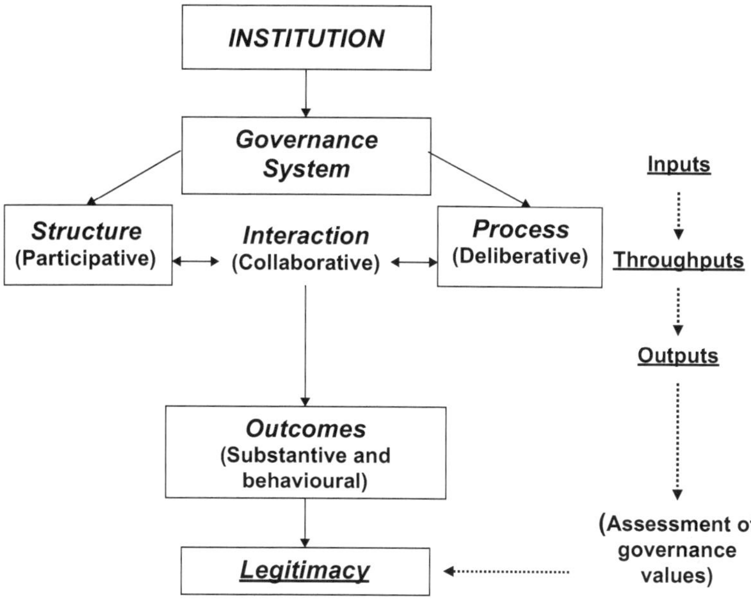

Figure 1.1 Theoretical model for the evaluation of contemporary global governance (Cadman 2011, adapted)

components.[6] This topic is explored in detail in Chapter 2; however, it is worth just sketching the broad ideas here.

Sampford theorises 'institutional integrity' as taking the same form as 'individual integrity'.[7] He starts with a version of Singer's fundamental ethical question – 'how are we to live our lives?'[8] *Individual ethics* involves individuals asking hard questions about their values, giving honest and public answers, and trying to live by those answers. If they do, they have integrity in the sense they are true to the values they claim to stand for. Similarly, *institutional ethics* involves an institution/organisation asking hard questions of its value and values, giving honest and public answers and living by them. This more complex institutional process starts with the vital questions that must be asked of any institution or organisation:

6 Hugh Breakey and Timothy Cadman, "A Comprehensive Framework for Evaluating the Integrity of the Climate Regime Complex," in *Ethical Values and the Integrity of the Climate Change Regime*, ed. Hugh Breakey, Vesselin Popovski, and Rowena Maguire (Surrey: Ashgate, 2015).

7 Developed during the 1990s but first set down at some length by Noel Preston and Charles Sampford, "Institutionalising Ethics," *Encouraging Ethics and Challenging Corruption*, Charles Sampford, Noel Preston, and Carmel Connors (Annandale: Federation Press, 2002).

8 Peter Singer, *How Are We to Live? Ethics in an Age of Self-Interest* (Melbourne: Text Publishing, 1993).

'What is it for?' 'What justifies the institution to the community in which it operates (recognising that the community gives it the right to operate and takes the risk that the concentration of power in that institution may be used against that community)?' Asking those questions involves an institutional and collective effort under the institution's formal and informal constitutional processes. An institution has integrity if it lives by its answers – by creating mechanisms which make it more likely that the institution keeps to the values to which it is publicly committed. These mechanisms might be seen as an institutional 'integrity system'. Such institutional integrity systems need to be articulated with external institutions and mechanisms designed to ensure that entrusted power is not abused for personal gain (Transparency International's definition of corruption) but is used for its claimed purposes (Sampford's definition of integrity).

Breakey and Cadman elucidate and expand, seeing integrity involving a process of reflecting on and developing one's values, giving a public profession of those values, and then working to achieve them.[9] An institution sets down its values in its *Public Institutional Justification* (PIJ). The PIJ stands as a public record (a code, charter, constitution, mission statement, etc.) of the institution's values, commitments, principles, and goals. When the institution's existence, activities, or privileges are challenged on social or moral grounds, the PIJ's role is to provide a justification for the institution.

Not all institutions will possess a PIJ, having never undertaken the process of developing one. If an institution does have a PIJ, then its actions, processes, and outcomes can be evaluated from the perspective of that PIJ. If the institution succeeds in living up to the PIJ's values, goals, and purposes, then the institution possesses 'consistency-integrity', as its actions are *consistent with* its values.

The institution may have succeeded in achieving consistency-integrity because its internal organisational arrangements are well designed to facilitate its living up to its PIJ. In this case, the institution can be said to boast 'coherence-integrity' because its internal constitution *coheres with* achieving its PIJ. An institution with coherence-integrity can be expected to go on achieving consistency-integrity in the future because its parts work together effectively to promote the desired outcomes.

Alternatively, the institution might achieve consistency-integrity because its external environment facilitates this outcome. In this happy arrangement, other organisations, regulations, and laws tend to support the institution when it is achieving its PIJ, but resist it when it betrays its PIJ.[10] In this case, the institution

9 See Hugh Breakey, Timothy Cadman, and Charles Sampford, "Conceptualizing Personal and Institutional Integrity: The Comprehensive Integrity Framework," *Research in Ethical Issues in Organizations* 14 (2015). For a fuller exploration of personal morality, see Hugh Breakey, "Compromise Despite Conviction: Curbing Integrity's Moral Dangers," *Journal of Value Inquiry* (2016 (forthcoming)).

10 The separation of powers is a good example of this dynamic, where institutions outside one organisation support it when it is playing its mandated role, but stymie it when it extends beyond its proper ambit. See Hugh Breakey, "Dividing to Conquer: Employing the Separation of Powers to Structure Institutional Inter-Relations," *Research in Ethical Issues in Organizations* 12 (2014).

enjoys 'context-integrity' because its external context facilitates its ongoing achievement of its PIJ.[11]

Governance values, on the account developed in Chapter 2, apply to the structures, processes, and mechanisms that steer an institution. Rather than focusing on institutional *integrity*, governance values confer *legitimacy* on the institution. The level of legitimacy conferred depends on the comprehensiveness – the 'thickness', as we will call it – of the governance values. 'Thin' governance values, like accountability and transparency, secure only basic oversight. 'Thickish' governance values capture broader concerns with the institution's effective functioning, including questions about who makes decisions and how those decisions are made. Finally, 'thick' governance values incorporate a rich portfolio of ethical values, including equality, democracy, and inclusiveness. If present, thick governance values confer legitimacy on the institution's structures and processes.

These two qualities of legitimacy and integrity differ: the former confers a social and moral status, while the latter relates purely to how the institution measures up to its own standards (as it sets down in the PIJ). Nevertheless, because governance values require attention to the way information, authority, feedback, and decision-making flow through the institution, there is often a fertile and mutually facilitative process between governance values and (especially context-integrity).

Integrity (evaluated through consistency, coherence, and context dimensions) and legitimacy (assessed through increasingly thick governance values) are not the only moral evaluations possible. An institution can boast both integrity and legitimacy without fulfilling everything that an objective moral theory – or a particular moral community – might demand of it. This is because integrity only asks whether the institution is being true to *its own* standards, while legitimacy applies only to the *structures and mechanisms* that steer an institution (at least in the theoretical terms of input, throughput, and output legitimacy, as Figure 1.1 conceptualises legitimacy). A comprehensive moral evaluation will draw in all of the substantive moral values and goals (for example, human rights, sustainable development) that an institution may be morally asked to uphold. This leads us to the idea of 'governing values', which we discuss in Chapter 2.

The following chapters will explore in more detail the various implications for the climate regime of these differences between institutional legitimacy-based governance values, PIJ-based integrity values, and morality-based governing values. (On the relationship between the governance values and the governing value of human rights touched on previously, see especially Chapter 9 by Bridget Lewis.) In general though, it is important to bear in mind the distinction between governing and governance values when discussing the climate regime. In particular, an expansion in the governing values that the climate regime pursues (or professes that it pursues) does not necessarily equate to an expansion in its governance values. Through outputs like the Paris Agreement, the climate regime

11 See also Breakey and Cadman, "Integrity of the Climate Regime Complex."

can declare its commitment to myriad different governing values. Whether those values can be achieved, however, depends upon the development of appropriate institutional machinery and processes to operationalise them. In short, it is easy enough for the climate regime to profess idealistic governing values, but much harder to put in place the governance values that alone can achieve those laudable goals.

Background to the Paris Agreement

By November 2015, as the twenty-first Conference and Meeting of the Parties (COP and CMP) to the UNFCC approached, the endless round of pre-negotiations referred to as the Ad Hoc Working Group on the Durban Platform for Enhanced Action (ADP), the two work streams focusing on the 2015 agreement, and the post 2020 ambition had not brought anything substantive to the table.[12] There was still no indication as to whether the largely expired Kyoto Protocol (KP) would be superseded by a 'protocol, another legal instrument or an agreed outcome with legal force'.[13] There were some significant obstacles to market mechanisms in the negotiations. Like-Minded Developing Countries (LMDCs), a somewhat volatile sub-grouping within the G77[14] and containing some of those countries most impacted by climate change, including Bangladesh, Sri Lanka, and Sudan, as well as China and Bolivia and various oil countries, had previously pushed the developed countries (including the US and UK) to acknowledge their historical responsibility for climate change.[15] This view found favor with Brazil in Warsaw at COP 19 in 2013.[16] It argued that the 1990 baseline for measuring country emissions reductions agreed to in the KP should also take emissions dating back to 1850 into consideration.[17] These discussions were connected to discussions around who should pay for current impacts (around a negotiating stream related to loss and damage under the Warsaw International Mechanism – WIM). While this position had some validity,[18] developed countries

12 More information on these bodies may be found by visiting the Convention website: http://www.unfccc.int. For COP 21: http://www.cop21.gouv.fr/; COP/CMP: http://unfccc.int/meetings/paris_nov_2015/meeting/8926.php. Background to the Convention: https://unfccc.int/essential_background/convention/items/6036.php. ADP: https://unfccc.int/bodies/body/6645.php. The work streams: http://unfccc.int/bodies/body/6645.php#ws1.

13 For KP see https://unfccc.int/kyoto_protocol/items/2830.php.

14 The Group of 77 is a grouping of developing countries within the UN; see https://unfccc.int/parties_and_observers/parties/negotiating_groups/items/2714.php.

15 See: http://unfccc.int/files/documentation/submissions_from_parties/adp/application/pdf/adp_lmdc_ws1_20140309.pdf.

16 For COP 19 see http://unfccc.int/meetings/warsaw_nov_2013/meeting/7649.php.

17 See http://unfccc.int/files/na/application/pdf/substa_submission_by_brazil_-_brazilian_proposal_final_corrected.pdf.

18 For a measured evaluation see, for example, Jonathan Pickering and Christian Barry, "On the Concept of Climate Debt: Its Moral and Political Value," *Critical Review of International Social and Political Philosophy* 15, no. 5 (2012).

proved consistently resistant. As a consequence, some developing countries opposed market mechanisms for emissions trading in a tit-for-tat strategy.

At the Paris meeting, market-based mitigation activities such as emissions trading became inextricably linked to the Nationally Determined Contributions (NDCs) that the Parties brought to the table (the more activities, the greater the opportunity for carbon 'offsets' and other schemes). The developed countries had pushed for many years to see a price on carbon so that the third Kyoto mechanism (international emissions trading) could be used to stimulate investment in renewable energy and enable countries that could not meet their emissions reduction targets to purchase 'carbon offsets' from countries that could.[19] REDD+, the 'poster child' of recent negotiations, which aims to provide payments to developing countries from developed countries to reduce emissions from deforestation and forest degradation, was entangled in, and somewhat impacted by, this debate.[20] REDD+ is a logical mechanism to generate offsets from avoided deforestation. But in order to be credible, offsets need to be tied to certified emissions reductions (popularly known as carbon credits) and verification and accounting (otherwise the credibility of the offsets generated is questionable). Only the Clean Development Mechanism (CDM) offered these methodologies, but by Paris was largely moribund due to the expiration of the KP.[21] As the climate talks entered their last few days, a compromise emerged for the LMDCs and others to drop their position on historical responsibilities, and for developed countries to take the lead on financing adaptation and increasing their mitigation activities. It also appeared that while there would be no protocol equivalent to Kyoto, an agreement with some legally binding provisions would emerge. The US delegation, which back home faced a hostile Congress, could not directly agree to legally binding emissions reduction targets, but with sixty-five lawyers in the delegation was prepared to support elements with legal requirements that did not require Congressional approval. Negotiators began furiously toing and froing over critical issues such as the transparency of reporting and how, when, and if the NDCs would be reviewed. The 'Comité de Paris', formed with the intention of keeping the text negotiations moving along and comprising some of the key national players, appeared to assist the last few hours of negotiations. In the end, compromise language emerged that recognised a target temperature of 'well below two degrees with an effort to achieve one point five degrees'. National reporting of commitments under the NDCs would be on a five-year basis, and increased funding would be made available to developing countries for adaptation.

And so the Paris Agreement (PA) came into being, subject to ratification by 55 per cent of signatories to the Convention from April 2016. It largely

19 For emissions trading see http://unfccc.int/kyoto_protocol/mechanisms/emissions_trading/items/2731.php.

20 For REDD+ see http://unfccc.int/land_use_and_climate_change/redd/items/7377.php.

21 For CDM see http://cdm.unfccc.int/.

replaced the Kyoto Protocol (although various implementation commitments remained, such as those included in the Doha Amendment).[22] The Ad Hoc Working Group for Advancing the Durban Platform (ADP, the main negotiating body prior to and during the Paris Conference of Parties (COP 21)) was disbanded and replaced by the Ad Hoc Working Group on the Paris Agreement (APA). The CMA (Conference of Parties Meeting as Parties to the Agreement) was created to complement the CMP (Conference of Parties Serving as the Meeting of the Parties to the Kyoto Protocol) as the main negotiating forum for the agreement.[23] Many of the administrative, oversight, and decision-making powers of the new agreement were linked to this body. The WIM will continue (with a review in 2016) and will act a clearing house for insurance and risk transfer to help Parties create and put risk management strategies in place.[24] WIM was also requested to establish a task force to develop recommendations to address displacement as a consequence of climate change. The Standing Committee on Finance (SCF) was given a renewed mandate to serve the agreement under responsibilities identified by COP.[25] The Green Climate Fund (GCF) and the Global Environment Facility (GEF), previously potentially in conflict over who was to prevail, were both deemed relevant and mandated to act as operational entities for climate finance under the agreement.[26] REDD+ continued to find favor with a majority of Parties and was identified as being eligible for funding under the GCF, as well as other existing and relevant funding sources. There was also some movement on the market mechanisms front. The door for a 'mechanism to contribute to greenhouse gas emissions' (code for emissions trading) was also left open, and notice was given of the need to ensure that there is a system of accounting to avoid double accounting of emissions reductions in future mechanisms. Efforts were also made to encourage Parties and other stakeholders to voluntarily cancel certified emissions reductions (CERs – 'carbon credits'/'carbon offsets') under Kyoto and to report on transfers of credits (possibly to allow them back into the trading system and maybe keep the CDM alive).

Importantly, given these commitments, and in relation to implementation specifically, a mechanism to facilitate implementation and compliance was also identified as being necessary. Currently there is a compliance committee and related entities that sit under the Kyoto Protocol, but not more broadly.[27] It was also

22 For the Doha Amendment see http://unfccc.int/kyoto_protocol/doha_amendment/items/7362.php.

23 For CMP see https://unfccc.int/bodies/body/6397.php.

24 For WIM see http://unfccc.int/adaptation/workstreams/loss_and_damage/items/8134.php.

25 For SCF see https://unfccc.int/cooperation_and_support/financial_mechanism/standing_committee/items/6877.php.

26 For GCF see http://www.greenclimate.fund/home; for the GEF: https://www.thegef.org/gef/whatisgef; for climate finance https://unfccc.int/focus/climate_finance/items/7001.php.

27 For the Compliance Committee see: https://unfccc.int/bodies/body/6432.php.

agreed that an accelerated implementation process from 2016 should be put in place. In an effort to ensure that Parties deliver on their commitments, it was also decided that a transparency framework (under the COP and based on the existing Cancun Agreement) should also be created and that a Capacity Building Initiative for Transparency should be established. From a governance perspective, it is interesting to note the implied linkages between capacity building, transparency, compliance, and implementation. Presumably, there is some understanding within the regime that if agreements are to be implemented effectively, transparency of actions taken is essential from a reporting viewpoint, and that Parties will need to provide additional resources (financial, technical, institutional) to improving and/or ensuring compliance.

A walk through the collection

This collection analyses the governance arrangements of the UNFCCC and an emerging, rudimentary, global carbon integrity regime. Part one of the collection explores the links between governance and integrity in the climate regime and comprises two chapters. This first introductory chapter provides some context on work carried out in a related collection exploring integrity issues and a method for examining integrity. The second chapter develops a conceptual framework for assessing governance arrangements which is employed in various ways across the chapters. The chapter starts by exploring the contested issues surrounding governance studies such as definitions and coverage and takes the position of adopting a value-neutral definition which focuses on systems and processes. The chapter establishes a methodology for evaluating governance by classifying governing values as thin, thickish, and thick. Thin governance is defined as providing minimum conditions for basic oversight, while thick governance is seen as having systems and processes in place to ensure legitimacy. The method established in this chapter allows for the contributors to explore governance issues within the UNFCCC using a common reference of analysis.

Part two of the collection, 'Evaluating governance arrangements within the UNFCCC', explores the current governance arrangements in operation within the UNFCCC. Philipp Pattberg, Lisa Sanderink, and Oscar Widerberg open this part by providing a meta-analysis of the institutions, players, and discourses involved in global climate governance. This empirical analysis considers the degree of fragmentation, or coherence, within global climate governance. The authors find that the global climate regime is operating in a polycentric form with two dominant discourses of green governmentality and ecological modernisation evident across the various institutions.

The fourth chapter by Eero Palmujoki continues to explore issues of fragmentation in the context of the Reducing Emissions from Deforestation and Forest Degradation (REDD+) initiative. The chapter finds that the basic institutions of sovereignty, the market, and justice define the borders of REDD+ governance across the various bodies involved in implementing REDD+ readiness or projects. The chapter unpacks the institutional bodies and instruments within the UNFCCC, which have steered the creation of REDD+ policy through an exploration of the Subsidiary Body for Scientific and Technological Advice (SBSTA) decisions and the more recent Paris Agreement. It is found that a failure to align domestic conservation measures with global carbon markets in a way that is compatible with the UNFCCC reduces the integrity of the REDD mechanism and as such suggests evidence of thin values only.

The fifth chapter by Emilie Bécault and Axel Marx explores how concepts of institutional integrity are highly relevant in the global governance of climate financing. This chapter carries out a meta-level assessment of climate finance governance by directing attention to the normative framework, overarching architecture, and geographical scope of the climate regime as a global governance system. It is shown that there are substantial disagreements about how the UNFCCC principles should be interpreted in order to constitute a fair, equitable, and effective provision of climate finance to developing countries. This lack of consensus has led to the development of a divergent finance architecture involving a myriad of institutions, including donor funds managed by UN agencies, donor funds managed by EU institutions, regional funds managed by regional development banks, bilateral funding initiatives, and domestic government financing which can be described as polycentric in operation. The Paris Agreement does not provide clarity on how fair, equitable, and effective climate finance should be delivered, nor does it establish specific monetary goals beyond 2025. This leaves uncertainty about what counts as official climate finance and the role that new actors and sources can play for the long-term mobilisation of climate finance.

The sixth chapter by Rowena Maguire continues to explore issues concerning finance and equity for developing countries by examining the governance arrangements within the UNFCCC to recognise people displaced by climate change. The chapter investigates the structures and processes within the regime that have been used to recognise the problem of climate displacement. While the regime does not holistically govern climate displacement, a range of norms and existing mechanisms within the UNFCCC are consistent with providing legal protection and assistance for persons displaced by climate change. The chapter suggests that the Loss and Damage Mechanism (WIM) should be used to assist those people who have to cross borders as a result of climate change and that adaptation finance should be implemented in such as way so as to identify and assist populations at risk of internal displacement. It is argued that thick governing values of stakeholder engagement and equality in decision-making is of vital importance when implementing strategies to assist vulnerable groups migrating as a result of climate change.

The seventh chapter by Kate Dooley and Gita Parihar continues the discussion about equity in the UNFCCC by examining the utility of human rights discourses and principles of equity within the UN Climate regime. The authors situate human rights and equity as substantive governing values, which move beyond procedural concerns to incorporate important ethical goals, including the demands of morality, justice, and legitimacy. They demonstrate that the principle of equity is operationalised through the principle of Common But Differentiated

Responsibilities (CBDR). The chapter shows the complementarity between human rights and equity principles and argues that it is not possible for a proposal on climate action to be compliant with human rights obligations if it does not address differential obligations in a manner that protects the most vulnerable. These two complementary frameworks are essential to the context-integrity of the regime, as they share an ethical basis appealing to different political actors and institutions and seek to inject justice and fairness into UNFCCC structures and processes.

The eighth and final chapter in part two by Tanvi Mani explores the governance arrangements of the Kyoto Protocol focusing on tradable permits. She finds that that the Kyoto Protocol focused on thin values such as compliance. She suggests that the integrity provisions within the Paris Agreement have the potential to increase the thickness of governance, but greater levels of collaboration are required if this is to eventuate.

The third part of the collection, 'Improving governance within the UNFCCC', considers how the governance arrangements of the UNFCCC can be enhanced. Chapter 9 by Bridget Lewis explores the contribution that human rights as a *governing* value can make to strengthen the *governance* values of the climate regime and thereby increase institutional legitimacy. Embedding human rights in such a way would help articulate the regime's PIJ and thereby fulfil its international commitments more effectively.

Chapter 10 by Lisa Elges from Transparency International (TI) discusses the challenges associated with mobilising and governing multilateral climate finance. The chapter discusses a study carried out by TI which assessed the six major climate funds: Least Developed Country Fund, Special Climate Change Fund, Climate Investment Fund, Forest Carbon Partnership Facility, UN REDD Programme, and the Adaptation Fund. These funds were assessed under three parameters: transparency, accountability, and integrity. She finds that the governance arrangements establishing multilateral climate finance bodies is 'thickish', but the different legal and non-legal responsibilities that multiple actors have makes it difficult to identify who is accountable for what, which rules apply, and where the decision-making lies. Many of the funds assessed in the chapter are in the early days of implementation and will require reassessment in the future once more funds are mobilised in order to track their transparency, accountability, and integrity.

Chapter 11 by Joseph Macharia and Hope Johnson examines how reforms are needed in the agricultural sector in order to ensure that food production is not threatened (part of the PIJ of the UNFCCC). The chapter explores how the concept of Climate-Smart Agriculture (CSA) can be used to account for both mitigation and adaptation activities in the agriculture sector. It explains that there is gap in knowledge between international institutional knowledge about CSA and community-level knowledge concerning CSA. The chapter argues that knowledge needs to be disseminated in new and creative ways and explores the role that social media can play in translating international climate knowledge into knowledge at the farm level. The online platform of Mkulima Young (which translates as 'young farmer' in Swahili) is used as a case study to show how knowledge concerning CSA can be classified as social learning and results in co-produced knowledge where local farmers use and share their knowledge in innovative ways in order to adapt techniques to local conditions. The chapter suggests that climate governance can be enhanced by the regime partnering with non-state actors with active online platforms, where co-produced knowledge can be shared and disseminated.

Chapter 12 by Peter Glynn explores the role that Civil Society Organisations (CSOs) have played within the UNFCCC. The chapter explains that CSOs have been active participants within the UNFCCC COP negotiations, but that their participation has been declining since the 2009 Copenhagen COP negotiations due to such groups feeling disenfranchised and the negotiating process being viewed as inefficient. The Paris COP saw a pledge from Laurent Fabius to hold forums in which corporations and civil society environmental organisations would be invited participants, signalling a big change in engagement. Glynn finds that the application of the integrity systems and governance values framework within the regime would have a beneficial impact on climate policy, and argues that there needs to be a higher level of participation of non-state actors in policymaking at both the domestic and international levels.

The final chapter by Klaus Radunsky and Tim Cadman serves as an afterword to the volume. Given their historical involvement in the regime, they explore the long road to Paris, adopting the perspectives of the 'insider' (governmental) negotiator and the 'outsider' (non-governmental) observer. They suggest that that the regime has moved from a 'thin' to a 'thickish' application of governing values. But as long as 'non-Party stakeholders' remain excluded from a deliberative role in the negotiations, and the nation states continue to make all the decisions, the legitimacy of the regime remains questionable.

References

Barnett, Jon. "Adapting to Climate Change: Three Key Challenges for Research and Policy – an Editorial Essay." *Wiley Interdisciplinary Reviews: Climate Change* 1, no. 3 (2010): 314–17.

Biermann, Frank, Michele Betsill, Aarti Gupta, Norichika Kanie, Louis Lebel, Diana Liverman, Heike Schroeder, et al. *Earth System Governance People, Places, and the Planet: Science and Implementation Plan of the Earth System Governance Project Report No. 1.* Bonn: International Human Dimensions Programme on Global Environmental Change, 2009. <http://www.ihdp.unu.edu/docs/Publications/ESG/IHDP_ReportNo20_ESG_ReportNo1.pdf> (accessed 15 June 2016).

Biermann, Frank, and Aarti Gupta. "Accountability and Legitimacy in Earth System Governance: A Research Framework." *Ecological Economics* 70, no. 11 (2011): 1856–64.

Breakey, Hugh. "Compromise Despite Conviction: Curbing Integrity's Moral Dangers." *Journal of Value Inquiry* 50, no. 3 (2016): 613–29.

———. "Dividing to Conquer: Employing the Separation of Powers to Structure Institutional Inter-Relations." *Research in Ethical Issues in Organizations* 12 (2014): 29–58.

Breakey, Hugh, and Timothy Cadman. "A Comprehensive Framework for Evaluating the Integrity of the Climate Regime Complex." Chap. 2 In *Ethical Values and the Integrity of the Climate Change Regime*, edited by Hugh Breakey, Vesselin Popovski and Rowena Maguire, 17–27. Surrey: Ashgate, 2015.

Breakey, Hugh, Timothy Cadman, and Charles Sampford. "Conceptualizing Personal and Institutional Integrity: The Comprehensive Integrity Framework." *Research in Ethical Issues in Organizations* 14 (2015): 1–40.

Cadman, Timothy. *Quality and Legitimacy of Global Governance: Case Lessons from Forestry*. New York and Houndmills, Basingstoke, Hampshire: Palgrave Macmillan, 2011.

Corbera, Esteve, and Heike Schroeder. "Governing and Implementing Redd+." *Environmental Science & Policy* 14, no. 2 (2011): 89–99.

Elliott, Christopher. *Forest Certification: A Policy Perspective*. Bogo: Center for International Forestry Research, 2000.

Kjaer, Anne Mette. *Governance*. Cambridge and Malden, MA: Polity Press, 2004.

Pickering, Jonathan, and Christian Barry. "On the Concept of Climate Debt: Its Moral and Political Value." *Critical Review of International Social and Political Philosophy* 15, no. 5 (2012): 667–85.

Preston, Noel, and Charles Sampford. "Institutionalising Ethics." *Encouraging Ethics and Challenging Corruption*, Charles Sampford, Noel Preston and Carmel Connors, 32–68. Annandale: Federation Press, 2002.

Sampford, Charles. "Global Lessons from Fitzgerald: From State and National to Global Integrity Systems." Chap. 13 In *The Fitzgerald Legacy: Reforming Public Life in Australia and Beyond*, edited by Colleen Lewis, Janet Ransley and Ross Homel, 210–33. Brisbane: Australian Academic Press, 2010.

———. "Global Transparency: Fighting Corruption for a Sustainable Future: From National Integrity Systems to Global Integrity Systems." *TI Commissioned Conference Overview Discussion Paper for 13 IACC*. Athens, 2008.

Scharpf, Fritz W. *Games Real Actors Play: Actor-Centered Institutionalism in Policy Research*. Boulder: Westview Press, 1997.

Schmidt, Vivien A. "Democracy and Legitimacy in the European Union Revisited: Input, Output and 'Throughput'." *Political Studies* 61 (2012): 2–22.

Singer, Peter. *How Are We to Live? Ethics in an Age of Self-Interest*. Melbourne: Text Publishing, 1993.

Thompson, Mary, Manali Baruah, and Edward Carr. "Seeing Redd+ as a Project of Environmental Governance." *Environmental Science and Policy* 14, no. 2 (2011): 100–10.

2 Governance values and institutional integrity

Hugh Breakey, Tim Cadman and Charles Sampford

Background and introduction

The UN Framework Convention on Climate Change (UNFCCC) aims to tackle the consequences of 'dangerous' climate impacts.[1] The Convention entered into effect in 1994, influenced by the dominant discourses of what might be termed 'the technological fix' and neo–liberalism, epitomized by market-based approaches, including the 'flexible mechanisms' of the 1997 Kyoto Protocol (KP). These mechanisms consisted of the Clean Development Mechanism (CDM), joint implementation (JI) and international emissions trading (IET). Emissions trading aims to 'offset' emissions through the purchase of carbon 'credits' generated through project activities in developing countries.[2] The Intergovernmental Panel on Climate Change (IPCC), established prior to Rio in 1991, provides the scientific information underlying global climate deliberations. The IPCC produced its Fifth Assessment Report (FAR) in 2014, summarizing the results of all the working group reports, and two special reports that came out in 2011. It is now 95 per cent certain that human activities are responsible for climate change. Global population and our current model of economic growth and land use[3] are the main sources of rising fossil fuel combustion leading to increasing greenhouse gases (GHGs). The more compelling the scientific evidence for climate change and its likely effects, the greater the need to address this by radically reducing GHGs.[4] International efforts have concentrated on

1 These impacts, however, are not specifically elucidated. See United Nations Framework Convention on Climate Change (UNFCCC) (1992), available at https://unfccc.int/files/essential_background/background_publications_htmlpdf/application/pdf/conveng.pdf (accessed 25 January 2016).

2 Timothy Cadman, "Introduction: Global Governance and Climate Change," in *Climate Change and Global Policy Regimes: Towards Institutional Legitimacy*, ed. Timothy Cadman (London: Palgrave Macmillan, 2013), pp. 1–2.

3 Responsible for 24 per cent of emissions (GHG).

4 Rajendra Pachauri, "Chairperson of the Intergovernmental Panel on Climate Change, Presentation to the Sixth International Forum for Sustainable Asia and the Pacific (ISAP)," in *Sixth International Forum for Sustainable Asia and the Pacific*, ed. Institute for Global Environmental Strategies (Yokohama: ISAP (Reproduced with permission), 2014).

negotiations for the 'protocol, another legal instrument or an agreed outcome with legal force', now referred to in the wake of the 2015 climate negotiations as the Paris Agreement, which largely replaces the Kyoto Protocol.[5] However, like most other 'wicked' problems (global or otherwise), legal instruments are only part of the answer. Lawyers may emphasize the formal agreements, but ethicists emphasize the importance of values and ethical standards, economists emphasize the importance of economic incentives and the absence of perverse incentives and political scientists emphasize the importance of institutions – while some insist that a combination of those measures at international, regional, national, sub-national, corporate and professional levels form a global 'ethics regime' or an 'integrity system'.[6]

Understanding these laws, values and institutions and the way they can effectively interact to secure the necessary reduction in GHGs requires a great deal of research – not least in exploring the dimensions of governance quality in climate mechanisms.[7] Governance, rather than government, is a more useful framework for understanding the political and social interactions that occur in the intergovernmental realm and the related contexts of state, society and the market. It acknowledges the importance of state and non-state actors' relationships within contemporary systems of global policymaking, and the structures and processes which underpin them.[8]

In the light of these observations, this chapter recognizes that structures and processes are fundamental to understanding the integrity of contemporary governance and, ultimately, institutional legitimacy. Legitimacy is a core analytical preoccupation for governance scholars, and its study is still evolving.[9] Legitimacy, we suggest, is the end point of activity within an institution. The successful interaction between the structural and procedural components of an institution's governance system (the rules, standards, etc.) plays an important role in the institution's longer-term behavioural and problem-solving impacts. To put it another way, an institution has integrity when it structures itself to fulfil the values it claims to further in its Public Institutional Justification (see section three). This view of legitimacy incorporates

5 United Nations Framework Convention on Climate Change (UNFCCC), "Report of the Conference of the Parties on Its Seventeenth Session, Held in Durban from 28 November to 11 December 2011" (Durban: UN, 2012).

6 See Charles Sampford, "Adam Smith's Dinner," in *The Future of Financial Regulation*, ed. Iain Macneil and Justin O'Brien (Oxford, UK: Hart Publishing, 2010); Charles Sampford, "Towards a Global Carbon Integrity System: Learning from the Gfc1 and Avoiding a GCC2," *Low Carbon Economy* 2, no. 4 (2011).

7 Esteve Corbera and Heike Schroeder, "Governing and Implementing Redd+," *Environmental Science & Policy* 14, no. 2 (2011).

8 Jan Kooiman, "Societal Governance: Levels, Models, and Orders of Social-Political Interaction," in *Debating Governance: Authority, Steering and Democracy*, ed. Jon Pierre (Oxford: Oxford University Press, 2000).

9 Frank Biermann and Aarti Gupta, "Accountability and Legitimacy in Earth System Governance: A Research Framework," *Ecological Economics* 70, no. 11 (2011).

the views of a number of existing governance theorists and is discussed in more detail below.[10]

This chapter consequently explains the conceptual governance framework that will underpin the collection. Section one introduces some key conceptual issues, describing governance as a value-neutral term that can apply to institutions, governments or complex networks. It distinguishes governance values – applying to mechanisms of steering and coordinating – from substantive 'governing values' more generally. Section two outlines a continuum of governance values, ranging from Thin Governance Values (covering only the most basic oversight arrangements) all the way through to Thick Governance Values (capturing all the morally desirable qualities able to be possessed by internal steering and coordinating processes). The second section reflects on the nuanced relationship between governance values and institutional integrity, before the final section – in line with the collection's area of focus – considers the main links between governance values and environmental values.

Section one: Conceptual issues

Governance as a value-neutral term

Controversy plagues attempts to define 'governance'.[11] Even within the single discipline of political science, theorists have observed the term's shifting meaning.[12] Some theorists view governance as referring to a *certain type of regime-complex*, namely a decentralized, fragmented, non-hierarchical, dynamic, cross-sectoral network with complex and smeared vertical and horizontal boundaries.[13] Others conceive the term to connote a *certain type of policy mechanism* – new, flexible, non-state instruments, as contrasted with traditional state command-and-control regulation.[14] Still others understand the term as a *larger rubric*, capturing all

10 See, for example, Fritz W. Scharpf, *Games Real Actors Play: Actor-Centered Institutionalism in Policy Research* (Boulder: Westview Press, 1997); Timothy Cadman, *Quality and Legitimacy of Global Governance* (London: Palgrave Macmillan, 2011); Vivien A. Schmidt, "Democracy and Legitimacy in the European Union Revisited: Input, Output and 'Throughput'," *Political Studies* 61 (2012).

11 See, for example, Economic and Social Council (ECOSOC), "Definition of Basic Concepts and Terminologies in Governance and Public Administration," (New York: United Nations, 2006), pp. 2–3; Marc F. Plattner, "Reflections on 'Governance'," *Journal of Democracy* 24, no. 4 (2013).

12 Andrew Jordan, Rüdiger K.W. Wurzel, and Anthony Zito, "The Rise of 'New' Policy Instruments in Comparative Perspective: Has Governance Eclipsed Government?" *Political Studies* 53 (2005): pp. 479–80. See also Plattner, "Reflections on 'Governance'."

13 See Economic and Social Council (ECOSOC), "Definition of Basic Concepts," p. 2; Andrew Crane, "From Governance to Governance: On Blurring Boundaries," *Journal of Business Ethics* 94 (2010): p. 18; Aaron Wachhaus, "Governance Beyond Government," *Administration & Society* 46, no. 5 (2014): pp. 574–77.

14 See, e.g., Jordan, Wurzel, and Zito, "Has Governance Eclipsed Government?"

aspects of steering and regulating – whether by state or non-state actors, or tangled institutional networks.

In this chapter we focus on the specific conception of 'governance' that appears when expert and lay discourse speaks of its normative aspects in phrases like 'good governance', 'governance values' and 'governance failures'. Such expressions apply to governments, institutions, organizations and networks alike. This usage of the term 'governance' accords with dictionary definitions, which define the term simply as '[t]he action or manner of governing a state, organization, etc.'[15]

We thus understand governance as referring to the *structures and processes used to steer and coordinate interactions within an institution or institutional complex (or in the case of UNFCCC a regime complex)*. We use the term 'regime complex' as an extension of existing regime theory in international relations to describe the various institutions and sub-institutions that make up the climate change policy regime (UNFCCC), co-opting terminology originally developed by Keohane and Victor,[16] but reflecting the approach to global integrity systems developed by Sampford.[17]

We also apply the term to the intersection between UNFCCC and the broader political economy of sustainable development, including the Conventions on Biological Diversity (CBD) and Combatting Desertification (CCD); as well as other conventions such as those covering wetlands (Ramsar); as well as health, migration and food availability – all of which are affected both by the climate regime and climate change itself.[18] These intersections are important, as they can result in trade-offs across policy regimes that can positively or negatively impact

15 http://www.oxforddictionaries.com/definition/english/governance (accessed 27 May 2015). See also the term's etymology at http://www.etymonline.com/index.php?term=govern&allowed_in_frame=0 (accessed 27 May 2015). A similar notion is captured by Thakur and Weiss in their work on global governance: ' "Governance" is the sum of laws, norms, policies, and institutions that define, constitute, and mediate relations among citizens, society, market, and the state . . .' Thomas G. Weiss and Ramesh Thakur, *Global Governance and the UN: An Unfinished Journey* (Bloomington: Indiana University Press, 2010), pp. 5–6. See similarly Wachhaus, "Governance Beyond Government," p. 574.

16 Robert O. Keohan and David G. Victor, "The Regime Complex for Climate Change." *Perspectives on Politics* 9, no. 1 (2011): pp. 7–23.

17 See Charles Sampford, "Law, Institutions and the Public Private Divide," *Federal Law Review* 185 (1990, 1992); Charles Sampford, "Global Transparency: Fighting Corruption for a Sustainable Future: From National Integrity Systems to Global Integrity Systems," in *TI Commissioned Conference Overview Discussion Paper for 13 IACC* (Athens, 2008); Charles Sampford, "Global Lessons from Fitzgerald: From State and National to Global Integrity Systems," in *The Fitzgerald Legacy: Reforming Public Life in Australia and Beyond*, ed. Colleen Lewis, Janet Ransley, and Ross Homel (Brisbane: Australian Academic Press, 2010) for this applied state-based approach to ethics regimes and national integrity systems (Jeremy Pope's term on which he and Sampford worked). See application to global issues: Sampford, "Global Transparency: Fighting Corruption for a Sustainable Future: From National Integrity Systems to Global Integrity Systems"; Sampford, "Global Lessons from Fitzgerald: From State and National to Global Integrity Systems."

18 Cadman, "Introduction: Global Governance and Climate Change."

governance.[19] This leads into a broader discussion of the institutional integrity of the climate regime, notably issues of context, consistency and coherence, which we discuss in a previous volume.[20]

Understood in this way, governance focuses on what we might call an institution's 'internal organizational arrangements' regarding how the institution identifies its constituencies, makes decisions and encourages or polices compliance with those decisions; and how information, critical feedback and authority flow within the institution and outward into the public realm.[21] Colloquially, governance refers to 'how things are done' in a given institution or complex.[22]

'Governance arrangements' thus refer to any and all instruments and practices organizations employ to facilitate steering and coordination.[23] On this footing, 'governance' emerges as a value-neutral term. It thus differs from concepts like 'justice' or 'integrity',[24] which carry an intrinsic moral loading. Instead, the normative status applying to governance arises only when we explicitly include value-laden terms, such as when we refer to 'good' and 'bad' governance, to governance 'principles' or 'values' or to governance 'failures'.

Governance can apply to discrete institutions, politics or complex institutional networks

Understood in this fashion, governance can apply to a myriad of different sorts of entities, from discrete institutions (corporations, NGOs, community groups, sporting clubs), to nation states and other political entities wielding great power,[25] to institutional regimes and more amorphous complexes (so that we might speak of

19 Jamie Pittock, "Climate Change and Sustainable Water Management," in *Climate Change and Global Policy Regimes*, ed. Tim Cadman (London: Palgrave Macmillan, 2013).

20 Hugh Breakey and Timothy Cadman, "A Comprehensive Framework for Evaluating the Integrity of the Climate Regime Complex," in *Ethical Values and the Integrity of the Climate Change Regime*, ed. Hugh Breakey, Vesselin Popovski and Rowena Maguire (Surrey: Ashgate, 2015).

21 See Hugh Breakey, Timothy Cadman, and Charles Sampford, "Conceptualizing Personal and Institutional Integrity: The Comprehensive Integrity Framework," *Research in Ethical Issues in Organizations* 14 (2015).

22 On this approach, governance emerges as a broader concept than government. See Economic and Social Council (ECOSOC), "Definition of Basic Concepts," p. 3.

23 The term has been similarly used by other scholars to refer to a range of specific mechanisms influencing 'the interaction between various actors pursuing common goals'. See Mathias Koenig-Archibugi, "Introduction: Institutional Diversity in Global Governance," in *New Modes of Governance in the Global System: Exploring Publicness, Delegation and Inclusiveness*, ed. Mathias Koenig-Archibugi and Michael Zürn (Basingstoke and London: Palgrave Macmillan, 2006), p. 24, footnote 13.

24 Discussed in section three.

25 Weber saw states as enjoying a monopoly on the legitimate use of force, but this is either palpably false for all real states or trivially true by defining legitimate power as state power – see Charles Sampford, "Legality and Legitimacy: A Dozen Years after Goldstone," in *Responsibility to Protect and Sovereignty*, ed. Charles Sampford and Ramesh Thakur (Farnham: Ashgate, 2013).

'global governance' in general terms). For all of these entities we can enquire into what capabilities for steering and coordinating each possesses or can possess. In what follows, for expository ease, we will refer to the governance of *institutions*, but this term should be understood to also cover institutional regimes and complexes.

We noted earlier that some political theorists reserve the term governance for decentralized, non-state networks, thereby distinguishing it from *government* studies which apply to state agencies. So understood, government studies centre on traditional state authority and command-and-control regulatory policies.[26] While recognizing that the distinction is useful for some, we adopt the broader sense of governance covering state agencies, corporations, professions and NGOs. This allows us to recognize the continuities between governance challenges and solutions for different kinds of institutions and to avoid some of the false dichotomies drawn between state and non-state actors and between 'public' and 'private' institutions.[27] The broader term of governance can capture styles, qualities and mechanisms of steering and coordination that depart from such traditional hierarchies – not least within state agencies that were previously imagined in those hierarchical forms. Since non-hierarchical, cross-sectoral networks are unlikely to manifest the full scope of virtues and outputs we might demand of traditional governments, concerns for good governance – rather than good government – take centre stage.[28]

The fact that concepts like 'good governance' can apply to different sorts of entities (institutions, professions, polities, networks) does not mean that the term's substance remains stable as we shift from one entity to the next. To the contrary, what a single thinker means by 'governance values' may shift in content and priority as we move from speaking of the governance of a corporation to an institution of government and then again to a global regime such as the UNFCCC. This can reflect the fact that some values are more important to some institutions than are others.[29] For example, the governance values of democratic

26 Even so, so-called 'new', 'soft' governance mechanisms often involve some state involvement (Jordan, Wurzel, and Zito, "Has Governance Eclipsed Government?"). More broadly, while authoritative command and coercive enforcement of legislated rules are the hallmark of state activity, states routinely engage in subtler forms of steering – and are themselves subject to such steering, for example through the separation of powers. See Hugh Breakey, "Dividing to Conquer: Employing the Separation of Powers to Structure Institutional Inter-Relations," *Research in Ethical Issues in Organizations* 12 (2014).

27 See Sampford, "Law, Institutions and the Public Private Divide." Indeed, Sampford wrote about corporate governance and professional governance before becoming involved in public governance reform in Queensland in 1991. See Charles Sampford and David Wood, "The Future of Business Ethics? Legal Regulation, Ethical Standard Setting and Institutional Design," *Griffith Law Review* 1, no. 1 (1992).

28 Note Weiss and Thakur: 'In the domestic context . . . governance is government plus the additional mechanisms required to ensure order and predictability in problem-solving. For the planet, however, governance is the whole story because there is no central authority'. Weiss and Thakur, *Global Governance and the UN*, p. 29.

29 Breakey, "Dividing to Conquer," pp. 38–39.

inclusiveness are more important in legislative bodies compared with its role in the military or courts (juries are not a 'democratic' initiative but a critical check and process on attributing criminal guilt). Of course, the degree to which declared governance values are built into the governance of the relevant institution is another variable. Those that do can be seen as practising 'values-based governance'[30] and at least one sense of 'integrity'.[31]

A further important consideration that should also be taken into account is the view amongst some governance scholars that greater attention should be paid to evaluating the success of climate change policy on the basis of the social processes by which decisions are made.[32] Social-political dynamics play a significant role in the climate negotiations. It has been contended, for example, that the more powerful the country, the greater influence that country has on the policy positions adopted, resulting in problems of (in)justice and (in)equality.[33] Historical conflicts between developed and developing countries, dating back to the original 1992 Rio Conference, have been identified as one of the contributing factors to what has been referred to as institutional ossification within the climate regime.[34] Climate change governance has also been described as a prime example of arrested development.[35]

Governance values versus 'governing values'

Governance values enhance the institution's legitimacy by capturing qualities that improve – from a moral and social perspective – the way the institution performs its *steering and coordinating* efforts.

Because of its focus on steering and coordinating, governance values need not capture everything one might morally want from an institution. For example, one might (for excellent moral and social reasons) believe an institution should cleave to substantive moral values such as promoting human rights and nurturing environmental value. Rather than evaluating *who* was involved and *how* the process

30 See, Sampford, "Law, Institutions and the Public Private Divide."

31 Where institutions ask hard questions about their values, give honest and public answers and live by them: see, Noel Preston and Charles Sampford, "Institutionalising Ethics," in *Encouraging Ethics and Challenging Corruption*, ed. Charles Sampford, Noel Preston and Carmel Connors (Annandale: Federation Press, 2002).

32 Jon Barnett, "Adapting to Climate Change: Three Key Challenges for Research and Policy – an Editorial Essay," *Wiley Interdisciplinary Reviews: Climate Change* 1, no. 3 (2010).

33 Chukwumerije Okereke, "Climate Justice and the International Regime." *Wiley Interdisciplinary Reviews: Climate Change* 1, no. 3 (2010): pp. 462–74.

34 Timothy Cadman, Eastwood Lauren, Federico Lopez-Casero Michaelis, Tek Narayan Maraseni, Jamie Pittock, and Tapan Sarker, *The Political Economy of Sustainable Development: Policy Instruments and Market Mechanisms* (Cheltenham, UK: Edward Elgar Publishing, 2015); Joanna Depledge, "The Opposite of Learning: Ossification in the Climate Change Regime," *Global Environmental Politics* 6, no. 1 (2006).

35 Oran R. Young, *Institutional Dynamics: Emergent Patterns in International Environmental Governance* (Cambridge, MA: MIT Press, 2010).

occurred, substantive values open the question of what ethical goals the institution has sought to further. A combination of good substantive and good governance values increases the likelihood that the *right* decisions were made and implemented, producing *just* or *good* outcomes.[36]

In the sense that one might want a given system of governance – such as the global climate regime – to pursue these sort of substantive values, one might term them 'governance values'.[37] Sampford provides a good example of this approach. He lists the governance values for the global regime as including rule of law, liberty, equality, fraternity (citizenship in sovereign states), human rights, democracy and the environment.[38]

For our purposes here we distinguish these values from our understanding of governance values, as they extend beyond the way the institution conducts its steering and coordinating efforts and into the rich, substantive values that can shape decision-making. Leo Huberts marks this distinction by separating the *ethics* of the governance process and the *results* of the process.[39] Governance is about the *how*, not the *why* or the *what*. Thus, we will speak of these types of substantive values as 'governing values' – as they govern the goals and

36 Note that the distinction between procedural and substantive values differs from the distinction between instrumental and intrinsic values. 'Instrumental' values are those that are valuable only insofar as they enhance another sort of value. 'Intrinsic' values are valuable in and of themselves, irrespective of their effects. Procedural values often possess instrumental value, as they can enhance and illuminate other types of values – for example, transparency encourages decisions to be made on a more public-spirited basis. (For an illustrative discussion of the instrumental value of transparency in the EU context, see Deirdre Curtin and Albert Jacob Meijer, "Does Transparency Strengthen Legitimacy? A Critical Analysis of European Union Policy Documents," *Information Polity* 11 (2006). However, procedural values (and governance values more generally) can also carry intrinsic value: treating people with due process and non-discrimination under the rule of law respects their rights and so is intrinsically valuable.

37 Applied to polities and global regimes, Crane terms this 'Big-G' governance as 'how societal goals are set, what mechanisms are put in place to achieve them, and how these are monitored and enforced . . .' Crane, "From Governance to Governance," p. 17. Crane contrasts this with 'small-g' governance, which maps onto our thinner versions of governance values (thin and targeted).

38 Charles Sampford, "Sovereignty and Intervention," in *Human Rights in Theory and Practice*, ed. Tom Campbell and Burton M. Leiser (London: Ashgate, 2001). Naturally, different institutions would attract different moral expectations, so corporations would be required to live up to a different list of moral demands (though they should still be seen as having responsibilities for furthering human rights) – see Charles Sampford, "Business and Human Rights: Reflections and Observations," in *Business and Human Rights*, ed. Wesley Cragg (Cheltenham: Edward Elgar, 2012).

39 Leo Huberts, *The Integrity of Governance: What It Is, What We Know, What Is Done, and Where to Go* (Hampshire: Palgrave Macmillan, 2014). Huberts's focus is on compliance with relevant rules and codes. Which rules are 'relevant' is set by the general public. This makes his account closer to what we will later term Thick Governance Values, albeit that he does not rely on a set list but allows the content to shift according to the local community's values. This aligns with the approach in Preston and Sampford, "Institutionalising Ethics."

priorities of executive decision-making. In contradistinction, we will reserve the term 'governance values' (our focus in this chapter) for values that bear directly on the question of how the institution is steered and its collective action coordinated.[40]

Because they refer to everything that can morally be desired of a given institution, governing values differ conceptually and normatively from both integrity and governance values. However, there are overlaps, in the sense of there being cases when the qualities required by morality-based governing values parallel those demanded by governance values aimed at delivering institutional legitimacy. Often this overlap will be contingent and instrumental, inasmuch as governance values will prove a crucial resource for the sorts of 'higher-order' outcomes that an institution's governing values might aspire to. For example, improved participation (through inclusiveness of interests represented in a given policy community) and deliberation (through democratic decision-making of those interests) are governance values that will increase legitimacy – and thereby enhance an institution's ability to achieve the morally desired goals of sustainable development and poverty alleviation (governing values under Article 1.1 of the Paris Agreement).

In other cases, the overlap between governing and governance values can be tighter, especially when the former explicitly includes procedural requirements, impacting an institution's internal organizational arrangements. A good example is human rights. As well as setting down desirable policy outcomes (that is, the promotion and protection of key freedoms and human interests), the governing value of human rights also lays down procedural constraints on how state-institutions can pursue these goals, such as requirements to follow due process, the rule of law and so on. In this way, for certain sorts of organizations (namely, state institutions) the governing values of human rights may dictate requirements that overlap with governance values. An example would be that of free, prior and informed consent (FPIC), which requires the application of such governance values as inclusiveness, accountability and transparency. While the governing value in this case (FPIC) remains different conceptually and normatively from governance values, there are areas of direct overlap as to how it is operationalized. Another example is provided by the courts. The rule of law is a central governance value about how courts should (and in most countries hopefully do) operate, but it is also a substantive value that they seek to further through their judgements and the imposition of the rule of law on government agencies and other actors.

In the context of climate change and the UNFCCC, these 'governing values' also include substantive claims about climate justice and equity. Of course, some accounts of equity include procedural values focused on decision-making

40 That said, the two sorts of values do interrelate. Arguably, Thick Governance Values (see section two of this chapter) are a key way of facilitating – and perhaps even in some cases a necessary condition for – the types of rich governing values of which Sampford speaks.

processes.[41] By including concern for participation, transparency, accountability and deliberation, such 'procedural equity' would lay down Governance Values (rather than substantive governing values).[42] However, most accounts of climate justice, and the usual meanings put forward for climate equity as it emerges in agreements and decisions under the UNFCCC, focus on substantive outcomes and distributive burden-sharing rather than on decision-making processes. For example, staple meanings given to climate equity include intergenerational justice, intra-generational justice, need-based entitlements, capacity-based burden-sharing and historical responsibility – all principles for fair allocations of resources rather than procedural values.[43] Even theorists aiming to highlight the importance of procedural equity (and therefore of 'Governance Values') stress how much the existing understanding of equity hones in on substantive rather than procedural justice.[44]

In summary, sustainable development, poverty alleviation, climate equity and climate justice usually refer to substantive principles for distributing entitlements and burdens. As governing values, these principles can be distinguished from the three different types of governance values outlined in the following section.

41 See Jonathan Pickering, "Top-Down Proposals for Sharing the Global Climate Policy Effort Fairly: Lost in Translation in a Bottom-up World?" in *Ethical Values and the Integrity of the Climate Change Regime*, ed. Hugh Breakey, Vesselin Popovski, and Rowena Maguire (Surrey: Ashgate, 2015).

42 This account of 'procedural equity' is drawn from ibid., pp. 94–99.

43 See, for example, Henry Shue, *Climate Justice: Vulnerability and Protection* (Oxford: Oxford University Press, 2014); Stephen M. Gardiner, Simon Caney, Dale Jamieson, and Henry Shue, eds., *Climate Ethics: Essential Readings* (Oxford: Oxford University Press, 2010); Rowena Maguire, "Incorporating International Environmental Legal Principles into Future Climate Change Instruments," *Carbon and Climate Law Review* 6, no. 2 (2012); Aaditya Mattoo and Arvind Subramanian, "Equity in Climate Change: An Analytical Review," *World Development* 40, no. 6 (2012); Jonathan Pickering, Steve Vanderheiden, and Seumas Miller, "'If Equity's in, We're Out': Scope for Fairness in the Next Global Climate Agreement," *Ethics & International Affairs* 26, no. 4 (2012); International Law Association (ILA), "Legal Principles Relating to Climate Change," in *Report of the International Law Association's Committee on Legal Principles Relating to Climate Change*, ed. Shinya Murase (Washington: ILA, 2014). Intergovernmental Panel on Climate Change (IPCC), "Fifth Assessment Report – Synthesis Report," in *Climate Change 2014: Synthesis Report. Contribution of Working Groups I, II and III to the Fifth Assessment Report of the Intergovernmental Panel on Climate Change*, ed. R.K. Pachauri and L.A. Meyer (Geneva: IPCC, 2014), 151 pp. Lange, Vogt and Ziegler note 'process-based' criteria in their discussion of equity, only to set such criteria aside and focus on 'allocation-based' and 'outcome-based' criteria (Andreas Lange, Carsten Vogt, and Andreas Ziegler, "On the Importance of Equity in International Climate Policy: An Empirical Analysis," *Energy Economics* 29 (2007): p. 547). This focus on substantive/distributive justice aligns with the term's use in the Framework Convention itself, as well as its most recent employment in the Paris Agreement and Decision. See United Nations Framework Convention on Climate Change (UNFCCC), "United Nations Framework Convention on Climate Change," (Rio: United Nations, 1992), Art.3(1); "Conference of the Parties: COP21 Paris Agreement," (FCCC/CP/2015/L.9: UNFCCC, 2015).

44 See, for example, Pickering, "Top-Down Proposals."; Idil Boran, "Principles of Public Reason in the UNFCCC: Rethinking the Equity Framework," *Science and Engineering Ethics* 22, no. 4 (2016): pp. 1–19.

Section two: Thin, Thickish, Thick Governance Values

This section plots different notions of, and institutional instantiations of, Governance Values on a continuum from Thin Governance Values, through 'Thickish' Governance Values, up to the most inclusive and comprehensive 'Thick' Governance Values. We use the terms 'thick' and 'thin' in their figurative meanings,[45] where 'Thin' Governance Values connote a limited and basic (though tightly focused) set of specific values, while 'Thickish' and 'Thick' Governance Values expand the richness and substance of the values covered.[46]

Thin Governance Values: Minimum conditions for basic oversight

Thin Governance Values refer to specific aspects of the structures and processes that determine how things are done within an institution. In particular, they refer to various qualities that tend to ensure reliability and basic oversight. These Thin Governance Values include transparency, accountability and compliance. Thin Governance Values tend to be the focus of staple governance mechanisms such as accountability measures, compliance officers, reporting duties and regulatory bodies.

Governance values can also include a concern for *efficacy*: that is, the impact of the institution as it pursues its goals.[47] Many would agree that efficacy comprises one part of ordinary notions of governance.[48] It would indeed seem odd to say, 'This institution has good governance, but it rarely gets the job done.' Equally, the classic governance activities of 'steering' and 'coordinating' both suggest concern for the institution putting agreements or decisions into practice, making efficacy central to attributions of good governance. Efficacy can itself be understood in thicker and thinner ways. A thin understanding of efficacy, which we can place under the rubric of Thin Governance Values, amounts to a narrow concern for at least *avoiding gridlock*. This value looks to whether the institution is internally paralyzed, either because it is incapable of making decisions or incapable of acting on those decisions in a timely and productive manner. A gridlocked institution suffers from fatal governance flaws. In the areas where it is gridlocked, the institution cannot be steered at all.

45 Merriam-Webster's dictionary defines *thin* as (inter alia) 'lacking substance or strength', and *thick* as including 'having great depth or extent'.

46 Noting that, just as in Rawls 'thick' 'conceptions' of justice, the thicker conceptions are, the more controversial and potentially culturally, temporally and institutionally specific they may be.

47 Efficacy is not a procedural value, as it refers to *whether* things get done and not *how* they get done. But nor is efficacy a substantive value in the usual sense, as it does not refer to a particular quality that is morally or politically desirable, but rather to whether the institution or network succeeds in making decisions, implementing its policies and realizing its objectives (whatever these objectives are).

48 See, for example, Plattner, "Reflections on 'Governance'," pp. 20–23. Notions of efficiency, effectiveness, competence, implementation and service-delivery permeate several of the definitions surveyed in Economic and Social Council (ECOSOC), "Definition of Basic Concepts.", p. 4.

A concern for Thin Governance Values often captures what people have in mind when they say that an institution needs 'better governance' or that there have been 'governance failures'.[49] Such complaints typically refer to the need for improved (internal or external) regulation and accountability and other measures to ensure compliance with rules and codes.

Thickish Governance Values: Functional realism

Thickish Governance Values expand the portfolio of qualities regarding how the institution is steered and how its various elements are coordinated. Alongside (thin) staples such as accountability, transparency and compliance, Thickish Governance Values adds elements of deliberativeness and democratic decision-making. Moving beyond mere accountability, Thickish Governance Values incorporates qualities regarding how the institution goes about making its decisions and whom it involves in this process. So, too, more robust concerns for efficacy emerge as the governance values thicken. Rather than a concern for simply avoiding internal gridlock, Thickish Governance Values can include reference to whether the institution is *functional.* This efficacy-quality looks to whether the institution's internal organization arrangements are effective in getting things done *given what is realistically possible in terms of the institution's actual powers, resources and limitations.* We can say an institution is functional (or competent) if it does all that it realistically can do in terms of making and implementing decisions – even if it does not significantly impact the final result because of limits on its raw strength. Here we could point to the 'REDD+' mechanism under the Convention, which uses various 'safeguards' to ensure the thin values of accountability and transparency as well as thicker values such as inclusiveness.[50] However, more challenging values, such as democracy, which implies non-state actors have decision-making agency on par with sovereign states, are largely absent in most sub-institutional components under the initiative.[51]

Vivien Schmidt's notion of 'throughput legitimacy' presents an example of Thickish Governance Values.[52] Schmidt distinguishes 'throughput legitimacy'

49 For example, Justin O'Brien and Olivia Dixon, "The Common Link in Failures and Scandals at the World's Leading Banks," *Seattle University Law Review* 36 (2013): p. 941.

50 We note that inclusiveness has two dimensions – inclusiveness of external stakeholders and the inclusiveness of members of the institution. The key tasks of asking questions about the institution, giving honest and public answers and living up to those values can generally only be achieved with the active involvement of members of the institution at all levels (see Charles Sampford, "Institutionalising Public Sector Ethics," in *Ethics for the Public Sector: Education and Training*, ed. N. Preston (Sydney: Federation Press, 1994)).

51 Timothy Cadman, Tek Narayan Maraseni, Hwan Ok Ma, and Federico Lopez-Casero, "Five Years of Redd+ Governance: The Use of Market Mechanisms as a Response to Anthropogenic Climate Change," *Forest Policy and Economics* 4 (2016), http://dx.doi.org/10.1016/j.forpol.2016.03.008. REDD+ is an acronym for: Reducing emissions from deforestation and forest degradation and the role of conservation, sustainable management of forests and enhancement of forest carbon stocks in developing countries.

52 Schmidt, "Input, Output and 'Throughput'."

from input and output legitimacy. While output legitimacy connotes government *for* the people, and input legitimacy stems from government *by* the people, throughput legitimacy facilitates government *with* the people, focusing on the efficacy, accountability and transparency of an institution's governance.[53] Throughput legitimacy hones in on the institution's internal organizational arrangements – both in terms of who is represented and how deliberation occurs.[54]

Schmidt's account of throughput legitimacy therefore extends beyond Thin Governance Values, as it includes aspects of inclusiveness/openness/democracy, deliberativeness/discursive qualities and efficacy/productiveness. However, Schmidt's list remains thinner than the full scope of Thick Governance Values, to which we now turn.[55]

Thick Governance Values: Governance for legitimacy

Thick Governance Values capture the full gamut of social and moral qualities that can be demanded of an institution's mechanisms for steering and coordinating. If an institution possesses all of these qualities, then (while this may not cover everything morality might demand of the institution)[56] the institution will possess legitimacy in the ways it goes about making, authorizing, implementing and appraising its decisions. Thick Governance Values thus constitute a plausible answer to the question of what sorts of internal organizational arrangements would make institutions and institutional complexes legitimate.[57] Institutions and networks gain in legitimacy the more that their actions are transparent and accountable, the more that they employ constructive deliberation, the more that they provide affected stakeholders with the capacity to provide input into their workings and the more that they produce effective results.

In this way Thick Governance Values expand the portfolio of required qualities to cover structural and process values regarding who participates in the governance systems and how decisions are made and implemented. Examples of thick governance values, such as the UNDP Principles and the Cadman Framework, stretch beyond the 'usual suspects' of transparency, accountability and rule of law to cover enhanced criteria of participation (including the inclusiveness of all

53 Ibid., p. 2.
54 Ibid., p. 5.
55 For example, Cadman incorporates factors impacting input and output legitimacy. As Schmidt focuses on distinguishing throughput legitimacy from input and output legitimacy, she employs narrower interpretations of her notions of (for example) inclusiveness and democracy.
56 As noted earlier, governance values focus on the arrangements for steering and coordinating. While Thick Governance Values constitute the most we could ask, out of concern for institutional legitimacy, they do not constitute the most we could ask for the institution as a whole. As section one observed, we can have moral concerns (characterized as 'governing values') about the institution's substantive values and goals distinct from moral concerns about its methods of steering and coordinating.
57 Allen Buchanan and Robert O. Keohane, "The Legitimacy of Global Governance Institutions," *Ethics and International Affairs* 20, no. 4 (2006).

stakeholders and equality between them) and deliberation (including the mode of decision-making: democratic).[58] Thick Governance Values also include stronger demands for efficacy. As well as an institution being effective in the sense of avoiding gridlock (from Thin Governance Values) and being functional (from Thickish Governance Values), Thick Governance Values will look to whether the institution is *effective*. An institution is effective if it succeeds in performing its stated function – if it impacts its target problem in a substantial and meaningful way and is the best realistic alternative for doing so.

To provide one illustrative example of Thick Governance Values, Cadman identifies two governance principles and then incorporates an array of criteria and indicators specifying the principles. See Table 2.1 below.[59]

These values remain the same as we shift from one institution (or institutional complex) to the next. Hence Thick Governance Values comprise a generally settled and sufficiently agreed list.[60] Of course, it may be that different types of institutions can and should implement these values in different ways, with some of the values prioritized by some institutions more than others: for example, democratic participation would be implemented in different ways by courts, armed forces, corporations and community groups. Similarly, a failure to be democratic might not be a problem for law courts making decisions about criminal guilt, but could prove fatal to a community group. In other words, while the governance values present in the objective list remain stable, the way that they instantiate, and their importance in certain contexts, could shift from one type of institution to the next. Table 2.2 below provides a list of the staple members of each type of Governance Values, illustrating how the thicker types encompass and add to the thinner ones.

Scholars have made a number of observations regarding the governance quality of the climate regime in the years leading up to the end of Kyoto and the new arrangement (now referred to as the Paris Agreement). Given the role of market mechanisms under the protocol, it was acknowledged that there had certainly been an increase in cooperative governance structures between governmental agencies and non-state actors over the years, even if state actors remained the ultimate decision makers.[61] Institutional arrangements outside the formal negotiations increasingly favoured 'partnerships', employing network-type models of

58 See also the accounts described in the ECOSOC Note, most of which would be characterized as 'Thick Governance Value' approaches: Economic and Social Council (ECOSOC), "Definition of Basic Concepts," 4. For another example of Thick Governance Values: Weiss and Thakur see good governance as incorporating: '"Good governance" incorporates peoples' participation and empowerment with respect to public policies, choices, and offices; the rule of law and an independent judiciary . . . ; and standards of probity and incorruptibility, transparency, accountability, and responsibility.' Weiss and Thakur, *Global Governance and the UN*, p. 6.

59 Cadman, *Quality and Legitimacy of Global Governance*, p. 17. The Cadman Framework was developed in the context of global forest policy post Rio (1992), and has also been applied to the UNFCCC.

60 For a fuller exploration see ibid., pp. 1–18.

61 Liliana B. Andonova, Michele M. Betsill, and Harriet Bulkeley, "Transnational Climate Governance," *Global Environmental Politics* 9, no. 2 (2009): pp. 57–58.

Table 2.1 Principles, criteria and indicators in the Cadman framework

Principle	*Criterion*	*Indicator*
Meaningful Participation	Interest Representation	Inclusiveness Equality Resources (capacity building)
	Organizational responsibility	Accountability Transparency
Productive Deliberation	Decision-making	Democracy Agreement Dispute settlement
	Implementation	Behavioural change Problem-solving Durability

Source: Cadman, 2011 (reproduced courtesy of Palgrave Macmillan).

Table 2.2 This table expresses in an indicative way the expansion of governance values, from Thin, to Thickish, to fully Thick

Types of factors	*Thin Governance Values*	*Thickish Governance Values*	*Thick Governance Values*
Procedural and/ or Structural Factors	Accountability	Accountability	Accountability
	Transparency	Transparency	Transparency
	Compliance	Compliance	Compliance
	Rule of law	Rule of law	Rule of law
		Deliberative	Deliberative
		Democratic	Democratic
		Degrees of capacity building	Extensive capacity building
		Inclusion of some non-state actors	Inclusion of all stakeholders
			Stakeholders' equality in decision-making
			Strategies for dispute settlement
Efficacy Factors	Avoid gridlock	Avoid gridlock	Avoid gridlock
		Functional	Functional
			Effective and problem-solving

governance.[62] But within the regime itself, it was argued that there were very few clearly articulated policies on how non-state actors could participate in the negotiations under existing rules and procedures.[63] Further concerns were also expressed regarding the divergent capabilities between countries, as well as the ineffectiveness and inefficiencies of climate funding mechanisms.[64] Questions were also raised about the legitimacy and effectiveness of the climate regime in view of the ever-rising numbers of non-state actors actively governing a range of initiatives beyond the traditional UNFCCC institutions.[65] Some commentators pointed to an increasing lack of clarity around national as opposed to non-state sovereignty, as well as how sustainable development was being put into practice under the regime, noting conflicts between national and institutional priorities, the demands of the market and the extent to which non-state stakeholders were (or were not) involved in different initiatives.[66] This led to calls for greater consistency across formal and informal systems, and the recognition of non-state stakeholders directly and indirectly impacted by climate change based on the principles of social justice and human rights.[67] Governments appear to have listened to some of these calls, as the Paris Agreement has specific provisions regarding the role of 'non-Party stakeholders' in reducing global emissions.[68] But if such provisions are to be effective, non-state actors will require a role in decision-making. They are still formally excluded from playing any part in reaching agreements under the Convention. If a 'call' were to be made on the current thickness of the governance values underpinning the regime, we estimate somewhere between 'thin' and 'thickish' – but not 'thick'.

Section three: Relation between governance values and institutional integrity

Governance values interrelate with concerns about institutional integrity. In some cases the terms are used almost interchangeably, with one commentator describing governance problems, where another would speak of a breakdown in

62 Karin Bäckstrand, "Accountability of Networked Climate Governance: The Rise of Transnational Climate Partnerships," *Global Environmental Politics* 8, no. 3 (2008).

63 Chukwumerije Okereke, "Climate Justice and the International Regime," *Climate Change* 1, no. 3 (2010): p. 466.

64 Adrian Macey, "Climate Change: Governance Challenges for Copenhagen," *Global Governance: A Review of Multilateralism and International Organizations* 15, no. 4 (2009).

65 Harriet Bulkeley, "Climate Policy and Governance: An Editorial Essay," *Wiley Interdisciplinary Reviews: Climate Change* 1, no. 3 (2010).

66 Adam G. Bumpus and John C. Cole, "How Can the Current CDM Deliver Sustainable Development?," ibid., no. 4.

67 Timothy Cadman, ed. *Climate Change and Global Policy Regimes* (London: Palgrave Macmillan, 2013), pp. 253–54.

68 United Nations Framework Convention on Climate Change (UNFCCC), "Conference of the Parties (COP): COP21 Paris Decisions," (Paris: UNFCCC, 2015), Decisions 122–24 and section V, Decisions 134–137.

(institutional) integrity.[69] This section provides a framework for thinking about institutional integrity, and then considers the relationship between integrity and governance values.

There are two reasons why it is helpful to have a strong understanding of both governance values and institutional integrity and the interrelations between them. First, for theorists, critics and institutional reformers, a clear understanding of these qualities allows a sharper isolation of exactly where institutional problems lie and better knowledge of how discrete issues might contribute to larger failings. Rather than just saying an institution suffers from a general governance-cum-integrity issue, we can describe where that issue arises in a framework that lets us consider how the issue creates further ramifications, what measures might resolve it and what other parts of the larger system might contribute to mitigating it. For example, as we will see further on, an institution struggling to achieve its goals by purely internal reforms might invest in improving its governance values in order to facilitate more fruitful interactions with external forces (such as media and civil society).

Second, since integrity (as we will see) involves staying true to the institution's values, most institutions and institutional leaders have a straightforward internal reason to work towards enhancing institutional integrity. (This analogizes to the individual case, where integrity offers significant personal benefits to its holder.)[70] While institution leaders may feel they have no equivalent reason to pursue governance values, a fuller understanding of the relationship between good governance and integrity can help them see the cases where pursuing governance initiatives will, in fact, enhance their own agendas for improved institutional results.

On our reckoning, the concepts of 'integrity' and 'governance values' relate, yet hold distinct meanings. We understand institutional integrity as paralleling personal integrity, in the sense that both types of integrity involve an agent asking hard questions about their values, giving honest and public answers and striving to live by those answers.[71] In the institutional context, this requires an institution (by asking hard questions about its values and giving honest and public answers) to develop its 'Public Institutional Justification' (PIJ).[72] Usually including an

69 See, for example, the use of the terms in Justin O'Brien, "The Future of Financial Regulation: Enhancing Integrity through Design," *Sydney Law Review* 32, no. (2010).

70 Hugh Breakey, "Compromise Despite Conviction: Curbing Integrity's Moral Dangers," *Journal of Value Inquiry* 50, no. 3 (2016): pp. 613–29.

71 This understanding of institutional integrity was developed by Sampford in the 1990s and outlined in Preston and Sampford, "Institutionalising Ethics." It was further refined in a series of articles over the following decade and most fully theorized in Breakey, Cadman, and Sampford, "Comprehensive Integrity Framework," and applied to the climate regime in Breakey and Cadman, "Integrity of the Climate Regime Complex."

72 Many accounts of institutional integrity – and also of institutional legitimacy, governance and ethics – employ a device resembling the public institutional justification. See, e.g., Preston and Sampford, "Institutionalising Ethics"; Daniel E. Wueste, *We Need to Talk . . . About Institutional Integrity* (New York: RIT Press, 2005); Buchanan and Keohane, "The

account of the *raison d'être* of the institution (what it is 'for'), the institution's members and representatives use the PIJ to justify the institution and to show the public it deserves their support or at least tolerance.[73]

While the PIJ stands as the central organizing concept for understanding institutional integrity, different notions of integrity attach to the PIJ in different ways. The Comprehensive Integrity Framework (developed by Breakey, Cadman and Sampford)[74] provides a terminological system for speaking about these different types of integrity.

Coherence-integrity: An institution enjoys coherence-integrity if its internal qualities drive its successful pursuit of its PIJ – that is, the institution is constituted so as to live up to its public persona. Two types of internal factors contribute to coherence-integrity. The first factor we touched upon earlier: the institution's 'internal organizational arrangements'. Recall that these arrangements consist of the processes governing how the institution makes its decisions, the methods it uses for policing and encouraging compliance with those decisions, its policies on transparency, accountability and critical feedback and so on. If these internal organizational arrangements direct and empower the institution to live up to its PIJ, then this enhances its coherence-integrity. Members' values constitute the second factor. These refer to the virtues, values, commitments and ideals driving the behaviour of the institution's members (including employees, executives and participants). While uniform consensus on values across all personnel is unlikely, if most members believe in playing their role in pursuing the PIJ, this greatly enhances coherence-integrity.

Context-integrity: Factors outside the institution can also contribute to it living up to its PIJ – these factors make up the institution's context-integrity. One crucial factor is the institution's legal context, including the laws that regulate the institution's actions, and the agencies that police those rules. The institution's organizational context – that is, the larger institutions and networks of which it forms one part – can also play a strong role in facilitating its pursuit of its PIJ as well. Other relevant external factors include other (competing or supporting) organizations, media and journalistic practices and the local community's

---

Legitimacy of Global Governance Institutions."; Johan J. Graafland and Bert V. van de Van, "The Credit Crisis and the Moral Responsibility of Professionals in Finance," *Journal of Business Ethics* 103 (2011); Geoff Moore, "The Virtue of Governance, the Governance of Virtue," *Business Ethics Quarterly* 22, no. 2 (2012). Note, though, that others understand related concepts (such as corporate integrity) quite differently – see, for example, Thomas Maak, "Undivided Corporate Responsibility: Towards a Theory of Corporate Integrity," *Journal of Business Ethics* 82 (2008).

73 This approach requires that the institution or institutional complex is sufficiently coherent to have its own Public Institutional Justification, which can limit its applicability to amorphous polycentric networks. In the context of the global climate regime on this point, see Anne Schwenkenbecher, "Polycentric Systems and the Integrity Approach," in *Ethical Values and the Integrity of the Climate Change Regime*, ed. Hugh Breakey, Vesselin Popovski, and Rowena Maguire (Surrey: Ashgate, 2015).

74 Breakey, Cadman, and Sampford, "Comprehensive Integrity Framework."

values, expectations and levers of power.[75] In general, the institution has context-integrity when these external forces support the institution when it is working towards its PIJ, but resist it when it does not.

Consistency-integrity: An institution displays consistency-integrity if its actions consistently accord with its PIJ. The activities the institution performs, and the impact of those actions, further the values, goals and ideals set down publicly in its PIJ.

Comprehensive-integrity: An institution secures the ultimate prize of comprehensive-integrity when it enjoys coherence-integrity (it is internally well constituted), context-integrity (it nests in a supportive environment) and consistency-integrity (it consistently lives up to its PIJ). Obviously, an institution that possesses both coherence- and context-integrity will tend to live up to its PIJ, and so will also display consistency-integrity. However, this is only a tendency: unexpected and contingent events (such as natural disasters, civil conflicts or economic meltdowns) may derail the consistency-integrity of an otherwise well-positioned institution.

While the PIJ is the central organizing idea for institutional integrity, some institutions may not have a PIJ at all. Such institutions will have never been through the process of developing and publicizing a statement of their organization's goals, ideals and values. Other institutions may possess a PIJ, but may not actually make any efforts to live up to it. In this case, the PIJ is mere rhetoric: a public relations façade disconnected from the organization's actual driving values and goals. In the environmental context, such an institution may be engaged in 'greenwashing': cloaking its damaging activities with a patina of environmental rhetoric.[76] In more flagrant failures of integrity the organization may even be a 'decoy' institution, set up to appear as if its fundamental goal is a morally worthy one precisely to undermine the achievement of that goal.

Having institutional integrity does not imply that the institution enjoys legitimacy or moral sanctity. This is because there is no guarantee that its PIJ itself will be morally worthy. Of course, since the institution uses the PIJ to justify its existence to stakeholders and local community, the PIJ will not usually be totally evil or rapaciously self-interested – for if it was, then the PIJ would fail one of its primary functions, namely to serve as a justification for the institution's existence and activities.[77] But to say that the PIJ will not be evil or rapacious does not mean that it will be everything that an objective morality – or a given local

75 For a discussion of the ways in which external and competing organizations can work to keep an institution conforming to its role, see Hugh Breakey, "Dividing to Conquer: Employing the Separation of Powers to Structure Institutional Inter-Relations," ibid.12 (2014).

76 This can sometimes evolve over time into a more genuine engagement with environmental values: see Cristina Besio and Andrea Pronzini, "Morality, Ethics, and Values Outside and Inside Organizations: An Example of the Discourse on Climate Change," *Journal of Business Ethics* 119 (2014).

77 In this way the PIJ differs from an individual person's statement of values – because the institution requires justification for its existence in a way that a person herself does not. See Breakey, Cadman, and Sampford, "Comprehensive Integrity Framework," pp. 28–30.

community – might wish of the organization. Institutions may aim to advance specific political goals, corporations will evince a strong concern for profit and organizations may further specific cultural goods and ways of life. In any given case, some of these goals (and the PIJs that express them) may be more or less morally worthy, and more or less socially acceptable to local stakeholders.

To make the terminology of the Comprehensive Integrity Framework more concrete, consider as an example the UN Framework Convention on Climate Change (UNFCCC).[78] The UNFCCC's Public Institutional Justification is given in the convention's Article 2, which publicly sets down its goal of stabilizing greenhouse gas concentrations at a level that would prevent dangerous anthropogenic interference with the climate system.[79] The UNFCCC's coherence-integrity thus includes any internal factors that impact it living up to this goal. This includes its members' values (whether its executive and employees are committed to playing their role in achieving this objective) and its internal organizational arrangements (the UNFCCC's decision-making bodies, myriad sub-institutions and the interrelations between them).[80] The UNFCCC's context-integrity looks to the institution's larger environment: its legal environment (international law, including the UN Charter and other important treaties),[81] its organizational context (as a sub-institution of the United Nations itself) and its other external relations (such as to civil society, academia, scientific bodies and so on).

With these terms and examples laid out, we can now turn to the key question at hand: *what is the relationship between governance values and institutional integrity?* As an initial point, consider the relationship between governance values and institutions that *fail* to achieve any institutional integrity. These institutions either do not have a PIJ at all, or their alleged PIJ is actually just a facile marketing ploy. Because such an institution is not committed to pursuing its PIJ, it is bound to have low coherence-integrity. In either case, the institution will still require mechanisms for steering and coordination. These mechanisms will require some forms of internal accountability and measures for ensuring the compliance of low-level actors – for example, to make sure they are pursuing the corporation's profits[82] and not abusing their position to enrich themselves at the expense of the

78 For more detail, see Breakey and Cadman, "Integrity of the Climate Regime Complex."

79 "United Nations Framework Convention on Climage Change," http://unfccc.int/essential_background/convention/background/items/1349.php. For discussion, see Rowena Maguire, "Mapping the Integrity of Differential Obligations within the UNFCCC," in *Ethical Values and the Integrity of the Climate Change Regime*, ed. Hugh Breakey, Vesselin Popovski, and Rowena Maguire (Surrey: Ashgate, 2015).

80 For a close analysis of these, see Timothy Cadman, "Stakeholder Perspectives on the Integrity of the Climate Regime," ibid.

81 See Felicity Deane, "The Context-Integrity of the Global Carbon Regime: The Relevance and Impact of the World Trade Organization," ibid.

82 Of course, some corporations may seek to justify themselves solely on the basis that they maximize profits for their shareholders. However, if this really is the only goal of the corporation, irrespective of its effects on the community, it is a claim to a form of corporate sociopathy that few corporations would make and few communities would tolerate (see Charles

corporation. It is thus possible to speak of the 'governance' of such organizations and to consider how well their internal organizational arrangements facilitate the institution's actual goals. However, such institutions would be unlikely to possess any substantial governance values, as we have defined the term; while they might have some level of transparency, accountability and rule-compliance (as per Thin Governance Values), these will all be internal to the institution: they will not be transparent, accountable and compliant with respect to external agencies. That said, mechanisms promoting governance values may be forced upon the institution by an external agent – perhaps through the coercive force of law in the state itself, or perhaps through oversight from other market or civil society actors (such as market-based standards agencies).

We turn now to consider institutions that are genuinely pursuing their PIJ and hence trying to achieve institutional integrity. As a general rule of thumb, the more morally laudable and socially endorsed the institution's PIJ, the thicker the governance values that it will require in its efforts to live up to that PIJ. Qualities of inclusiveness, democratic decision-making and deliberation will only enhance its integrity. On the other hand, an institution with a PIJ that departs from what common-sense morality and the local community would demand will tend to avoid Thick Governance Values. Practices like including local stakeholders in decision-making, ensuring they have a strong voice equal in strength to other participants and enhancing deliberation may all work to frustrate the institution in achieving its PIJ. In the terms defined earlier, because such an institution has a low context-integrity (its surrounding environment is hostile to its pursuit of its PIJ), it will not want to degrade its coherence-integrity with Thick Governance Values that would only hamper its pursuit of its PIJ. Of course, such an institution may see the wisdom in incorporating at least Thin Governance Values – transparency, accountability and rule-compliance may well help the institution achieve its publicly stated goals and demonstrate to both internal and external stakeholders that it is doing so. As such, institutions with a less-than-socially-laudable PIJ may be expected to rank only at the thin end of the governance-values spectrum.

Contrariwise, an institution with a laudable PIJ will usually have little to fear from Thick Governance Values because the institution can expect strong agreement and solid support from most stakeholders.[83] (Indeed, the institution's PIJ may be socially laudable precisely because the institution pays attention to its impact on stakeholders, and perhaps was even formed with their input and

Sampford and Virginia Berry, "Shareholder Values, Not Shareholder Value: The Role of 'Ethical Funds' and 'Ethical Entrepreneurs' in Connecting Shareholders' Values with Their Investments," *Griffith Law Review* 13, no. 1 (2004)). Accordingly, corporations generally claim to further the interests of the communities in which they operate as part of their PIJ. The issue is that we have to deal with the fact that there are very strong governance measures in some corporations to pursue profit, and some are very upfront about it.

83 The one qualification is if the governance values are not worked into a coherent and workable regime within the institution, or if governance values are pursued as ends in themselves, losing sight of the PIJ or merely failing to achieve it.

interests in mind. When this occurs, good governance helps enhance the PIJ – 'enhance' in the sense of improving the PIJ's capacity to fulfil its signature role of nurturing social support for the institution.) As well as having little to fear, the institution will have much to gain from pursuing Thick Governance Values. Since the institution has a legitimate PIJ, Thick Governance Values will serve to demonstrate that legitimacy to all internal and external stakeholders, enhancing the institution's 'brand' and building support from local actors, civil society agencies and (perhaps) governmental actors. As a general rule, then, institutions genuinely pursuing a laudable PIJ have good reason to aim for Thick Governance Values. In this case we can think of governance values as qualities regulating the crucial mediating layer sitting between coherence-integrity (regarding internal organizational arrangements) and context-integrity (regarding the external elements such as public opinion and civil society groups). In particular, Thick Governance Values allow the institution to marshal resources external to it (media, stakeholders, civil society) and to structure their interaction with the institution's internal workings – through transparency, inclusiveness, deliberativeness and so on – so as to better achieve its institutional goals (the PIJ).

Still, to say such institutions have good reason to pursue Thick Governance Values does not mean that that pursuit will be effective. Implementing improved governance practices requires time, expertize, leadership and investment of resources. Since these are finite resources, even the most morally upright institutions will need to make hard trade-offs between their governance ambitions and their other goals. (For example, an international charity will want to have enough internal oversight to make sure its benefactions reach those who most need them, but will not want to spend so much money on its governance that it has precious little funds left over to distribute.)

The situation becomes more complex in an institution that has (what we might evaluate as) a morally upright PIJ – but one that *does not* have the support of the local community or government, such as a human rights organization taking a stand against a community or government that is breaching an unpopular minority's human rights. In this case, Thick Governance Values may serve only to allow institutional capture by local actors aiming to undermine the institution's work.

Can an institution with a laudable PIJ achieve consistency-integrity (that is, live up to its aspirations) without developing Thick Governance Values? We have already observed that Thick Governance Values will enhance the institution's legitimacy and work as a key strategy in achieving its PIJ. However, the quality and structure of the institution's internal organizational arrangements are – as we noted earlier – not the only internal factor that contributes to its integrity. Coherence-integrity also includes reference to member's values. If the institution's members hold strong values aligning with its PIJ, then the institution may be able to achieve integrity even without pursuing Thick Governance Values. The virtues and discretionary decision-making of its members, who act individually and in combination with internal and external actors, may work quite reliably in furthering the PIJ. Outside the institution, the institution's context-integrity might also help facilitate its pursuit of its PIJ. That is, its legal and cultural environment may support the institution in pursuing its PIJ and work to constrain it when it is not. In these ways an institution may achieve consistency-integrity – perhaps even across significant periods and in the face of significant challenges – without Thick Governance Values. However, even in such cases the pursuit of Thick Governance Values might well be worthwhile, as these would ensure the institution keeps to its purpose if its members' values shift over time, or if its external context changes.

Summing up, integrity goes beyond governance in considering all the elements that help an institution develop and pursue its values and goals. Governance goes beyond integrity by considering all the qualities that enhance the legitimacy of the institution's structures and processes for steering and governing. Understood in this way, the distinct ideas of institutional integrity and governance values furnish us with an array of distinct criticisms of a given institution:

1 The institution does not possess a PIJ, and so does not possess integrity.
2 The institution does not possess a morally and socially laudable PIJ. It may attain comprehensive-integrity, but it will tend to resist even Thin Governance Values and so struggle to achieve social credibility.
3 The institution possesses a laudable PIJ, but in fact the PIJ is just window dressing. The institution fails to have coherence-integrity and will usually possess (at best) Thin Governance Values.
4 The institution possess a laudable PIJ, and Thick Governance Values would enhance its capacity to pursue this PIJ. However, the institution has not (yet) achieved this standard of governance, suggesting it possesses weak coherence-integrity.

The UNFCCC, governance values, and equity and environmental values

The foregoing discussion provides several ways that environmental and equity values, such as inter-generational justice in mitigating/adapting to climate change or 'sustainable development' more generally, might inject into discussions of good governance.[84] As we noted earlier, these types of 'governing' values focus on outcomes or distributive entitlements.[85]

84 In the context of environmental ethics, one difference between governing (substantive) values and governance (procedural) values warrants remark. Substantive values, containing claims about intrinsic value, can directly announce the value of non-human entities – they can lay down a non-anthropocentric ethic (See, for example, Aldo Leopold, *A Sand County Almanac: With Sketches Here and There* (New York: Oxford University Press, 1949)). And compare Bryan G. Norton, "Why I Am Not a Nonanthropocentrist: Callicott and the Failure of Monistic Inherentism," *Environmental Ethics* 17 (1995). Procedural values, insofar as they regulate human institutions and interactions, necessarily apply only to the human agents populating such institutions and can only indirectly provide respect to animals, species and ecologies. (See n. 87.)

85 See n. 43.

Environmental objectives of various kinds might feature in the Public Institutional Justification (PIJ) of a given institution. We saw this earlier in our discussion of the UNFCCC, which contains an array of environmental goals and equity values in its objectives (Art. 2).[86] The UNFCCC's Public Institutional Justification is also closely interwoven with the norm of sustainable development. The UNFCC was developed as a consequence of the Earth Summit, which focused on sustainable development. This is reflected in many of the UNFCCC policy mechanisms (including REDD+ and the CDM, which are supposed to deliver sustainable development). This is why there is a sustainable development discourse within the UNFCCC, while its explicit Public Institutional Justification is to reduce 'harmful' emissions.

In cases like the UNFCCC, where environmental and equity values feature in the PIJ, then any mechanisms that help secure this institutional objective will bolster the institution's integrity. These mechanisms might be ones to improve the institution's internal organizational arrangements (and its coherence-integrity). But they also might be mechanisms to improve the quality of the interface between internal and external forces – that is, to improve the layer mediating between coherence-integrity and context-integrity. Such efforts will often need to aim at improving the institution's ranking on the governance continuum; Thick Governance Values will enhance the institution's legitimacy even as they facilitate its pursuit of the environmental values it prizes. So, too, the efficacy with which the institution implements and achieves those results will be reflected in measures of its 'problem-solving' and 'behavioural change', captured in Thick Governance Values. Finally, the UNFCCC's environmental and equity values play an enhanced role when key governance values are improved. This is especially so for the value of deliberation (as well as inclusiveness and equality of stakeholders, in addition to equitable distribution of resources), as improved deliberation allows the prized substantive values (rather than private and parochial agendas) to guide decision-making.[87]

In a separate case, in circumstances where the local culture and community possess strong environmental and equity commitments, an institution aiming to improve its legitimacy by thickening its governance values will come to engage with environmental values. While the institution itself may not (initially) possess environmental or equity values within its PIJ, its improved governance will tighten its social and moral connections to stakeholders, and their values will begin to impact the institution's activities.[88] Here the institution's Thick

86 United Nations Framework Convention on Climate Change (UNFCCC), "United Nations Framework Convention on Climate Change." See Maguire, "Mapping the Integrity of Differential Obligations within the UNFCCC."

87 Even ethical and political theories of justice and equity can play a role in the deliberations of 'bottom-up' decision-making regimes: see, for example, Hugh Breakey, "COP20's Ethical Fallout: The Perils of Principles without Dialogue," *Ethics, Policy & Environment* 18, no. 2 (2015); Pickering, "Top-Down Proposals."

88 An example is where initial greenwashing gives way to genuine environmental engagement. See Besio and Pronzini, "Discourse."

Governance Values, when connected to an environmentally aware community, work to inject environmental values into its functioning – and perhaps even into its PIJ.

Conclusion

This chapter has developed a continuum ranging from Thin Governance Values of basic oversight through to Thick Governance Values that furnish institutional legitimacy. A given institution's governance ambitions and achievements (these two may differ, as implementation often falls sort of aspiration) may fall at different points on this continuum. We saw that the thickness of an institution's governance values interrelate with its pursuit of institutional integrity in complex ways – with a key factor being the substance of the institution's Public Institutional Justification and external stakeholders' opinion of that justification. The more external stakeholders are on the same moral page as the institution, the more that Thick Governance Values can work not only to increase institutional legitimacy but also to enhance institutional integrity.

References

Andonova, Liliana B., Michele M. Betsill, and Harriet Bulkeley. "Transnational Climate Governance." *Global Environmental Politics* 9, no. 2 (2009): 52–73.

Bäckstrand, Karin. "Accountability of Networked Climate Governance: The Rise of Transnational Climate Partnerships." *Global Environmental Politics* 8, no. 3 (2008): 74–102.

Barnett, Jon. "Adapting to Climate Change: Three Key Challenges for Research and Policy – an Editorial Essay." *Wiley Interdisciplinary Reviews: Climate Change* 1, no. 3 (2010): 314–17.

Besio, Cristina, and Andrea Pronzini. "Morality, Ethics, and Values Outside and Inside Organizations: An Example of the Discourse on Climate Change." *Journal of Business Ethics* 119 (2014): 287–300.

Biermann, Frank, and Aarti Gupta. "Accountability and Legitimacy in Earth System Governance: A Research Framework." *Ecological Economics* 70, no. 11 (2011): 1856–64.

Boran, Idil. "Principles of Public Reason in the UNFCCC: Rethinking the Equity Framework." *Science and Engineering Ethics* 22, no. 4 (2016): 119.

Breakey, Hugh. "Compromise Despite Conviction: Curbing Integrity's Moral Dangers." *Journal of Value Inquiry* 50, no. 3 (2016): 613–29.

———. "COP20's Ethical Fallout: The Perils of Principles without Dialogue." *Ethics, Policy & Environment* 18, no. 2 (2015): 156–69.

———. "Dividing to Conquer: Employing the Separation of Powers to Structure Institutional Inter-Relations." *Research in Ethical Issues in Organizations* 12 (2014): 29–58.

Breakey, Hugh, and Timothy Cadman. "A Comprehensive Framework for Evaluating the Integrity of the Climate Regime Complex." Chap. 2 In *Ethical Values and the Integrity of the Climate Change Regime*, edited by Hugh Breakey, Vesselin Popovski and Rowena Maguire, 17–27. Surrey: Ashgate, 2015.

Breakey, Hugh, Timothy Cadman, and Charles Sampford. "Conceptualizing Personal and Institutional Integrity: The Comprehensive Integrity Framework." *Research in Ethical Issues in Organizations* 14 (2015): 1–40.

Buchanan, Allen, and Robert O. Keohane. "The Legitimacy of Global Governance Institutions." *Ethics and International Affairs* 20, no. 4 (2006): 405–37.

Bulkeley, Harriet. "Climate Policy and Governance: An Editorial Essay." *Wiley Interdisciplinary Reviews: Climate Change* 1, no. 3 (2010): 311–13.

Bumpus, Adam G., and Cole. John C. "How Can the Current CDM Deliver Sustainable Development?" *Wiley Interdisciplinary Reviews: Climate Change* 1, no. 4 (2010): 541–47.

Cadman, Timothy, ed. *Climate Change and Global Policy Regimes.* London: Palgrave Macmillan, 2013.

———. "Introduction: Global Governance and Climate Change." In *Climate Change and Global Policy Regimes: Towards Institutional Legitimacy*, edited by Timothy Cadman, 1–16. London: Palgrave Macmillan, 2013.

———. *Quality and Legitimacy of Global Governance.* London: Palgrave Macmillan, 2011.

———. "Stakeholder Perspectives on the Integrity of the Climate Regime." Chap. 4 In *Ethical Values and the Integrity of the Climate Change Regime*, edited by Hugh Breakey, Vesselin Popovski and Rowena Maguire, 43–59. Surrey: Ashgate, 2015.

Cadman, Timothy, Lauren Eastwood, Federico Lopez-Casero Michaelis, Tek Narayan Maraseni, Jamie Pittock, and Tapan Sarker. *The Political Economy of Sustainable Development: Policy Instruments and Market Mechanisms.* Cheltenham, UK: Edward Elgar Publishing, 2015.

Cadman, Timothy, Tek Narayan Maraseni, Hwan Ok Ma, and Federico Lopez-Casero. "Five Years of Redd+ Governance: The Use of Market Mechanisms as a Response to Anthropogenic Climate Change." *Forest Policy and Economics* 4 (2016): Table 2. http://dx.doi.org/10.1016/j.forpol.2016.03.008.

Corbera, Esteve, and Heike Schroeder. "Governing and Implementing Redd+." *Environmental Science & Policy* 14, no. 2 (2011): 89–99.

Crane, Andrew. "From Governance to Governance: On Blurring Boundaries." *Journal of Business Ethics* 94 (2010): 17–19.

Curtin, Deirdre, and Albert Jacob Meijer. "Does Transparency Strengthen Legitimacy? A Critical Analysis of European Union Policy Documents." *Information Polity* 11 (2006): 109–22.

Deane, Felicity. "The Context-Integrity of the Global Carbon Regime: The Relevance and Impact of the World Trade Organization." Chap. 6 In *Ethical Values and the Integrity of the Climate Change Regime*, edited by Hugh Breakey, Vesselin Popovski and Rowena Maguire, 73–86. Surrey: Ashgate, 2015.

Depledge, Joanna. "The Opposite of Learning: Ossification in the Climate Change Regime." *Global Environmental Politics* 6, no. 1 (2006): 1–22.

Economic and Social Council (ECOSOC). *Definition of Basic Concepts and Terminologies in Governance and Public Administration.* New York: United Nations, 2006.

Gardiner, Stephen M., Simon Caney, Dale Jamieson, and Henry Shue, eds. *Climate Ethics: Essential Readings.* Oxford: Oxford University Press, 2010.

Graafland, Johan J., and Bert V. van de Van. "The Credit Crisis and the Moral Responsibility of Professionals in Finance." *Journal of Business Ethics* 103 (2011): 605–19.

Huberts, Leo. *The Integrity of Governance: What It Is, What We Know, What Is Done, and Where to Go.* Hampshire: Palgrave Macmillan, 2014.

Intergovernmental Panel on Climate Change (IPCC). "Fifth Assessment Report – Synthesis Report." In *Climate Change 2014: Synthesis Report. Contribution of Working Groups I, II and III to the Fifth Assessment Report of the Intergovernmental Panel on Climate Change*, edited by R.K. Pachauri and L.A. Meyer, 151 pp. Geneva: IPCC, 2014.

International Law Association (ILA). "Legal Principles Relating to Climate Change." In *Report of the International Law Association's Committee on Legal Principles Relating to Climate Change*, edited by Shinya Murase, 1–19. Washington: ILA, 2014.

Jordan, Andrew, Rüdiger K.W. Wurzel, and Anthony Zito. "The Rise of 'New' Policy Instruments in Comparative Perspective: Has Governance Eclipsed Government?" *Political Studies* 53 (2005): 477–96.

Koenig-Archibugi, Mathias. "Introduction: Institutional Diversity in Global Governance." In *New Modes of Governance in the Global System: Exploring Publicness, Delegation and Inclusiveness*, edited by Mathias Koenig-Archibugi and Michael Zürn, 1–30. Basingstoke and London: Palgrave Macmillan, 2006.

Kooiman, Jan. "Societal Governance: Levels, Models, and Orders of Social-Political Interaction." In *Debating Governance: Authority, Steering and Democracy*, edited by Jon Pierre, 138–65. Oxford: Oxford University Press, 2000.

Lange, Andreas, Carsten Vogt, and Andreas Ziegler. "On the Importance of Equity in International Climate Policy: An Empirical Analysis." *Energy Economics* 29 (2007): 545–62.

Leopold, Aldo. *A Sand County Almanac: With Sketches Here and There*. New York: Oxford University Press, 1949.

Maak, Thomas. "Undivided Corporate Responsibility: Towards a Theory of Corporate Integrity." *Journal of Business Ethics* 82 (2008): 353–68.

Macey, Adrian. "Climate Change: Governance Challenges for Copenhagen." *Global Governance: A Review of Multilateralism and International Organizations* 15, no. 4 (2009): 443–49.

Maguire, Rowena. "Incorporating International Environmental Legal Principles into Future Climate Change Instruments." *Carbon and Climate Law Review* 6, no. 2 (2012): 101–13.

———. "Mapping the Integrity of Differential Obligations within the UNFCCC." Chap. 3 In *Ethical Values and the Integrity of the Climate Change Regime*, edited by Hugh Breakey, Vesselin Popovski and Rowena Maguire, 31–42. Surrey: Ashgate, 2015.

Mattoo, Aaditya, and Arvind Subramanian. "Equity in Climate Change: An Analytical Review." *World Development* 40, no. 6 (2012): 1083–97.

Moore, Geoff. "The Virtue of Governance, the Governance of Virtue." *Business Ethics Quarterly* 22, no. 2 (2012): 293–318.

Norton, Bryan G. "Why I Am Not a Nonanthropocentrist: Callicott and the Failure of Monistic Inherentism." *Environmental Ethics* 17 (1995): 341–58.

O'Brien, Justin. "The Future of Financial Regulation: Enhancing Integrity through Design." *Sydney Law Review* 32, no. 1 (2010): 63–85.

O'Brien, Justin, and Olivia Dixon. "The Common Link in Failures and Scandals at the World's Leading Banks." *Seattle University Law Review* 36 (2013): 941–72.

Okereke, Chukwumerije. "Climate Justice and the International Regime." *Climate Change* 1, no. 3 (2010): 462–74.

Pachauri, Rajendra. "Chairperson of the Intergovernmental Panel on Climate Change, Presentation to the Sixth International Forum for Sustainable Asia and the Pacific (ISAP)." In *Sixth International Forum for Sustainable Asia and the Pacific*, edited by Institute for Global Environmental Strategies (IGES), 7–10. Yokohama: ISAP (Reproduced with permission), 2014.

Pickering, Jonathan. "Top-Down Proposals for Sharing the Global Climate Policy Effort Fairly: Lost in Translation in a Bottom-up World?" Chap. 7 In *Ethical Values and the Integrity of the Climate Change Regime*, edited by Hugh Breakey, Vesselin Popovski and Rowena Maguire, 89–104. Surrey: Ashgate, 2015.

Pickering, Jonathan, Steve Vanderheiden, and Seumas Miller. " 'If Equity's in, We're Out': Scope for Fairness in the Next Global Climate Agreement." *Ethics & International Affairs* 26, no. 4 (2012): 423–43.

Pittock, Jamie. "Climate Change and Sustainable Water Management." In *Climate Change and Global Policy Regimes*, edited by Tim Cadman, 138–56. London: Palgrave Macmillan, 2013.

Plattner, Marc F. "Reflections on 'Governance'." *Journal of Democracy* 24, no. 4 (2013): 17–28.

Preston, Noel, and Charles Sampford. "Institutionalising Ethics." In *Encouraging Ethics and Challenging Corruption*, edited by Charles Sampford, Noel Preston and Carmel Conners, 32–68. Annandale: Federation Press, 2002.

Sampford, Charles. "Adam Smith's Dinner." In *The Future of Financial Regulation*, edited by Iain Macneil and Justin O'Brien, 23–40. Oxford, UK: Hart Publishing, 2010.

———. "Business and Human Rights: Reflections and Observations." In *Business and Human Rights*, edited by Wesley Cragg, 315–32. Cheltenham: Edward Elgar, 2012.

———. "Global Lessons from Fitzgerald: From State and National to Global Integrity Systems." Chap. 13 In *The Fitzgerald Legacy: Reforming Public Life in Australia and Beyond*, edited by Colleen Lewis, Janet Ransley and Ross Homel, 210–33. Brisbane: Australian Academic Press, 2010.

———. "Global Transparency: Fighting Corruption for a Sustainable Future: From National Integrity Systems to Global Integrity Systems." *TI Commissioned Conference Overview Discussion Paper for 13 IACC.* Athens, 2008.

———. "Law, Institutions and the Public Private Divide." *Federal Law Review* 20, no. 2 (1990, 1992): 185–222.

———. "Legality and Legitimacy: A Dozen Years after Goldstone." In *Responsibility to Protect and Sovereignty*, edited by Charles Sampford and Ramesh Thakur, 143–70. Farnham: Ashgate, 2013.

———. "Sovereignty and Intervention." In *Human Rights in Theory and Practice*, edited by Tom Campbell and Burton M. Leiser, 335–51. London: Ashgate, 2001.

———. "Towards a Global Carbon Integrity System: Learning from the GFC1 and Avoiding a GCC2." *Low Carbon Economy* 2, no. 4 (2011): 201–19.

———. "Institutionalising Public Sector Ethics." In *Ethics for the Public Sector: Education and Training*, edited by N. Preston, 14–34. Sydney: Federation Press, 1994.

Sampford, Charles, and Virginia Berry. "Shareholder Values, Not Shareholder Value: The Role of 'Ethical Funds' and 'Ethical Entrepreneurs' in Connecting Shareholders' Values with Their Investments." *Griffith Law Review* 13, no. 1 (2004): 115–23.

Sampford, Charles, and David Wood. "The Future of Business Ethics? Legal Regulation, Ethical Standard Setting and Institutional Design." *Griffith Law Review* 1, no. 1 (1992): 56–72.

Scharpf, Fritz W. *Games Real Actors Play: Actor-Centered Institutionalism in Policy Research.* Boulder: Westview Press, 1997.

Schmidt, Vivien A. "Democracy and Legitimacy in the European Union Revisited: Input, Output and 'Throughput'." *Political Studies* 61 (2012): 2–22.

Schwenkenbecher, Anne. "Polycentric Systems and the Integrity Approach." Chap. 2 In *Ethical Values and the Integrity of the Climate Change Regime*, edited by Hugh Breakey, Vesselin Popovski and Rowena Maguire, 31–38. Surrey: Ashgate, 2015.

Shue, Henry. *Climate Justice: Vulnerability and Protection.* Oxford: Oxford University Press, 2014.

"United Nations Framework Convention on Climate Change." <http://unfccc.int/essential_background/convention/background/items/1349.php>.

United Nations Framework Convention on Climate Change (UNFCCC). "Conference of the Parties (COP): COP21 Paris Decisions." Paris: UNFCCC, 2015.

———. "Conference of the Parties: COP21 Paris Agreement." FCCC/CP/2015/L.9: UNFCCC, 2015.

———. "Report of the Conference of the Parties on Its Seventeenth Session, Held in Durban from 28 November to 11 December 2011." Durban: UN, 2012.

———. "United Nations Framework Convention on Climate Change." Rio: UN, 1992.

Wachhaus, Aaron. "Governance beyond Government." *Administration & Society* 46, no. 5 (2014): 573–93.

Weiss, Thomas G., and Ramesh Thakur. *Global Governance and the UN: An Unfinished Journey.* Bloomington: Indiana University Press, 2010.

Wueste, Daniel E. *We Need to Talk . . . About Institutional Integrity.* New York: RIT Press, 2005.

Young, Oran R. *Institutional Dynamics: Emergent Patterns in International Environmental Governance.* Cambridge, MA: MIT Press, 2010.

Part II

Evaluating governance arrangement within the UNFCCC

3 The climate governance regime complex

Institutions, actors and discourses[1]

Philipp Pattberg, Lisa Sanderink and Oscar Widerberg

Introduction

Over the past twenty years, global efforts to combat climate change have become an increasingly complex matter. The central forum for multilateral, state-led climate governance, the United Nations Framework Convention on Climate Change (UNFCCC), has been complemented by numerous cross-border initiatives comprising both state and non-state actors, including non-governmental organizations (NGOs), firms, academia, cities, sub-national regions and international organizations.[2]

The broader institutional structure has developed from a single regime to a regime complex,[3] showing increasing signs of fragmentation and functional overlaps that threaten coherence and overall effectiveness.[4] In Ostrom's words, global climate governance is best described as a polycentric system with 'multiple governing authorities at different scales rather than a monocentric unit'.[5] However, while fragmentation is largely accepted in theory as a ubiquitous phenomenon in global climate governance, few empirical studies exist that map institutional complexity and consequently attempt to measure degrees of fragmentation or coherence.[6]

1 We would like to acknowledge support from the Netherlands Organization for Scientific Research (CONNECT project 'Assessing and Reforming the Current Architecture of Global Environmental Governance', contract number 2012/15789/MaGW) and research assistance by Flavia Dias Guerra.
2 Kenneth Abbott, "The Transnational Regime Complex for Climate Change," Environment & Planning C: Government & Policy 30, no. 4 (2012); Robert Keohane and David Victor, "The Regime Complex for Climate Change," *Perspectives on Politics* 9, no. 1 (2011); Harriet Bulkeley, Liliane Andanova, Michele Betsill, Daniel Compagnon, Thomas Hale, Matthew Hoffman, Peter Newell, Matthew Paterson, Charles Roger, and Stacy VanDeveer, *Transnational Climate Change Governance* (Cambridge: Cambridge University Press, 2014).
3 Keohane and Victor, "The Regime Complex for Climate Change."
4 Frank Biermann, Philipp Pattberg, Harro Van Asselt, and Fariborz Zelli, "The Fragmentation of Global Governance Architectures: A Framework for Analysis," *Global Environmental Politics* 9, no. 4 (2009).
5 Elinor Ostrom, "Polycentric Systems for Coping with Collective Action and Global Environmental Change," *Global Environmental Change* 20, no. 4 (2010): p. 552.
6 Fariborz Zelli and Harro Van Asselt, "Introduction: The Institutional Fragmentation of Global Environmental Governance," *Global Environmental Politics* 13, no. 3 (2013).

This chapter consequently presents new empirical data on three related measures of institutional coherence or fragmentation in global climate governance. First, we present a mapping of all governance institutions that constitute the climate change regime complex, departing from Abbott and Snidal, Abbott and Keohane and Victor.[7] We include both international and non-state public institutions (for example, the UNFCCC and transnational municipal networks) as well as the broad range of bottom-up initiatives constituted by various mixes of actors, including firms, civil society, governments and international organizations. Second, we add the organizational, agent-based dimension of climate governance by showing how the institutional meta-structure is constituted by actors and their connections. To this end we have collected data on membership in all climate governance institutions that make up the regime complex, in total more than 10,000 unique organizations. As a third analytical step, we report on advances in measuring discursive structures in the climate change regime complex. Here we analyse how the 80 climate governance institutions relate to 4 pre-identified meta-discourses in environmental policy via their mission statements. In the discussion section we revisit the three complementary assessments of the current climate change regime complex to establish whether the overall institutional structure is fragmented, that is, polycentric, or rather centralized, that is, integrated. In the conclusions we reflect on how to advance this research agenda further.

Regime complexes, fragmentation and polycentric governance: How to analyse governance systems?

While previous research of regime complexes in global climate governance has been preoccupied with theoretically assessing the benefits and shortcomings of fragmented/diversified governance architectures,[8] thereby analyzing predominantly dyadic relations between institutions or suggesting management options for improved effectiveness and efficiency,[9] few attempts have been made to map and measure the overall institutional structure of the climate regime complex along clearly defined variables and with clearly defined methodologies beyond listings of individual institutional components. This chapter therefore offers a preliminary mapping-and-measuring approach – preliminary because questions about appropriate methodologies and measurement methods remain – based on an interpretation of the climate change regime complex as a multi-dimensional structure that includes actors and their linkages as well as discourses and framings

7 Kenneth Abbott and Duncan Snidal, "The Governance Triangle: Regulatory Standards Institutions and the Shadow of the State," in *The Politics of Global Regulation*, ed. Walter Mattli and Ngaire Woods (Princeton: Princeton University Press, 2009); Kenneth Abbott, "The Transnational Regime Complex for Climate Change," *Environment & Planning C: Government & Policy* 30, no. 4 (2012); Keohane and Victor, "The Regime Complex for Climate Change."

8 For example, Biermann et al., "The Fragmentation of Global Governance Architectures: A Framework for Analysis."

9 Fariborz Zelli, "The Fragmentation of the Global Climate Governance Architecture," *Wiley Interdisciplinary Reviews: Climate Change* 2, no. 2 (2011).

in addition to institutional arrangements and their relations vis-à-vis. The question that consequently motivates this chapter is this: how is the climate governance architecture structured in terms of institutions, actors and discourses and how fragmented is global climate governance along these three dimensions? This is a first necessary step towards a better theory of climate governance. This section therefore first introduces the main concepts used in this chapter before we propose a framework for measuring fragmentation/integration along a continuum across the three dimensions of institutions, actors and discourses.

The dominant theoretical approach to analyzing the performance and effectiveness of institutions such as the UNFCCC is regime theory or neoliberal institutionalism. The focus of most empirical work to date is either on one international environmental institution (for example, the ozone regime, the climate regime or the regime to combat desertification) or dyadic relations between two international institutions. As Raustiala and Victor have noted:

> Most empirical studies focus on the development of a single regime, usually centered on a core international agreement and administered by a discrete organization.[10]

Only recently have researchers started to map and partially measure the overall institutional structure of global environmental governance, that is, the clusters of norms, principles, regimes and other institutions generally referred to as the 'governance architecture' of an issue area.[11]

An important step towards analyzing the macro-level of governance institutions was Raustiala's and Victor's conceptualization of *regime complexes* as 'an array of partially overlapping and nonhierarchical institutions governing a particular issue-area'.[12] While firmly rooted in a state-based, international ontology, their interest was in understanding institutionalization beyond single regimes. Keohane and Victor apply this conceptualization to the realm of climate governance (also with a focus on international cooperation exclusively).[13] On this account, institutions can be placed on a continuum ranging from fully integrated institutions that regulate via top-down authority to collections of loosely coupled institutions (regime complexes) to fragmented institutional structures that lack coordination and linkages among constituent parts. To better map what Keohane and Victor have described as a 'Cambrian explosion', the emergence of a wide array of new institutional forms,[14] Abbott has developed the transnational regime complex of climate change and displayed participating institutions in a governance triangle based on the dominant actors in decision-making and their respective

10 Kal Raustiala and David Victor, "The Regime Complex for Plant Genetic Resources," *International Organization* 58, no. 2 (2004): p. 279.
11 Biermann et al., "The Fragmentation of Global Governance Architectures: A Framework for Analysis."
12 Raustiala and Victor, "The Regime Complex for Plant Genetic Resources," p. 279.
13 Keohane and Victor, "The Regime Complex for Climate Change."
14 Ibid., p. 12.

mixes.[15] We use the heuristic of the governance triangle[16] to map the overall institutional architecture of climate governance. However, in addition to simply mapping institutions, we are interested in offering a more robust measurement of the institutional structure along a continuum ranging from highly fragmented to integrated.

How then should fragmentation be conceptualized? Rather than relying on a single measure for fragmentation/integration, we argue that fragmentation is a multi-dimensional concept that includes not only the relations between institutions, but also the linkages of actors (organizations which are members to governance institutions) and the specific framings used to describe themselves. Consequently, in this chapter we provide a measurement for institutional structure, the underlying actor-network and the related discursive configurations. An operationalization of each analytical step is provided in the respective following sections (methodology and approach).

Mapping the climate governance regime complex: Institutions

The first step in our analysis is to build a database of institutions that make up the climate change regime complex.

Methodology and approach

In order to compile such a database, clear and unambiguous criteria are required. We therefore include (i) international and transnational institutions, which not only have the (ii) intentionality to steer policy and the behaviour of their members or a broader community, but also explicitly mention the (iii) common governance goal, accomplishable by (iv) significant governance functions. The common governance goal can also be considered to be the Public Institutional Justification (PIJ), as explained in the conceptual background of this book. With nearly universal membership, the UNFCCC and its legally binding Kyoto Protocol set the overall goal of global climate change governance – stabilizing greenhouse gas concentrations in the atmosphere at a level that will prevent dangerous human interference with the climate system. The institutions' members and representatives use this broad governance goal as the PIJ to justify the existence of the institutions. A comprehensive mapping of the institutional architecture of global climate change must, however, include both international regimes and transnational arrangements. Since both have already been mapped with some accuracy, this research relies on a review and update of previous case studies according to the criteria set out.[17]

15 Abbott, "The Transnational Regime Complex for Climate Change."

16 See also Abbott and Snidal, "The Governance Triangle: Regulatory Standards Institutions and the Shadow of the State."; Kenneth Abbott and Duncan Snidal, "Strengthening International Regulation through Transnational New Governance: Overcoming the Orchestration Deficit," *Vanderbilt Journal of Transnational Law* 42, no. 2 (2009).

17 Abbott, "The Transnational Regime Complex for Climate Change."; Bulkeley et al., *Transnational Climate Change Governance.*

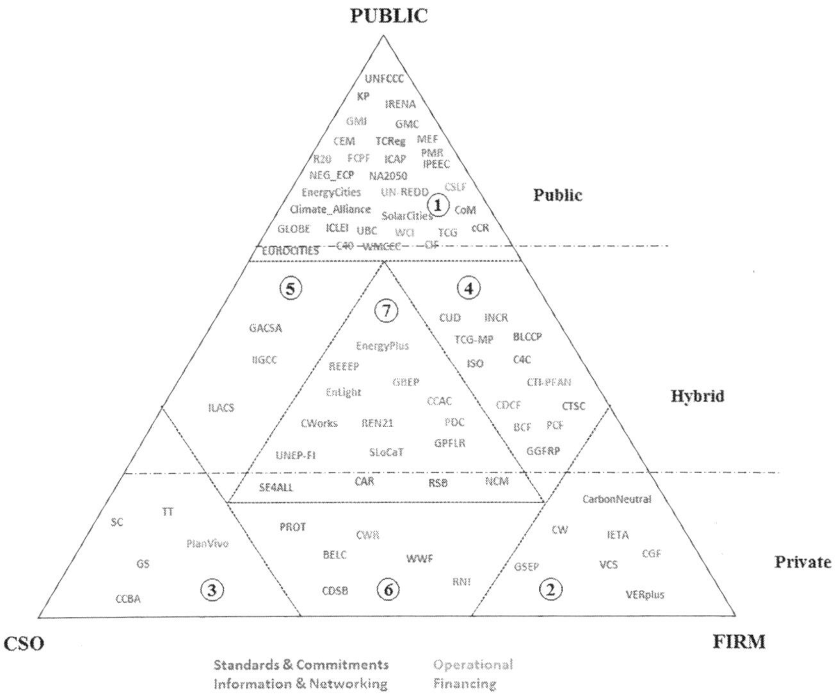

Figure 3.1 Global climate change governance triangle (2015); based on Abbott and Snidal 2009a, 2009b and Abbott 2012; own data

Institutions that meet our criteria are placed in the governance triangle according to the identity of their founding members and members with decision-making power – *State, Firm and Civil Society Organization* (CSO).[18] The triangle is divided into seven *zones*, which represent the major combinations of actor *types*; it also highlights *role*, or governance function, of each institution by means of colour-coding – *standards and commitments* (red), *operational activities* (orange), *information and networking* (green) and *financing* (blue).

The extent and nature of the global climate change governance architecture is thus represented in Figure 3.1 using an adapted form of the transnational governance triangle combined with the international regime complex approach.

An alternative way to display the global climate change governance architecture is based on the institutions' governance functions and is displayed in Figure 3.2. The triangle shape highlights multiple forms of governance within climate change, while the decagon provides a basis to discuss functional differentiation.

18 Abbott and Snidal, "The Governance Triangle: Regulatory Standards Institutions and the Shadow of the State."

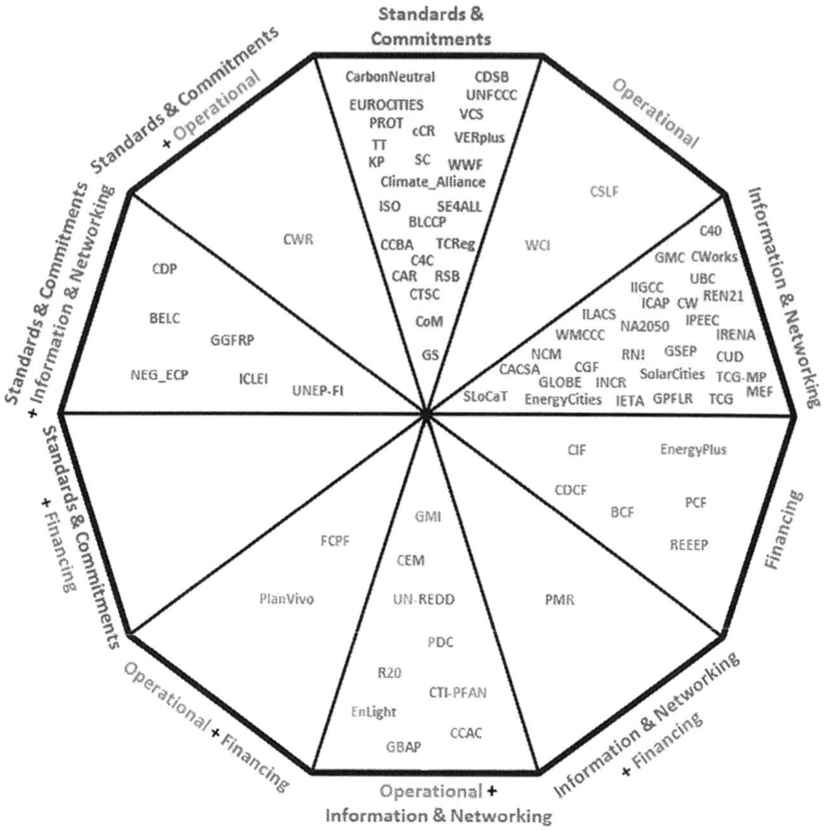

Figure 3.2 Global climate change governance decagon (2015)

Results and discussion

The current global climate change governance architecture (data collection ended April 2015) comprises 80 governance institutions; states participate in 61 arrangements (76 percent), 31 of which remain solely state-led, while 30 constitute public-private partnerships, whereas the private tier (both profit and non-profit) is represented by 19 institutions in total. There is a significant expression of business actors (51 percent) taking on governance functions across the institutional structure of climate change governance, even though the majority of these arrangements partner with states (zone 4 and 7). Information and networking is overall the most common role (also when coupled with operational activities or standards and commitments), closely followed by exclusive standard-setting functions. It is interesting to see that most of the institutions that engage in standards and commitments comprise private actors or public-private partnerships instead of just states. Financing is the least represented among climate governance arrangements, but the few that exist tend to be hybrid institutions and exclusively devoted to this role. Table 3.1 summarizes our findings.

Overall, the institutional structure of global climate change governance is characterized by the shared relevance between state and non-state actors, particularly firms, engaged in information and networking and standards and commitments.

To assess whether the institutional structure can be described as fragmented or integrated, we assess the distribution of institutions across the seven zones in the governance triangle. We understand the overall institutional structure to be integrated if institutions are concentrated in few zones (with the strongest form of integration being present in cases where all institutions fall into just one zone) and consequently to be fragmented if institutions are distributed widely across all zones (with the strongest form of fragmentation being present in cases where institutions are spread evenly across all seven zones). To examine the variability in the distribution of institutions in different zones of the triangle, we calculate the standard deviation (SD) of the distribution of institutions per zone in percentages. We can hence control for population size when comparing with other issue areas in which the number of institutions might differ. A high concentration of institutions to any one zone, meaning a low level of fragmentation, is shown by a relatively high SD. In contrast, low concentration of institutions results in a relatively low SD. Calculating the SD for our data shows a slightly fragmented institutional structure (SD 0.11 out of .035 for full concentration compared to 0.00 for ideal equal distribution; see Table 3.2).

Actors and agency in the climate governance regime complex

The next step is to explore the features of the institutional structure. The structure of the climate regime complex can be conceptualized as a network consisting of the institutions, which were mapped in the previous section, and the participants in these institutions. For this we use network analysis to compute the topological features of the regime complex and to suggest measures of fragmentation, following a method described by Widerberg.[19]

Methodology and approach

Network analysis is increasingly used within international relations studies[20] and has been applied to understand the structure of, for instance, private carbon

19 Oscar Widerberg, "Mapping Institutional Complexity in the Anthropocene: A Network Approach," in *Environmental Politics and Governance in the Anthropocene: Institutions and Legitimacy in a Complex World*, ed. Philipp Pattberg and Fariborz Zelli (London: Routledge, 2016).

20 Emilie Hafner-Burton, Miles Kahler, and Alexander Montgomery, "Network Analysis for International Relations," *International Organization* 63, no. 3 (2008); Miles Kahler, ed. *Networked Politics: Agency, Power and Governance* (Ithaca and London: Cornell University Press, 2009).

Table 3.1 Total of global climate change governance institutions per zone and per role (2015)

	Standards & Commitments	*Operational*	*Information & Networking*	*Financing*	*Standards & Commitments + Operational*	*Operational + Information & Networking*	*Information & Networking + Financing*	*Standards & Commitments + Information & Networking*	*Standards & Commitments + Financing*	*Operational + Financing*	*Total (Zone)*	*%*
1	7	2	13	1	0	4	1	2	0	1	31	38.8%
2	3	0	4	0	0	0	0	0	0	0	7	8.8%
3	4	0	0	0	0	0	0	0	0	1	5	6.3%
4	4	0	3	3	0	1	0	1	0	0	12	15.0%
5	0	0	3	0	0	0	0	0	0	0	3	3.8%
6	3	0	1	0	1	0	0	2	0	0	7	8.8%
7	3	0	5	2	0	4	0	1	0	0	15	18.8%
Total (Role)	24	2	29	6	1	9	1	6	0	2	**80**	

Table 3.2 Distribution of institutions across seven zones: results for sample and ideal types (low and high fragmentation)

	Our sample	*Low fragmentation*	*High fragmentation*
1	0.39	1.00	0.14
2	0.09	0.00	0.14
3	0.06	0.00	0.14
4	0.15	0.00	0.14
5	0.04	0.00	0.14
6	0.09	0.00	0.14
7	0.19	0.00	0.14
SD	0.11	0.35	0.00

accounting standards[21] and Multilateral Environmental Agreements.[22] To understand the global climate change regime complex, we construct a bipartite network consisting of the institutions in the governance triangle presented in the previous section and their members or participants. We define members 'as actors with the formal position to influence the rules, norms, operations or performance of an institution. They have access to the network an institution provides and benefit from the privileges it may accrue,'[23] excluding those actors that simply ascribe to the values, rules, norms or missions of an institution and do not have the ability to influence the governance of the institution. Bipartite, also called two-mode networks, connect two types of nodes – in our case institutions and their participants. It is assumed that participating in the same institutions increases the likelihood of sharing material and non-material resources. Resources thus flow through the network through shared members, which could foster resource and knowledge exchange, policy diffusion and policy innovations.[24] Bipartite networks can be transformed into one-mode networks through a method called projection. Two different one-mode networks are created. The first projection connects institutions that share members, that is, if an institution shares one member with another institution, a link is created between them.

The second projection connects the members by shared institution, that is, if two members participate in the same institution, a link is created. For understanding fragmentation we calculate measures of density and centrality. Density

21 Jessica Green, "Order out of Chaos: Public and Private Rules for Managing Carbon," *Global Environmental Politics* 13, no. 2 (2013).

22 Rakhyun Kim, "The Emergent Network Structure of the Multilateral Environmental Agreement System," *Global Environmental Change* 23, no. 5 (2013).

23 Oscar Widerberg, "Mapping Institutional Complexity in the Anthropocene: A Network Approach."

24 Andrew Jordan and Dave Huitema, "Innovations in Climate Policy: The Politics of Invention, Diffusion, and Evaluation," *Environmental Politics* 23, no. 5 (2014).

is measured by average degree in the network, that is, how many connections an actor has on average. Centrality signifies the structural 'importance' or 'prestige' of individual nodes vis-à-vis other nodes. Two different types of centrality are used: degree centrality measures the number of links a node has, and betweenness centrality measures to what extent a node lies between other nodes. In other words, we measure which node has the most friends and which is best at connecting different groups of friends.

Results and discussion

By screening the homepages of the 80 institutions that are part of the global climate change regime complex for their governing members, we find a diverse group of actors spread across five different categories. Approximately 10,700 unique organizations are part of the dataset, including 8,500 sub-national actors (cities and regions), 1,500 companies, 320 non-governmental organizations and 50 international organizations. A few institutions consisting of sub-national authorities dominate the field in terms of number of members. The Covenant of Mayors, for instance, have in our dataset nearly 5,700 members.[25] Note that we only include members with a governing function which tends to inflate the size of institutions with a networked governing structure. In some cases, such as with the CDP, a vast number of companies and sub-national authorities are rule-takers under the institutions; however, they are not formally included in governing the rule-framework. In the Covenant of Mayors, on the other hand, cities can actively take part in the governing board of the institution. If we project the networks and plot it as a one-mode network of institutions, a clearer structure of the regime complex emerges. After removing isolates, that is, nodes not connecting to other nodes, Figure 3.3 shows 73 institutions connected by shared members.

The nodes are quite densely connected, with an average degree of the network of around 15, and about 23 percent of all possible edges are present. The network also shows a clear tendency to form a tight cluster around institutions with states as members. This is perhaps not surprising since states are central actors in cross-border governance. However, it is interesting to note that the UNFCCC is deeply enmeshed with a host of other institutions in which its signatories engage in different climate-related governance activities. Also, the network shows how a few border institutions emerge, connecting state and non-state actors. For example, it looks like ICAP, REEEP and TCG_MP, for example, could be candidates for filling this bridging function. These observations are supported by computing the centrality of the nodes in terms of betweenness and degree scores and plotting these in a graph (see Graph 3.1).

The plot reveals that the majority of the institutions have a low centrality degree, meaning that few are connecting institutions in terms of membership. The betweenness centrality measure favors nodes which connect clusters of other

25 Since the data was collected in 2014, this number has increased to nearly 6,700.

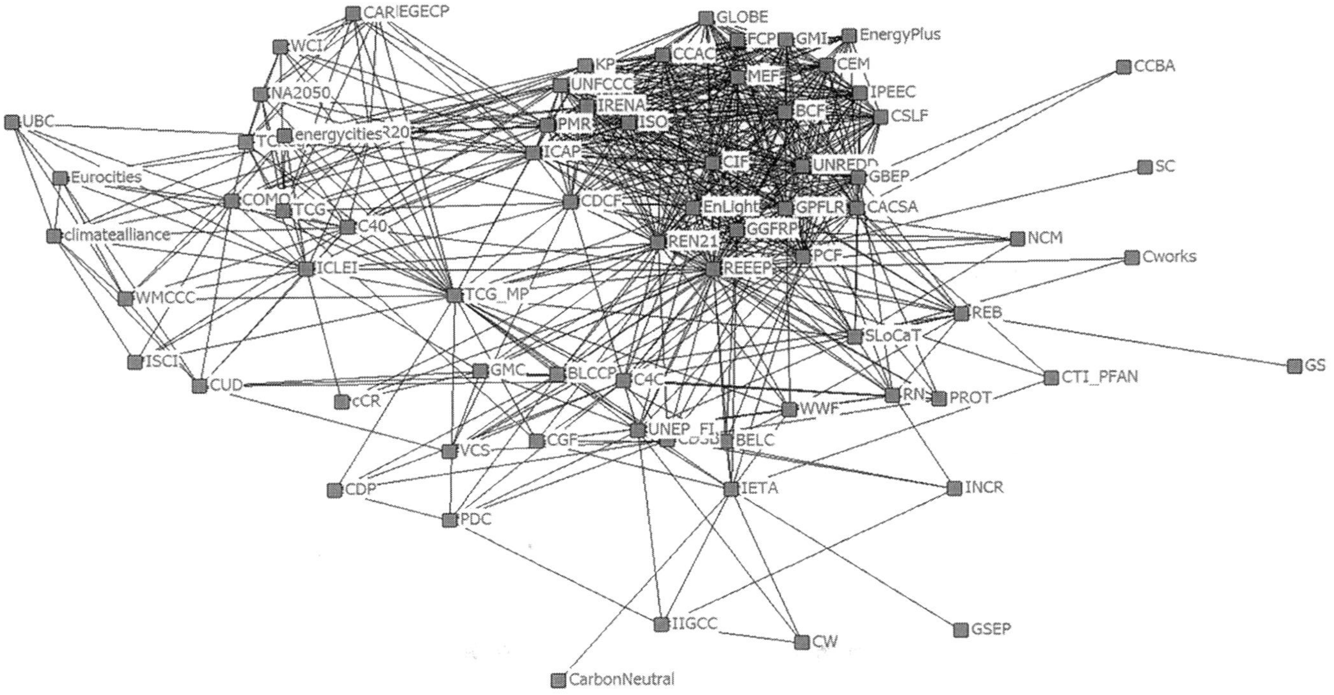

Figure 3.3 One-mode network of global climate change governance institutions (2015)

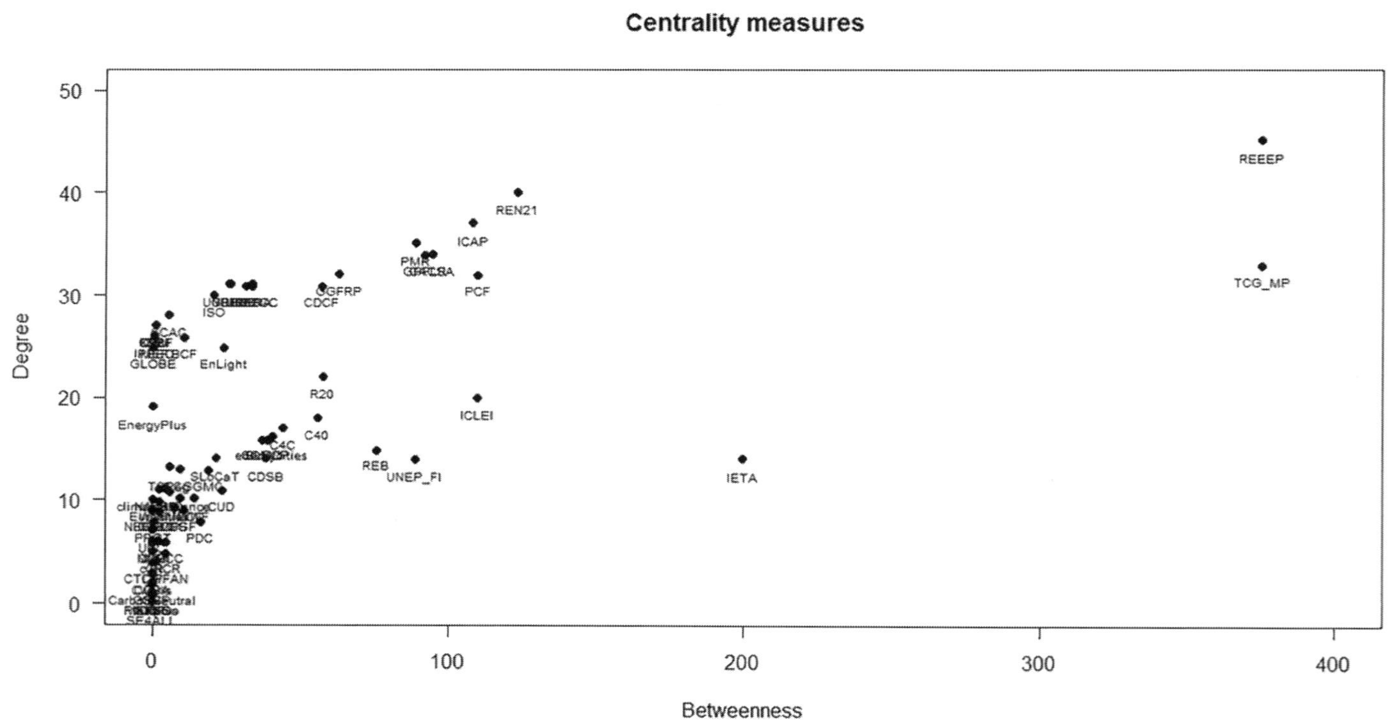

Graph 3.1 Centrality of global climate change governance institutions (2015)

nodes in our network state and non-state actors. Thus, large multi-stakeholder organizations such as REEEP and The Climate Group score high since they combine several types of actors. Using this quite simple network analysis, the structure emerging from overlapping memberships could be used to discuss the institutional meta-structure of global climate governance, namely whether it is a 'core-periphery' structure, 'loosely coupled regimes'[26] or 'polycentrism'.[27]

Meta-discourses and narratives in the climate governance regime complex

Moving towards measuring the degree of fragmentation, the third analytical step on discourses implies that the level of competition or overlap in discourses, or rather the fragmentation of discursive structure within an issue area, could play an important role in explaining the outcome and effectiveness of governance arrangements.

Methodology and approach

Our approach to study discursive structures is based on computer-assisted content analysis using Qualitative Data Analysis (QDA) Miner 4 and WordStat 7. Content analysis using QDA Miner and WordStat has been increasingly applied in Political Science, yet for much simpler analyses. Still no clear step-by-step protocol exists for applying it to map meta-discourses across global governance architectures, especially not one that ensures both replicability and reliability on large data samples.

In order to define the discursive structure of global climate governance architecture, we adopt a top-down approach where four meta-discourses are pre-identified: Green Governmentality, Ecological Modernization, Reformist Civic Environmentalism and Radical Civic Environmentalism.[28] The extent to which these meta-discourses are present in the governance structure is retrieved by automatic coding of words and phrases that are linked to specific sub-categories of the meta-discourses. These sub-categories are also adapted from literature. The data that is used for the coding process comprise the mission statements of all 80 governance institutions that are included in the governance triangle.

26 Keohane and Victor, "The Regime Complex for Climate Change."

27 Ostrom, "Polycentric Systems for Coping with Collective Action and Global Environmental Change."

28 Karin Bäckstrand and Eva Lövbrand, "Planting Trees to Mitigate Climate Change: Contested Discourses of Ecological Modernization, Green Governmentality and Civic Environmentalism," Global Environmental Politics 6, no. 1 (2006); Beth Caniglia, Robert Brulle, and Andrew Szasz, "Civil Society, Social Movements, and Climate Change," in *Climate Change and Society: Sociological Perspectives*, ed. Riley Dunlap and Robert Brulle (Oxford: Oxford University Press, 2015).

To measure fragmentation we propose two indicators that operate at different levels: first, the discursive structure of the overall climate governance architecture, and second, the discursive structure within each zone of the climate triangle. Both indicators are based on cross tabulation between meta-discourses and zones, which shows the distribution of each meta-discourse across the seven zones. The first indicator is the amount of dominant discourses in the overall governance structure (≥ 25.0 percent). The second indicator is the level of discursive homogeneity[29] or heterogeneity[30] within each zone, which allows us to study the discursive coherence across the governance triangle. The assumption is that if the overall governance structure is dominated by one meta-discourse while discursive homogeneity is high within individual zones of the triangle, discursive fragmentation is low. On the other hand, discursive fragmentation is high when the overall sample displays various equally strong meta-discourses and individual zones are discursively heterogeneous (that is, the discourses are equally strong).

In order to do so, we follow a four-step methodology. In the first step a sample of seven mission statements, produced by one representative institution out of each zone, is coded manually based on the pre-defined meta-discourses and sub-categories. After coding these, it is necessary to retrieve the coded text under each sub-category and run WordStat to collect keywords. These keywords serve as a basis to create an inclusion dictionary.[31] Alongside qualitative identification of narratives and their clustering into meta-discourses, we employ specific features of the software, such as keyword measurement, to identify additional narratives on the whole dataset, as it allows for greater reproducibility and faster coding of large datasets. Refining the inclusion dictionary is crucial to ensure both the quality and validity of computer coding, and it involves an iterative process until the saturation level is reached. After reaching saturation level, the inclusion dictionary can be applied to the whole dataset to automatically code all 80 mission statements and to retrieve results.

Results and discussion

The discursive structure of the global climate governance is studied by analyzing the cross tabulation between meta-discourses and zones. Considering that there are four meta-discourses permeating global climate change governance, our results show that there are two meta-discourses with values higher than 25 percent dominating the overall discursive structure (Table 3.3). Green governmentality emerges as the dominant meta-discourse, followed by ecological

29 High homogeneity consists of one strong meta-discourse per zone, and it can vary between 7 and 17.

30 Equally strong meta-discourses within each zone represent high heterogeneity, which ranges from 18 to 28.

31 Dictionaries are lists of words (and phrases) associated with each of the sub-categories and categories. These words are issue specific and serve as indicators of the identified meta-discourses within global climate governance.

Table 3.3 Indicator 1 – Overall discursive structure of global climate change governance (dominance of meta-discourses ≥ 25,0%) (2015)

Zone / *Discourse*	*1*	*2*	*3*	*4*	*5*	*6*	*7*	*Average*
Green Governmentality	53.7%	42.3%	45.8%	38.0%	29.2%	42.3%	35.0%	40.9%
Ecological Modernization	18.0%	52.6%	22.1%	39.8%	41.7%	45.6%	24.2%	34.8%
Reformist Civic Environmentalism	21.1%	3.6%	29.8%	19.0%	20.8%	9.3%	37.9%	20.2%
Radical Civic Environmentalism	7.2%	1.5%	2.3%	3.2%	8.3%	2.7%	2.9%	4.1%

Table 3.4 Indicator 2 – Discursive coherence (homogeneity/heterogeneity) within each zone of the climate governance triangle (2015)

Zone / *Discourse*	*1*	*2*	*3*	*4*	*5*	*6*	*7*	
Green Governmentality	53.7%	42.3%	45.8%	38.0%	29.2%	42.3%	35.0%	
Ecological Modernization	18.0%	52.6%	22.1%	39.8%	41.7%	45.6%	24.2%	
Reformist Civic Environmentalism	21.1%	3.6%	29.8%	19.0%	20.8%	9.3%	37.9%	
Radical Civic Environmentalism	7.2%	1.5%	2.3%	3.2%	8.3%	2.7%	2.9%	
≥25%	1	2	2	2	2	2	2	**13**

modernization. These findings indicate that global climate governance is predominantly a science-driven and centralized multilateral negotiation order with top-down climate monitoring and mitigation techniques implemented on a global scale. In addition, we can observe that a decentralized liberal market order that aims to provide flexible and cost-optimal solutions pervades climate change governance. Meanwhile, the meta-discourses that advocate a fundamental transformation of Northern consumption patterns, as well as a complete structural reorientation to eco-centric value systems, only account for one-third of the discursive structure of global climate governance architecture.

If we move to assessing discursive coherence across the governance triangle by looking at the level of discursive homogeneity or heterogeneity within each zone, the results indicate that the discursive coherence is relatively high (Table 3.4). Green governmentality and ecological modernization, with a value higher than 25 percent, again dominate in nearly all zones. However, such analysis allows us to detect variations in the discursive structure produced by different sets of actors, thus giving insight into the micro-level patterns of discourse production. It highlights the political dimension of global environmental governance by focusing on the role of actors and different discourses manifested in such processes and to look for similarities or differences in the way they are produced. We can observe that reformist civic environmentalism dominates zone 3 and zone 7 together with green governmentality. Zone 3 indicates the sole presence of civil society organizations, while zone 7 indicates a hybrid type of governance arrangements, namely public-private partnerships. It is surprising that the narratives such as civil society participation, local governance or the issue of the North-South divide are more present in zone 7 than they are in zone 3. Such a finding shows how global climate governance is a negotiated process, as different actors involved in the negotiating process draw from different dominant discourses.

The results also allow for a more detailed analysis of the exact key narratives in global climate action. The key narratives are the sub-categories that are coded most frequently and that occur in the greatest number of mission statements, and that therefore can be perceived as generally accepted (see Table 3.5). The narratives can also be seen as governance values, as these relate to the qualities that improve the steering and coordination by institutions or the regime complex as a whole. More specifically, the narratives are 'thin' governance values. They merely refer to the features of the institutions that determine how the institutions get things done. Accordingly, the narratives cannot be seen as 'thicker' or 'thick' governance values, as they do not answer the question whether the institution is functional, let alone legitimate or effective. In other words, the narratives solely tell us the way the climate regime 'avoids gridlock'.

The results suggest, first, that governing the climate issue area is pinned on the importance of science, knowledge and expertise to extend government control on climate. To support this governmental control, worldwide monitoring and reporting systems, and policy measures, target implementation, and bureaucratic structures are presumed to be required. Second, the results prove that market strategies, innovation and technocracy are seen as key for climate action in order to create a win-win situation for both the climate system and economic growth. More interesting is that even though the civic environmentalism discourses do

Table 3.5 Key narratives of global climate change governance based on high frequency and case-occurrence (2015)

Narrative	*Meta-discourse*	*Frequency*	*Narrative*	*Meta-discourse*	*Case-occurrence*
Science driven	GG	353	*Science driven*	GG	68.75%
Monitoring and reporting	GG	317	*Public regulatory governance*	GG	68.75%
Public regulatory governance	GG	303	*Market driven*	EM	61.25%
Innovative and technocratic	EM	223	*Monitoring and reporting*	GG	55.0%
Co-benefits	RefCE	175	*Innovative and technocratic*	EM	55.0%

Table 3.6 Key narratives derived from zone 1 of the global climate change governance triangle based on high frequency and case-occurrence (2015)

Narrative	*Meta-discourse*	*Frequency*	*Narrative*	*Meta-discourse*	*Case-occurrence*
Science driven	GG	241	*Science driven*	GG	83.87%
Monitoring and reporting	GG	200	*Public regulatory governance*	GG	77.42%
Public regulatory governance	GG	178	*Monitoring and reporting*	GG	64.52%
Innovative and technocratic	EM	137	*Stakeholder participation*	RefCE	61.29%
Local complexities	RefCE	98	*Innovative and technocratic*	EM	58.06%

not prevail in the overall architecture, there is one concept of the reformist version to which high value is attached: co-benefits. Consequently, global climate governance structure recognizes the potential of networked governance and flexible mechanism projects, if these stimulate, for example, clean energy and sustainable forest management, and improve local employment and capacity-building.

While Tables 3.3 and 3.4 show the relation between meta-discourses and the seven zones of the governance triangle, these tables also indirectly demonstrate the relation between meta-discourses and actors. Zone 1 represents institutions that are constituted by states, groups of states and international organizations. Green governmentality being the most important meta-discourse in this zones, confirms that governments are the main actors advocating state regulation, and associating climate change as an intergovernmental issue. Moreover, comparing this meta-discourse among the zones in the governance triangle shows that green governmentality is most advocated in precisely zone 1, which reinforces the argument even more.

An assessment of the key narratives (sub-categories) within zone 1 provides a more detailed view of the key concepts of state-actors' storylines (Table 3.6). In line with the dominant meta-discourse, governmental actors believe in science-based monitoring systems and bureaucratic top-down regulations to effectively govern the global climate system. However, state-actors also support innovative and technocratic ideas, related to ecological modernization. More unexpected are the narratives related to the reformist civic environmentalism meta-discourse. It shows that states advocate a reflexive form of green governmentality, meaning that state-actors acknowledge local complexities through the increased willingness to invite local actors to the creation of institutions,[32] and that they capture some democratic participation ideals.

32 Cathleen Fogel, "The Local, the Global and the Kyoto Protocol," in *Earthly Politics. Local and Global in Environmental Governance*, ed. Sheila Jasanoff (Cambridge and London: MIT Press, 2003).

Table 3.7 Key narratives derived from zone 2 of the global climate change governance triangle based on high frequency and case-occurrence (2015)

Narrative	*Meta-discourse*	*Frequency*	*Narrative*	*Meta-discourse*	*Case-occurrence*
Flexible mechanisms	EM	27	*Market driven*	EM	85.71%
Public regulatory governance	GG	26	*Flexible mechanisms*	EM	71.43%
Monitoring and reporting	GG	25	*Public regulatory governance*	GG	71.43%
Corporate involvement	EM	24	*Science driven*	GG	71.43%
Market driven	EM	22	*Monitoring and reporting*	GG	57.14%

Zone 2 comprises institutions that are formed by firms, businesses and industry associations. In this zone the ecological modernization storyline is the most significant. However, percentages in Table 3.3 show that these actors to some extent agree on green governmentality narratives. This is interesting as businesses entered the climate change arena once the deregulated and market-based agenda of ecological modernization was pushed forward, which opposed the state-centric and science-based multilateral negotiation order related to green governmentality.[33] It seems as if climate governance is dealing with a 'weaker' version of ecological modernization, which claims that traditional regulation by the state should merely be minimized in favor of more flexible approaches, instead of fundamentally transforming the role of states.[34]

Furthermore, the key narratives derived from zone 2 demonstrate the present link between the two meta-discourses in more detail (see Table 3.7). These confirm that businesses believe in flexible mechanisms to enable carbon markets, which in turn broadens the participation of other firms, businesses and industrial associations. However, the importance attached to public regulatory governance, monitoring and reporting and science underlines that these market mechanisms require the legal and bureaucratic structures facilitated by governmental actors.

Finally, zone 3 consists of institutions of which the constituent actors are CSOs. CSOs are expected to commonly advocate reformist and radical civic environmentalism and consequently to support access and openness for stakeholders

33 Karin Bäckstrand and Eva Lövbrand, "Climate Governance Beyond 2012: Competing Discourses of Green Governmentality, Ecological Modernization and Civic Environmentalism," in *The Social Construction of Climate Change: Power, Knowledge, Norms, Discourse*, ed. Mary Pettenger (Aldershot: Ashgate, 2007).

34 Arthur Mol and David Sonnenfeld, "Ecological Modernization around the World: An Introduction," *Environmental Politics Special Issue: Ecological Modernization around the World Perspectives and Critical Debates* 9, no. 1 (2000).

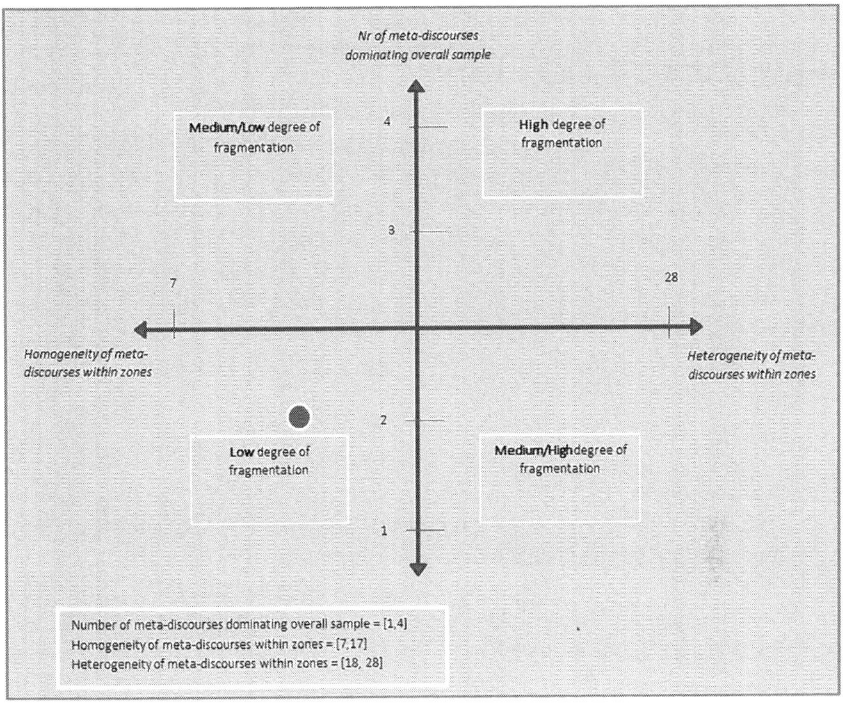

Figure 3.4 Discursive fragmentation in global climate governance architecture

and the transformation of consumption patterns, as well as to criticize expert dominance.[35] Instead, the Green Governmentality meta-discourse dominates zone 3 and is therewith most expressed by CSOs. Moreover, CSOs are expected to challenge the neoliberal approaches to climate governance found in ecological modernization.[36] Conversely, these approaches are slightly more supported than the ideals of reformist civic environmentalism. Therefore, the presumed link between CSOs and civic environmentalism has not been proven.

Based on the results discussed previously, we can conclude that the discursive structure of the global climate governance architecture experiences low degree of fragmentation, as presented in Figure 3.4. Green governmentality and ecological modernization are the two meta-discourses that dominate the overall structure, and are the most expressed by public as well as private actors. Furthermore, the evident link between the 'weak' version of ecological modernization and green

35 Bäckstrand and Lövbrand, "Planting Trees to Mitigate Climate Change: Contested Discourses of Ecological Modernization, Green Governmentality and Civic Environmentalism."
36 Bäckstrand and Lövbrand, "Climate Governance Beyond 2012: Competing Discourses of Green Governmentality, Ecological Modernization and Civic Environmentalism."

governmentality results in both meta-discourses reinforcing each other, while reformist and radical civic environmentalism are nowhere near to the same importance. Global climate change governance is therefore characterized as a science-driven centralized negotiation order associated with top-down regulations and monitoring systems. In addition, the idea of 'limits to growth' is opposed and market dynamics, innovation and technology are highly supported as long as these deliver co-benefits. It seems as if from the actors' different standpoints that institutions in global climate change action succeed to discern a common and clear message. Ultimately, that message explains the PIJ of the broader polycentric climate regime. The regime's existence is justified by pursuing a stabilization of greenhouse gas concentrations, derived from the UNFCCC, and more specifically through the use of science, top-down regulations, monitoring systems, economic growth, market dynamics, innovation and technology.

Conclusions

This chapter attempts to map institutional complexity and to measure degrees of fragmentation or coherence in global climate governance. Therewith, it presents new empirical data on three related dimensions of institutional integration or fragmentation. First, based on a mapping of all governance institutions that constitute the climate change regime complex, we provide a measurement of the distribution of institutions across all zones of the governance triangle as a proxy for overall institutional fragmentation/integration. Second, we add an organizational, agent-based dimension on climate governance by employing network analysis to explore the relational features of the institutional structure. The third analytical step is measuring discursive structure, assuming that the level of competition or overlap in discourses, or rather the fragmentation of discursive structure, could play an important role in explaining the outcome and effectiveness of governance arrangements.

Following the first step in our analysis, we built a database of institutions that make up the climate change regime complex. The institutions are mapped in a transnational governance triangle,[37] showing the extent and nature of global climate governance, and by ways of a decagon, based on the institutions' governance functions. What can be concluded is that the institutional structure of global climate governance is marked by the shared relevance between state and non-state actors, particularly business actors. Furthermore, information and networking and standard setting are the most common functions the institutions in the regime complex engage in. Our analysis of the actual distribution of institutions across the zones shows a slightly fragmented structure.

The network analysis, conceptualizing the climate change regime complex as a network consisting of institutions and the participants of the institutions, identified 10,700 unique organizations. These include 8,500 cities and regions, 1,500

37 Abbott and Snidal, "The Governance Triangle: Regulatory Standards Institutions and the Shadow of the State."

Table 3.8 Results of measurements for all three dimensions of institutional fragmentation/integration

Indicator	*Explanation*	*Value*
Institutional distribution	Showing variation using standard deviation in the distribution of institutions per zone in percentages	0.11
Density	The percent of total possible links that are present	23%
Clustering	The average of clustering coefficients for all nodes in the networks (ranging from 0–1)	0.717
Overall discursive structure	How many dominating discourse there are (ranging from 1–4)	2
Discursive coherence	Measuring homogeneity vs. heterogeneity among the discourses (ranging from 7–28)	13

companies, 320 NGOs and 50 international organizations. When these are plotted as a one-mode network that connects institutions which share members, a clear structure emerges, in which the institutions are quite densely connected. The network forms a tight cluster around institutions with states as members and includes a few border institutions that serve as a bridge between the state and non-state actor. Computing the centrality measures, degree and betweenness reveals that only few institutions are connected in terms of membership, but that some institutions connect clusters of institutions. This structure leaves us with new insights with regard to the discussion of whether the institutional meta-structure should be seen as a 'core-periphery structure', 'loosely coupled regimes'[38] or 'polycentrism'.[39]

Finally, the innovative way of mapping the discursive structure shows a low degree of fragmentation that global climate governance experiences. Green governmentality and ecological modernization are the most important meta-discourses in climate governance, meaning that the governance structure resembles a science-driven centralized negotiation order where top-down regulations and monitoring systems are seen as key. On top of that, the importance of economic growth is emphasized, as well as the potential of market dynamics, innovation and technology as long as co-benefits are being delivered. The discursive structure suggests that from the actors' different perspectives climate institutions do follow a common storyline. With the discursive structure we have identified the PIJ and the thin governance values of the broader climate change regime. Table 3.8 provides a combined overview of all measurements of fragmentation/integration that we have discussed in this chapter.

38 Keohane and Victor, "The Regime Complex for Climate Change."

39 Ostrom, "Polycentric Systems for Coping with Collective Action and Global Environmental Change."

With this research we introduced three innovative methods to measure institutional coherence or fragmentation in global climate governance and illustrated the applicability of the methodological approaches. However, questions about the methodologies and measurement methods remain. Therefore, future research should take up the challenge to further refine the methodologies. More specifically, since our discussion on new data is based on computer-aided analyses, one should continue to explore the features and functions of the used software, as it has high potential for the analysis of big datasets. On top of that, it is of high importance to continually update the mappings of climate change governance, as the regime complex is constantly evolving. Especially since the latest climate summit in December 2015 in Paris, new institutions emerge, new actors join and others disappear. Finally, to advance this research agenda even further, future research could explore the link between institutional complexity and measurements of effectiveness and means to improve the effectiveness of climate change governance. After all, mapping the institutional complexity of climate change governance is merely the first step towards a better theory of climate governance.

References

Abbott, Kenneth. "The Transnational Regime Complex for Climate Change." *Environment & Planning C: Government & Policy* 30, no. 4 (2012): 570–90.

Abbott, Kenneth, and Duncan Snidal. "The Governance Triangle: Regulatory Standards Institutions and the Shadow of the State." In *The Politics of Global Regulation*, edited by Walter Mattli and Ngaire Woods, 44–88. Princeton: Princeton University Press, 2009.

———. "Strengthening International Regulation through Transnational New Governance: Overcoming the Orchestration Deficit." *Vanderbilt Journal of Transnational Law* 42, no. 2 (2009): 1–80.

Bäckstrand, Karin, and Eva Lövbrand. "Climate Governance beyond 2012: Competing Discourses of Green Governmentality, Ecological Modernization and Civic Environmentalism." In *The Social Construction of Climate Change: Power, Knowledge, Norms, Discourse*, edited by Mary Pettenger, 123–48. Aldershot: Ashgate, 2007.

———. "Planting Trees to Mitigate Climate Change: Contested Discourses of Ecological Modernization, Green Governmentality and Civic Environmentalism." *Global Environmental Politics* 6, no. 1 (2006): 50–75.

Biermann, Frank, Philipp Pattberg, Harro Van Asselt, and Fariborz Zelli. "The Fragmentation of Global Governance Architectures: A Framework for Analysis." *Global Environmental Politics* 9, no. 4 (2009): 14–40.

Bulkeley, Harriet, Liliane Andanova, Michele Betsill, Daniel Compagnon, Thomas Hale, Matthew Hoffman, Peter Newell, Matthew J. Paterson, Charles Roger, and Stacy D. VanDeveer. *Transnational Climate Change Governance*. Cambridge: Cambridge University Press, 2014.

Caniglia, Beth, Robert Brulle, and Andrew Szasz. "Civil Society, Social Movements, and Climate Change." In *Climate Change and Society: Sociological Perspectives*, edited by Riley Dunlap and Robert Brulle, 235–68. Oxford: Oxford University Press, 2015.

Fogel, Cathleen. "The Local, the Global and the Kyoto Protocol." In *Earthly Politics: Local and Global in Environmental Governance*, edited by Sheila Jasanoff, 103–25. Cambridge and London: MIT Press, 2003.

Green, Jessica. "Order Out of Chaos: Public and Private Rules for Managing Carbon." *Global Environmental Politics* 13, no. 2 (2013): 1–25.

Hafner-Burton, Emilie, Miles Kahler, and Alexander Montgomery. "Network Analysis for International Relations." *International Organization* 63, no. 3 (2008): 559–92.

Jordan, Andrew, and Dave Huitema. "Innovations in Climate Policy: The Politics of Invention, Diffusion, and Evaluation." *Environmental Politics* 23, no. 5 (2014): 715–34.

Kahler, Miles, ed. *Networked Politics: Agency, Power and Governance*. Ithaca and London: Cornell University Press, 2009.

Keohane, Robert, and David Victor. "The Regime Complex for Climate Change." *Perspectives on Politics* 9, no. 1 (2011): 7–23.

Kim, Rakhyun. "The Emergent Network Structure of the Multilateral Environmental Agreement System." *Global Environmental Change* 23, no. 5 (2013): 980–91.

Mol, Arthur, and David Sonnenfeld. "Ecological Modernization around the World: An Introduction." *Environmental Politics Special Issue: Ecological Modernization around the World Perspectives and Critical Debates* 9, no. 1 (2000): 1–14.

Ostrom, Elinor. "Polycentric Systems for Coping with Collective Action and Global Environmental Change." *Global Environmental Change* 20, no. 4 (2010): 550–57.

Raustiala, Kal, and David Victor. "The Regime Complex for Plant Genetic Resources." *International Organization* 58, no. 2 (2004): 277–309.

Widerberg, Oscar. "Mapping Institutional Complexity in the Anthropocene: A Network Approach." In *Environmental Politics and Governance in the Anthropocene: Institutions and Legitimacy in a Complex World*, edited by Philipp Pattberg and Fariborz Zelli, 81–102. London: Routledge, 2016.

Zelli, Fariborz. "The Fragmentation of the Global Climate Governance Architecture." *Wiley Interdisciplinary Reviews: Climate Change* 2, no. 2 (2011): 255–70.

Zelli, Fariborz, and Harro Van Asselt. "Introduction: The Institutional Fragmentation of Global Environmental Governance." *Global Environmental Politics* 13, no. 3 (2013): 1–13.

4 Institutions, institutional practices, and global climate change governance

The case of REDD+

Eero Palmujoki[1]

Introduction

According to several assessments, climate governance under the United Nations Framework Convention on Climate Change (UNFCCC) has shifted from using a top-down model to a more pragmatic approach in which national and local initiatives, as well as public, private, and business ones, are able to mould the Convention.[2] This change stems from the failure of predominant climate politics in the past, which sought a comprehensive, uniform convention to mitigate climate change. In contrast, the new approach seeks to form a legally binding agreement from nationally determined contributions at the 2015 COP 21 in Paris. Although the new agreement has been hailed as a pragmatic approach that ushers in a new era of climate governance, the way in which national contributions are reached nevertheless very much depends on national governments. The Paris Agreement therefore highlights issues of integrity and fragmentation in global climate governance by a single regime from a new perspective. Indeed, climate governance already involves a mixture of states, international organizations (IOs), civil society organizations, businesses, and private actors, as well as a variety of development policies and market practices; the process also accommodates both multilateral and bilateral state relations along with public–private ones. Such a fragmented governance architecture raises issues for global attempts to mitigate climate change in that it challenges states' abilities to tackle climate change as a single institution.[3]

By extension, the fragmentation of governance raises questions about the relationship of governance to the UNFCCC regime. For one, how do different

1 The author would like to thank Kate Dooley for her insightful and valuable comments and formulations on earlier drafts, and Rowena Maguire for her suggestions on the final drafts of this article.

2 Steinar Andresen, "International Climate Negotiations: Top-Down, Bottom-up or a Combination of Both?," *The International Spectator* 50, no. 1 (2015); Oliver Geden, "Pragmatism in Climate Policy," *Project Syndicate* (23 November 2015), https://www.project-syndicate.org/commentary/paris-climate-change-pragmatic-approach-by-oliver-geden-2015–11.

3 Robert Keohane and David Victor, "The Regime Complex for Climate Change," *Perspectives on Politics* 9, no. 1 (2011): p. 8; Michael Zürn and Benjamin Faude, "On Fragmentation, Differentiation, and Coordination," *Global Environmental Politics* 13, no. 3 (2013).

policies and practices in climate governance constitute a climate regime and that regime's integrity? The characteristics of fragmentation have been set in relief in the context of Reducing Emissions from Deforestation and Forest Degradation (REDD),[4] the development of which also illustrates the development of the UNFCCC climate regime during the post-Copenhagen period up to the consensus that emerged at COP 21.

Second, it is justified to ask how bottom-up governance moulds the interpretation of norms and rules that regulate climate mitigation. Do local, private, and national practices that are now accepted as legitimate create incompatible rules and different interpretations of norms in climate mitigation, and will this lead to the disintegration of climate regimes? In the case of REDD+, one such important practice is MRV (Measuring, Reporting, and Verification), which is already part of the common standards and procedures among the UNFCCC. Will the bottom-up model change already established common standards of MRV? Therefore, one of the main issues in this volume – thin or thick governance[5] – is crucial in this chapter as well.

In response, the present chapter scrutinizes the multifaceted topic of climate change governance from the perspective of international society's fundamental institutions. Understood here in a broad sense, *institutions* constitute established practices in the international community.[6] The chapter discusses whether global climate change governance is a new practice in international relations or whether it should be considered as representing a more limited regime. In comparison to other established practices of interaction and cooperation among states, including sovereignty and justice as well as procedural practices such as diplomacy, trade, and development assistance, the fragile entrenchment of climate change mitigation as an international practice offers a new perspective. Established practices have developed into institutions with canonical norms and rules, all of which inform not only states' roles but also private actors, businesses, and individuals' actions. In that sense climate change governance arguably cannot be deemed a strong institution, despite its incipient institutionalization.[7] Rather, it is formed by international society's more established

4 Harro van Asselt, "Integrating Biodiversity in the Climate Regime's Forest Rules: Options and Tradeoffs in Greening Redd Design," *Review of European Community & International Environmental Law* 20, no. 2 (2011).

5 See, Hugh Breakey, Tim Cadman, and Charles Sampford, Chapter 2 in this volume.

6 Hedley Bull, *The Anarchical Society: A Study of Order in World Politics* (London: Macmillan, 1977); Barry Buzan, *From International to World Society? English School Theory and the Social Structure of Globalisation* (Cambridge and New York: Cambridge University Press, 2004); Kalevi Jaakko Holsti, *Taming the Sovereigns: Institutional Change in International Politics* (Cambridge and New York: Cambridge University Press, 2004).

7 Barry Buzan, *An Introduction to the English School of International Relations: The Societal Approach* (Cambridge: Polity Press, 2014), pp. 161–63; Robert Falkner, "Global Environmentalism and the Greening of International Society," *International Affairs* 88, no. 3 (2012); Eero Palmujoki, "Fragmentation and Diversification of Climate Change Governance in International Society," *International Relations* 27, no. 2 (2013).

institutions – sovereignty, market practices, and development assistance – which the present chapter investigates in terms of their influence on efforts to mitigate greenhouse gas (GHG) emissions and to curb global warming, both among multilateral political processes and in national and local implementations. Many initiatives nevertheless employ top-down approaches, and this chapter does not suggest that international institutions constitute local practices only. Rather, it investigates how international institutions are shaped by different regional, national, and local practices.

However, a more limited definition of *institution* – that is, a rational and purposeful body – refers to the area-specific regimes upon which this chapter focuses. In line with this volume's focus, this definition also coincides with regime integrity. Accordingly, *integrity* more precisely refers to consistency or how agents (that is, UNFCCC members) live according to the regime's principles.[8] The climate regime under the UNFCCC represents one such rational institution, which in the present context can be termed a *secondary institution* that has emerged to arrange collective action to mitigate climate change. The case of global and local climate change mitigation efforts that is studied here is REDD, an initiative designed to curb GHG emissions by way of forest conservation, carbon trade, and development assistance.

At the same time the regime's rationality is examined here in the context of more fundamental international institutions. Accordingly, established practices of international interaction provide a framework that all regime members accept as the way the regime works. The examination of the regime in the framework of established practices focuses on context integrity, or integrity among the institutions that are fundamental in the climate regime: sovereignty and market and development assistance. From this perspective, the issue of the ways in which fundamental institutions influence regimes reflects how broadly members share institutions and how similarly institutions understand the content and relations among the institutions. If members of international society share numerous institutions and an approximately similar understanding of their content, norms, and procedures, then governance is deemed 'thick', and the regime works harmoniously and effectively. In a situation where few common institutions offer basic procedures for interaction, governance is termed 'thin', regimes form the setting for power struggles, and the regime's integrity is constantly at risk. The thinness of governance, not its organizational fragmentation, poses risk to governance's integrity and the regime's effectiveness. Such concepts of thick and thin governance bear a close affinity with Breakey's, Cadman's, and Sampford's use of the

8 Hugh Breakey and Timothy Cadman, "A Comprehensive Framework for Evaluating the Integrity of the Climate Regime Complex," in *Ethical Values and the Integrity of the Climate Change Regime*, ed. Hugh Breakey, Vesselin Popovski, and Rowena Maguire (Surrey: Ashgate, 2015); Hugh Breakey, Tim Cadman, and Charles Sampford, "Conceptualizing Personal and Institutional Integrity: The Comprehensive Integrity Framework," in *The Ethical Contribution of Organizations to Society*, ed. Michael Schwartz, Howard Harris, and Debra Comer (Bingley, UK: Emerald Group Publishing Ltd, 2015), pp. 6–8.

terms in this volume, though for the observation of the nature of governance the terms lack such sophisticated criteria.[9]

The institutions of international society and climate change regime

The concept of regime that surfaced in discussions about international relations (IR) in the 1980s was designed to round out and bolster rationalist IR theories addressing international cooperation. The aim of the concept was to explain how self-interested actors – in a word, states – negotiate the problems of collective action posed by increased interstate interaction.[10] The mainstream approach, which Wendt and Duval (1989) called 'New Institutionalism', is based on a choice theoretic approach and individual ontology, where the role of regimes is to organize state practices in a given area.[11] Regimes – or institutions – are conscious constructions of the states. Accordingly, regime came to describe the tool of a power struggle of a state in an interdependent world.

Regimes may play some independent roles in light of explanations of their functions, according to which they reduce uncertainty, facilitate communication, promote learning, and disseminate information. In some studies the political nature of regime is obscured, as regimes are perceived as technical mechanisms for solving global problems. However, regime theory has generally turned to power-based theories in explaining where the rule of regime originates – that is, if the regime works, it serves the interests of the most powerful; if it does not, or if the regime is weak, the interests of the most powerful are not involved.[12]

In the scholarship addressing the institutions of international society, which Wendt and Duval (1989) called 'Old Institutionalism', the emergence of regimes and international institutions has been studied from a rather distinct perspective. Accordingly, international institutions are not based on the choices of individual state agents but are established practices that constitute and enable meaningful international interaction.[13] The relationship between different international institutions can be hierarchical; there are 'more' fundamental institutions that

9 Buzan, *From International to World Society? English School Theory and the Social Structure of Globalisation*; cf. Breakey, Cadman and Sampford in this volume.

10 Andrew Hurrell, "International Society and the Study of Regimes: A Reflective Approach," in *Regime Theory and International Relations*, ed. Volker Rittberger and Peter Mayer (Oxford and New York: Clarendon Press, 1993); Sabine Reinecke, Till Pistorius, and Michael Pregernig, "UNFCCC and the REDD+ Partnership from a Networked Governance Perspective," *Environmental Science and Policy* 35 (2014).

11 Alexander Wendt, and Raymond Duvall "Institutions and International Order," in *Global Changes and Theoretical Challenges: Approaches to World Politics for the 1990s*, ed. E.O. Czempiel and J.N. Rosenau (Lexington and Massachusetts, MA: Lexington Books, 1989).

12 Hurrell, "International Society and the Study of Regimes: A Reflective Approach."; Stephen D. Krasner, "Structural Causes and Regime Consequences: Regimes as Intervening Variables," *International Organization* 36, no. 2 (1982).

13 Wendt and Duvall, "Institutions and International Order."

make development possible for 'less' fundamental institutions (regimes). This scholarly discussion, which involves the English School of international relations (ES), constructivist-oriented IR scholars, and even traditional IR scholars, has not produced any simple explanation for rules set by the most powerful, nor has endorsed the view of regimes as technical mechanisms in problem solving, but rather has focused on the relationship between institutions.[14]

A regime's functional explanation is the same as in mainstream regime theories: regimes have emerged in certain areas that call for collective action. The central idea is that fundamental institutions of international society constitute international regimes (that is, secondary institutions) differently. At the same time, this does not suggest that institutions define regimes, but that in order to work regimes derive from the principles, norms, and procedures of institutions. These institutions provide broad frames of cooperation, yet not only demarcate limits but also enable interaction and cooperation.[15]

In mainstream IR theories the struggle for power is arguably a permanent construct that seems inevitable. Regime characteristics are thus more static than dynamic, and changes in regimes can be explained by changes in the capacities of their members. From the perspective of Old Institutionalism, international institutions, by contrast, are dynamic; they arise, evolve, and vanish against the backdrop of civilization's transformations. The content of institutions varies and develops; for example, the principle of sovereignty takes on new content in the context of the European Union. By the same token, the market as a fundamental institution of international society has persisted since the Cold War ended.[16] Accordingly, changes in institutions and in their mutual relationships constitute changes in regimes. The dynamism of international institutions thus reflects on the dynamism of regimes.

The power struggle is held in the context of these institutions.[17] In certain areas the bargaining takes place in international organizations and regimes, which in turn influence the development and change of more fundamental institutions. Therefore, not only do these fundamental institutions enable specific regimes, but also the development of regimes (as a result of political bargaining) leads to changes in these fundamental institutions.[18]

14 Buzan, *From International to World Society? English School Theory and the Social Structure of Globalisation*; Cornelia Navari, "Primary and Secondary Institutions: Quo Vadit," *Cooperation and Conflict* 51, no. 1 (2016); Laust Schouenborg, "The English School and Institutions: British Institutionalists?," in *Guide to the English School in International Studies*, ed. Cornelia Navari and Daniel M. Green (Oxford: Willey Blackwell, 2014); Kilan Spandler, "The Political International Society: Change in Primary and Secondary Institutions," *Review of International Studies* 41, no. 3 (2015); Wendt and Duvall, "Institutions and International Order."

15 Buzan, *From International to World Society? English School Theory and the Social Structure of Globalisation*; Spandler, "The Political International Society: Change in Primary and Secondary Institutions."; Wendt and Duvall, "Institutions and International Order."

16 Falkner, "Global Environmentalism and the Greening of International Society," p. 519.

17 Andrew Hurrell, "Power, Institutions, and the Production of Inequality," in *Power in Global Governance*, ed. M. Barnett and R. Duvall (Cambridge: Cambridge University Press, 2005).

18 Wendt and Duvall, "Institutions and International Order."

The question of which institutions are fundamental to international society remains open, even to ES scholars, though there is a common understanding that there are established practices of interaction between states. Despite a consensus of common practices among members of international society, there is no sure agreement about which institutions are fundamental. I am inclined to follow the interpretation that stresses their importance in certain contexts. In this chapter I discuss institutions that clearly enable the climate change regime. In this sense institutions are historical constructions, as are other conceptual structures of international relations.[19]

The climate regime emerged from the call to stop global warming, and it is also the original rationale for REDD. However, climate mitigation does not restrict all the countries in a similar way when its practices have developed in the interplay of more established international institutions. Sovereignty and justice have stood out as two important principles framing ethical approaches to climate change mitigation since the UN Conference on Environment and Development (UNCED) in 1992. Both guided the creation of the UN Framework Convention on Climate Change (UNFCCC) two years later. Sovereignty emphasizes that governments are the only sovereign agents responsible for fulfilling the duties that the Convention assigns. The ethical frame of justice is reflected in the climate regime by the norm of equity. Equity refers to the responsibilities borne by the Convention's different members and the principle of common but differentiated responsibilities. Justice is a weaker institution than is sovereignty because some developed countries have reservations about it. In REDD+ justice and equity are operationalized by the procedural practices of development assistance. UNCED's important principle, the precautionary principle, is a crucial environmental principle well established in international law,[20] but it cannot be deemed an institution. In climate change governance the precautionary principle is a contested norm because it is not completely accepted in actual climate mitigation practices by some important UNCED signatories.[21]

Aside from these principles, the UNFCCC negotiations, the Kyoto Protocol, and subsequent initiatives have established important practices in climate change mitigation: market practices and development assistance. Market practices related first to emissions trading and were originally adopted domestically in the United

19 Charlotta Friedner Parrat, "Reform of the United Nations: Between Great Power Management and Sovereign Equality," *International Studies Association Conference* (New Orleans 20 February 2015). Paper presented in the panel International Organization in the Anarchical Society.

20 Caroline E. Foster, "Precaution, Scientific Development and Scientific Uncertainty under the WTO Agreement on Sanitary and Phytosanitary Measures," *Review of European Community and International Environmental Law* 18, no. 1 (2009); Owen McIntyre and Thomas Mosedale, "The Precautionary Principle as a Norm of Customary International Law," *Journal of Environmental Law* 9, no. 2 (1997); Jacqueline Peel, "Interpretation and Application of the Precautionary Principle: Australia's Contribution," *Review of European Community and International Environmental Law* 18, no. 1 (2009).

21 Falkner, "Global Environmentalism and the Greening of International Society," p. 518.

States in order to create voluntary, market-based mechanisms for curbing GHG emissions. This idea was later adopted by the Kyoto Protocol's flexible mechanisms and became an established practice for curbing GHG emissions among EU members.[22] It calls for numerous concrete practices and technical standards intended to mitigate climate change.

Development assistance as a mitigation practice is based on justice and the concept of sustainable development. In the Kyoto Protocol, the Clean Development Mechanisms (CDMs) were the only way to include developing countries in GHG emissions mitigation and emissions trading. This partly market-based practice enables developed countries to reach emissions targets by helping developing countries in their voluntary attempts to curb GHG emissions. However, the CDMs were not exploited extensively in the least developed countries, and subsequent Conferences of Parties (COPs) introduced new aid-related initiatives to bind developing countries into mitigating climate change as well. Common to all was a tendency to deviate from UNFCCC contexts, and many actors, organizations, and institutions that were traditionally part of developmental cooperation business also become active in this field. In this way the practices of development assistance became part of the global climate change regime. In the Paris Agreement these three institutions and international practices – sovereignty, market, and justice through development assistance – are even more emphasized than before.

Positioning REDD in the global climate change regime

International institutions and the scope of REDD

The impetus of the REDD initiative was to connect forest conservation to emissions trading and to create permanent mitigation mechanisms for developing countries with tropical forests. The original idea, presented by Papua New Guinea and Costa Rica at the COP-11 in Montreal in 2005, consisted of a proposal to reduce emissions from tropical deforestation through different financial mechanisms, including market-based mechanisms and direct transactions from developed countries to REDD target countries. The establishment of national funds to allocate REDD payments was also discussed.[23] In order to prove that emission reductions are additional, the COP discussions suggest using national reference levels defined by international bodies to measure reductions in forest loss and to avoid national leakage problems that may occur if the reference level is based on project-oriented accounting. In COPs that followed, the original proposal was equipped with several proposals to the REDD+ initiative that connect first to sustainable forest management (forest degradation) and then to

22 Loren R. Cass, *The Failures of American and European Climate Policy: International Norms, Domestic Politics, and Unachievable Commitments* (Albany: State University of New York Press, 2006), pp. 174–77.

23 Charlotte Streck, "Financing Redd+: Matching Needs and Ends," *Current Opinion in Environmental Sustainability* 4, no. 6 (2012).

development goals for global and local environmental conservation. REDD+ was adapted to the UNFCCC as a pivotal building block of the post-Kyoto Protocol climate regime.[24] In UNFCCC discussions and pilot projects, however, the idea of market mechanisms has become obscured in the preparations of REDD+, while the general principles of financing forest conservation have remained undecided. Despite some market-based pilot projects, the majority of funding has thus far been donor-based, largely owing to the fact that a great part of funding under the rubric of REDD+ has gone to REDD preparedness projects, not to compensations of forest conservation.

Significant in the present state of affairs is the exceptionally complex process to prepare both the climate change regime and potential REDD+ target countries for REDD, the diversification of REDD+ beyond the UNFCCC context, the different models for funding, and the different organizations' entry into REDD+. Similarly, if forests are understood as carbon storage, then the question may create environmental issues concerning biodiversity, which then creates questions about the general regulations of earned carbon assets and new issues of their monitoring.[25]

At the organizational level of the UNFCCC process, REDD+'s technical issues and some important policy questions were assigned to the Subsidiary Body for Scientific and Technological Advice (SBSTA). Proposals and discussions from the SBSTA are important to the development of REDD+ practices, as they are part of the bargaining between UNFCCC members. Although no decisions have been made based on most of these proposals, they still constitute REDD+ practices. These proposals operationalize REDD+ governance according to global climate change governance in particular and connect it to international institutions in general. These apparently technical systems, such as MRV, have already been applied to working carbon trade schemes (such as the EU's) and have enforced the climate change regime's integrity. They, nevertheless, ended up being revised in the SBSTA discussions about REDD+.

The issues did not emerge in the REDD agenda simultaneously, but evolved during the UNFCCC discussions and process after the REDD was put to the UNFCCC agenda in the Montreal COP 2005. The examination of the development of REDD+ practices in the present chapter is based mostly on the SBSTA discussion. It should be noted that a remarkable number of concrete REDD practices developed in different REDD processes, including the UN REDD, the World Bank's Forest Carbon Partnership Facility (FCPF) programme and

24 Jan Willem den Besten, Bas Arts, and Patrick Verkooijen, "The Evolution of REDD+: An Analysis of Discursive-Institutional Dynamics," *Environmental Science & Policy* 35, no. 1 (2014); Constanze Haug and Joyeeta Gupta, "The Emergence of Redd on the Global Agenda," in *Climate Change, Forests and Redd: Lessons for Institutional Design*, ed. J. Gupta, N. van der Grijp, and O. Kuik (Abingdon and New York: Taylor & Francis, 2013); Till Pistorius, "From Red to REDD+: The Evolution of a Forest-Based Mitigation Approach for Developing Countries," *Current Opinion in Environmental Sustainability* 4, no. 6 (2012).

25 Haug and Gupta, "The Emergence of Redd on the Global Agenda."

different donor governments programmes, which apply their own development assistance practices. Various programmes also give rise to a proliferation of REDD funding institutions. Although these programmes and their concrete projects, or domestic developments in the target REDD countries, are not covered here, they are reflected in the SBSTA discussions.

For countries to be eligible for financial compensation for REDD, forest conservation must actually reduce total emissions without creating social harm, including problems for the livelihoods of locals and rights of indigenous populations. Simply put, these issues create a complex chain of problems that seemed to be difficult to solve separately. At least three chief issues can be identified, none of which are simple governance problems but which refer to deeper international institutions – namely, how funding is managed; on what REDD MRV is based and how it is organized; and how forest conservation is organized.

REDD+ practices and international institutions

All these issues refer, on the one hand, to the institution of the market and, on the other hand, to sovereignty through states' authority to manage carbon assets. The third chief issue, how forest governance is organized, is addressed towards development, both in the distribution of the incomes of conservation as well as in the issues of participation in the management of forest conservation. The discussions held by the UNFCCC SBSTA seem to be very technical. In that sense they seem to refer to 'thin' governance values. However, apparently technical discussions refer to certain practices that are very political in nature. They include concrete practices promoted by the states to address the aspects of REDD management, such as measuring the change in carbon stock stored in the forest or the evaluation of socio-economic impact of forest conservation in mitigation activities. These issues tend to lead to 'thicker' governance values, including the issues of decision-making, participation, and local and indigenous peoples' rights.

For the goal of curbing GHG emissions, target countries need to create credible MRV systems that ensure emissions reductions. The systems are apparently technical but, as a last resort, political and deeply attached to the international institution of sovereignty. Technical concerns appeared even in the first REDD proposal. First, the concept of additionality requires governments to establish national baseline rates. Accordingly, a mitigation measure 'is additional if anthropogenic emissions of GHGs by sources are reduced below those that would occur in the absence of" that measure.[26] Here baseline rates refer to the estimated emissions that would occur if no mitigation measures were taken. Second, carbon leakage can occur in numerous ways, both among states and domestically. In certain cases leakage occurs among different production sectors of a company operating in two countries with contrasting emissions policies. In the case of REDD

26 "CDM Rulebook: What Is Additionality?" http://www.cdmrulebook.org/84.html; United Nations Framework Convention on Climate Change (UNFCCC/CCNUCC), "Glossary CDM Terms," Version 08.0 (2 March 2012), https://cdm.unfccc.int/Reference/Guidclarif/glos_CDM.pdf.

the full protection of forests in a certain part of a country or district can cause deforestation in another part of the country, for the market demand for timber may remain high regardless of protection policies. Third, permanence refers to the fact that the emission reductions must be guaranteed to be permanent for a given period of time, especially if sold on a carbon market and thereby generating an offset. Therefore, carbon-trading mechanisms must evince an established institutional premise that sustainable mitigation is possible. Finally, efficient yet cost-effective monitoring of emissions is necessary for any mitigation process.[27] To activate all of these aspects, the governments' administrative capacities should be strengthened, which also cements the sovereignty of government.[28]

The third issue – that is, how forest conservation is organized – reflects both a country's administrative capacities and governance models offered by IOs. When MRV issues and definitions are developed in UNFCCC projects, the development of REDD+ governance models very much follows the models already applied in development assistance and sustainable forest management, including criteria and indicator (C&I) systems developed in IOs, such as in the International Tropical Timber Organization. The development practices are consistent with the high-level forums between donors and developing countries that resulted in a compilation of general practices listed in the Paris Declaration in 2005 and subsequently elaborated on in the Accra (2008) and Busan (2011) forums.

The IOs – the WB particularly – have used different assessment tools of governance in guiding and reviewing the governance quality of their development projects. These tools have been applied in the REDD projects too. IOs and donors endorse the models that try to realize good governance by aiming to consider civil society when making decisions and implementing development projects. The models are based on different kinds of stakeholder models first used in business and routinely applied in development projects since the 1990s. These civil society engagement models seek to lessen the problems of corruption and unworkable administration via the participation of stakeholders and by disseminating information about ongoing projects to local beneficiaries.[29] The problem, however, is that successful development cooperation calls for a sound, functioning relationship between donors and governments, and situating civil society actors between donors and governments does not work in most cases. The governance issue in REDD+ emerges when many mitigation measures require the strengthening of governments, while the IOs development model suggests more power for civil society. Substantial issues with structural governance models surface, for example, in different conservation models that can vary from partial conservation, which

27 United Nations Framework Convention on Climate Change (UNFCCC), "Report of the Conference of the Parties on Its Eleventh Session, Held at Montreal from 28 November to 10 December 2005," (FCCC/CP/2005/MISC.1 2005), Item 6 of the provisional agenda.

28 Jacob Phelps, Edward L. Webb, and Arun Agrawal, "Does REDD+ Threaten to Recentralize Forest Governance?," *Science* 328, no. 5976 (2010).

29 Eero Palmujoki and Pekka Virtanen, "Global, National, or Market? Emerging REDD+ Governance Practices in Mozambique and Tanzania," *Global Environmental Politics* 16, no. 1 (2016).

guarantees the economic conditions of locals, to fortress conservation, which disallows all exploitation of natural resources by local populations.[30]

REDD preparedness projects implemented by several IOs, including the UN (UNREDD) and the World Bank, show direct concern for the two latter issues, though the idea of the market was emphasized in the UNFCCC discussion. The experiences of REDD+ projects and their cooperation with REDD+ preparedness projects have not yet elucidated what the funding model of the REDD should be, while more established practices of MRV and governance models have already been established in REDD governance. Different REDD projects, owing to the procedural governance standards of the IOs and donors, may reflect on thick governance values at least on paper, but at national levels their implementation may be different. These experiences reflect back on UNFCCC discussions.[31]

The evolution of the REDD practices in the UNFCCC SBSTA conversations

When the original RED – Reducing Emissions from Forest Deforestation initiative – focused only on deforestation, the monitoring of changes in forest cover was believed to be easily done by GIS (Geographic Information System), based on satellite-based remote sensing technologies together with ground-truthing.[32] However, as it became obvious that forest cover does not indicate the size of the carbon stock or that the changes in forest cover do not indicate the changes in the size of emissions, forest degradation was added to RED, making it REDD. This made the verification process of the forests far more complicated and added sustainable forest management as a crucial part of REDD governance. Degradation and its monitoring, including the definition of forest degradation – whether this should be defined according to universal or local definitions and whether MRV should be based on international or national governance – became a dividing line in the SBSTA in two respects: general definitions and global surveillance are the preconditions for any meaningful market-based carbon trade system, but they also divide different country groups in the SBSTA. The issue was not necessarily a market or non-market-based system, but a country's ability to control its climate policies and manage its natural resources.[33] When forest cover can be monitored with remote sensing, forest degradation with the changes in carbon

30 Betsy A. Beymer-Farris and Thomas J. Bassett, "The REDD Menace: Resurgent Protectionism in Tanzania's Mangrove Forests," *Global Environmental Change* 22, no. 2 (2012).

31 Palmujoki and Virtanen, "Global, National, or Market? Emerging Redd+ Governance Practices in Mozambique and Tanzania."

32 United Nations Framework Convention on Climate Change (UNFCCC), "Report of the Conference of the Parties on Its Eleventh Session, Held at Montreal from 28 November to 10 December 2005."

33 United Nations Framework Convention on Climate Change (UNFCCC), *Views on the Range of Topics and Other Relevant Information Relating to Reducing Emissions from Deforestation in Developing Countries: Submissions from Parties* (Bonn Climate Change Conference FCCC/SBSTA/2007/MISC.2/Add.1: United Nations Office at Geneva, 2007), http://unfccc.int/documentation/documents/advanced_search/items/6911.php?priref=600004246.

stocks call for on-the-ground measurements.[34] Therefore, this question arises: to what extent is this monitoring a national question, and to what extent should the surveillance be the responsibility of an international body? In REDD, the submissions of the parties have not followed the general line between developing and developed countries. In fact, the majority of the REDD+ target countries have supported a market-based approach as well as universal definitions and common rules.[35] Similarly, as shown in Table 4.1, some developed countries have sympathized with the arguments for country-specific deviations and MRV systems.

Table 4.1 Approaches by some countries and country groups towards REDD[36]

	Scope	*Financing*	*Reference level*	*FVA*	*Safeguard SIS*
AOSIS[1]	REDD	F	HAB	B	
Australia	REDD+	MB+FVA	PB	M	International
Brazil	RED	F	HB		National
China	REDD+				
Colombia	REDD+	F	HAB		International
CRFN[2]	REDD	MB+FVA	HAB	B	International
EU	REDD+	ML	HAB	N	International
India	REDD+	MB	HB		
Indonesia	REDD+	MB+FVA	HB	B	National
Japan	REDD+			M	International
Mexico	REDD	MB+FVA	HAB	B	International
New Zealand	REDD	FVA		M	
Norway	REDD+	FVA	HAB		
Panama	REDD	MB	HAB		International
USA	REDD+			M	

Note: F = fund; MB=market-based; ML = marked-linked; PB = project-based; FVA(B) = broad, including Kyoto mechanisms, new market-based mechanisms, non-Kyoto market-based systems and non-market-based approaches; FVA(M) = same as previous, but without non-market-based approaches FVA(N) = only the Kyoto mechanisms with new market-based mechanisms.

1 Alliance of Small Island States (includes 38 of Caribbean, Indian Ocean and Pacific Island States).
2 Coalition of Rainforest Nations (Bangladesh, Belize, Chad, Cote d'Ivoire, Democratic Republic of Congo, Dominica, Dominican Republic, Fiji, Gabon, Guyana, Honduras, Kenya, Liberia, Nigeria, Panama, Papua New Guinea, Republic of Congo, Sierra Leone, Uganda).

34 United Nations Framework Convention on Climate Change (UNFCCC), "Background Paper for the Workshop on Reducing Emissions from Deforestation in Developing Countries," *Working Paper No. 1* (17 August 2006), https://unfccc.int/files/land_use_and_climate_change/lulucf/application/pdf/introduction_background_paper.pdf.

35 United Nations Framework Convention on Climate Change (UNFCCC), *Views on the Range of Topics and Other Relevant Information Relating to Reducing Emissions from Deforestation in Developing Countries. Submissions from Parties.*

36 The table is compiled from the following sources: Coalition of Rainforest Nations, "Submission of Views on Framework for Various Approaches" (30 August 2013), https://unfccc.

Naturally, concrete interests between the countries, surely in the context of international institutions, are obvious in the submissions. This is particularly apparent in the way that national reference levels (to which the forest conservation measures are compared) are defined. By this different governments argue for justice in REDD governance. The historical baseline (HB) represents the average level of historical deforestation that has taken place during the last decades in the country. The definition favours emerging countries that have had strong deforestation rates during recent decades. The historical adjusted baseline (HAB) tries to correct the distortion that HB creates for those countries where the forest cover is high and deforestation rates have been rather low, but which could potentially be higher in the future. The project-based (PB) reference level allows the most detailed set of guidance for creating baselines and focuses on local circumstances, but this is difficult to apply to international-level carbon trade.[37] Although much discussed, these reference levels have not been widely bargained, as the UNFCCC has been rather flexible in applying reference levels.

The emphasis on technical aspects in the implementation of REDD, including general accounting systems and the definition of baselines, indicates the aim for a market-based system. Though REDD originated partly as a market-based concept to connect developing countries to global emissions trading,[38] pilot projects have already revealed problems due to price variance in global carbon markets.[39] Such problems will obstruct longstanding conservation if there is no certainty of

int/files/documentation/submissions_from_parties/application/pdf/fva_coalition_for_rainforest_nations_15092013.pdf; M. Menton, C. Ferguson, R. Leimu-Brown, S. Leonard, M. Brockhaus, A.E. Duchelle, and C. Martius, *Further Guidance for REDD+ Safeguard Information Systems?: An Analysis of Positions in the UNFCCC Negotiations*, vol. 99 (CIFOR, November 2014); Brazil. Ministry of Environment, "Summary of Information on How the Cancun Safeguards Were Addressed and Respected by Brazil Throughout the Implementation of Actions to Reduce Emissions from Deforestation in the Amazon Biome between 2006 and 2010," (May 2015), http://redd.unfccc.int/files/brazil_safeguards_summary_final20150508.pdf; The Global Canopy Programme, "Reference Levels"; United Nations Framework Convention on Climate Change (UNFCCC) and Subsidiary Body for Scientific and Technological Advice (SBSTA), *Report on the In-Session Expert Meeting on Matters Relating to Non-Market-Based Approaches to Support the Implementation of the Activities in Decision 1/CP.16, Paragraph 70. Note by the Secretariat*; World Resources Institute, "Climate Finance".

37 The Global Canopy Programme, "Reference Levels," the REDD Desk, http://theredddesk.org/markets-standards/design-features/reference-levels.

38 Chukwumerije Okereke and Kate Dooley, "Principles of Justice in Proposals and Policy Approaches to Avoided Deforestation: Towards a Post-Kyoto Climate Agreement," *Global Environmental Change* 20, no. 1 (2010): pp. 82–95.

39 Godwell Nhamo and Alfred Bimha, "Carbon Pricing in Cap-and-Trade Systems: International Perspectives and Lessons for Africa," in *Green Economy and Climate Mitigation: Topics of Relevance to Africa*, ed. G. Nhamo (Pretoria: Africa Institute of South Africa, 2011), pp. 118–23; Willy Makundi, "African Forests and Trees in the Global Carbon Market," in *Climate Change and African Forest and Wildlife Resources*, ed. Emmanuel Chidumayo, David Okali, Goodwin Kowero and Muhamane Larwanou (Nairobi: African Forest Forum, 2011), pp. 194–96.

income for conservation. Though funding is either market or donor based, the target countries must construct a system for distributing assets among beneficiaries, which calls for a firm grip and the credible governance of governments. Interestingly enough, according to their submissions, the majority of the REDD target countries supported a market-based approach that connects their forest carbon stocks to the existing global carbon trade schemes. It should be noted, however, that the functioning carbon trade in the REDD system is expected to be reached after the different stages of REDD's development with diverse funding instruments. The EU's REDD has strictly followed the idea that REDD should be linked to carbon trade through the market instruments defined by the UNFCCC and not through private, national, and regional carbon trade schemes outside the UNFCCC system, which seems to be acceptable to most of the governments (see Table 4.1).

At the regime level the problem was how to harmonize the accounting of carbon assets of different conservation measures based on different kinds of funding mechanisms. The problem not only dealt with REDD but also actualized the importance of REDD as a mitigation measure, which took on more weight after the Copenhagen COP 15 in 2009. The issue has been addressed in the UNFCCC SBSTA for several years. Since 2012 the discussion has centred on the framework for various approaches (FVAs).[40] In this context the starting point is how carbon units created in domestic or private systems can be used in compliance with UNFCCC obligations. FVAs do not concern the units used exclusively in domestic accounting, but those in international transfers. FVAs include different kinds of market mechanisms and partly non-market-based approaches (NMAs). NMAs dealt with by SBSTA are accountable for an emissions reduction target of any contributor country and are thereby relevant to the limits agreed to in the UNFCCC. Market-based mechanisms include those agreed upon in Kyoto (International Emissions Trading, Joint Implementation, and Clean Development Mechanisms) and new market-based mechanisms (NMM) defined in the Durban COP 13 in 2011.

The SBSTA discussions are included in the Paris Agreement in Article 6. The article uses a little bit different vocabulary; instead of FVAs it refers to internationally transferred mitigation outcomes, and in Article 6.4 the agreement introduces a new sustainable development mechanism, which has similar content to NMM discussed in the SBSTA without mentioning it in the agreement.[41]

The crucial technical issue is double accounting, which can occur for different reasons. For example, it can be a result of carbon markets' having multiple parties, any of which can claim reductions in units for itself. Double accounting

40 United Nations Framework Convention on Climate Change (UNFCCC), *Outcome of the Work of the Ad Hoc Working Group on Long-Term Cooperative Action under the Convention. Revised Proposal by the President* (Doha Climate Change Conference – FCCC/CP/2012/L.14/Rev.1: United Nations Office at Geneva, November 2012).

41 United Nations Framework Convention on Climate Change (UNFCCC), "Conference of the Parties: COP21 Paris Agreement," (FCCC/CP/2015/L.9: UNFCCC, 2015), Articles 6.2 and 6.4.

can also occur when carbon credits are claimed both from financial contributions and mitigation purposes, or result from the fact that in combining carbon trade with different unit systems both systems count the same units. The discussion on FVAs attempted to solve these problems, but the parties have not been able to reach any concrete solution.[42] The Paris Agreement emphasizes that 'the use of internationally transferred mitigation outcomes towards nationally determined contributions . . . shall apply robust accounting to ensure, inter alia, the avoidance of double counting, consistent with the guidance adopted by the Conference of Parties serving as the meeting of the Parties to the Paris Agreement'.[43] According to the Paris Agreement, specific rules to avoid double accounting will be further elaborated on in the SBSTA.

The discussion of the UNFCCC SBSTA highlighted the different interests of country groups and their bargaining in the climate change regime. This bargaining and the arguments of the country groups followed the lines of international institutions, and as always there were different grouping-dependent issues involved. A rough division can be made according to the kind of approaches that should be taken under the rubric of FVAs. The EU's approach was the narrowest; it was ready to discuss under FVAs only market mechanisms defined in the UNFCCC (that is, the Kyoto Protocol and COP 13 in Durban). The situation differed with countries that have already used different accounting systems than the UNFCCC system – namely, domestic and regional systems, including those used by the United States, Japan, and Australia, all of which have advocated for a broader recognition of different market-based systems, which has also been included in the Paris Agreement.[44] The broadest interpretation of FVAs included developing countries that have adopted REDD+ non-market-based systems that did not involve carbon exchange units. The broadest interpretation has been adapted to the Paris Agreement, when non-market approaches have been included as Article 6.9. Accordingly, FVAs, or as the Paris Agreement referred to them, internationally transferred mitigation units, recognize the actions and mechanisms that include net reductions and the avoidance of GHG emissions by sources and removal of sinks due to forest conservation. Similarly, a group of developed and developing countries, the Environmental Integrity Group, supported a broad approach entailing voluntary activities to transfer some GHG emissions reductions to another country, all developed in and out of the UNFCCC in the case that they meet commonly defined requirements.[45]

42 United Nations Framework Convention on Climate Change (UNFCCC), *Report of the Subsidiary Body for Scientific and Technological Advice on Its Forty-Second Session, Held in Bonn from 1 to 11 June 2015* (FCCC/SBSTA/2015/2: United Nations Office at Geneva, 2015), http://unfccc.int/resource/docs/2015/sbsta/eng/02.pdf.

43 United Nations Framework Convention on Climate Change (UNFCCC), "Conference of the Parties: COP21 Paris Agreement," Article 6.2, see also Article 6.5.

44 Ibid., Article 6.2.

45 Environmental Integrity Group (EIG), Liechtenstein, Mexico, Monaco, Republic of Korea, and Switzerland, "Framework for Various Approaches SBSTA 39," (2013), https://unfccc.

There are more detailed proposals for dealing with the rules of measurement and reporting, how an independent verification process should proceed, how net emissions reductions are counted, and what kinds of institutional requirements are needed. As such, discussions in the UNFCCC SBSTA will most likely continue to follow those issues already set in relief amid different REDD preparedness projects carried out in the UNREDD under the World Bank's Forest Carbon Partnership Facility and in different donors' projects. Although the issues seem merely technical instead of political, their development in several COPs reflects their political nature and indicates how international institutions mould regimes via domestic experiences.

Several emergent issues in the REDD initiative in annual COPs address the influence of domestic experiences and practices in REDD+ target countries with the global regime. Owing to different social and environmental issues involved in REDD+ and its governance, UNFCCC COPs adopted safeguards to prevent environmental and social harms that implementing REDD+ may create. Many government and civil society actors have expressed their concern that the new wave of investment and competing interests for forest resources brought by REDD will further compromise the rights and resources of local communities. Safeguards constitute a mechanism developed by IOs to ensure that their investments do not create such negative outcomes. In 2010, parties to the UNFCCC agreed to a procedural practice consisting of such safeguards. Known as the Cancun Safeguards, these measures define broad criteria and principles intended to help developing countries create and implement their own national safeguards that can be used to enhance the role of the central government in the REDD+ process.[46]

From the perspective of justice, the way in which local communities share costs and benefits related to REDD+ projects is critical.[47] Although advocates of REDD+ argue that participation in the programme is likely to strengthen local

int/files/cooperation_support/market_and_non-market_mechanisms/application/pdf/fva_environmental_integrity_group.pdf; European Union, *Submission by Italy and The European Commission on Behalf of The European Union and Its Member States: Eu Submission 2014 on Strategies and Approaches for Scaling up Climate Finance* (Rome: Council of the European Union, 29 September 2014), http://ec.europa.eu/clima/policies/international/finance/docs/eu_strategies_and_approaches_en.pdf; United Nations Framework Convention on Climate Change (UNFCCC) and Subsidiary Body for Scientific and Technological Advice (SBSTA), *Report on the In-Session Expert Meeting on Matters Relating to Non-Market-Based Approaches to Support the Implementation of the Activities in Decision 1/CP.16, Paragraph 70. Note by the Secretariat* (Lima Climate Change Conference FCCC/SBSTA/2014/INF.13: United Nations Office at Geneva, 2014), http://unfccc.int/documentation/documents/advanced_search/items/6911.php?priref=600008078.

46 Stephanie Roe, Charlotte Streck, Luke Pritchard, and John Costenbader, *Safeguards in Redd+ and Forest Carbon Standards: A Review of Social, Environmental and Procedural Concepts and Application* (Amsterdam: Climate Focus, 2013).

47 Rowena Maguire, "Designing REDD+ to Be Just: Considerations for a Legally Binding Instrument," *Asian Journal of International Law* 4, no. 1 (2014).

access and property rights, both of which constitute a key institutional requirement for access to REDD+ financing[48] (COP 16, paragraph 72), past experiences have shown that administrative frameworks based on civil society engagement models to secure local rights are not enough if enforcement mechanisms are weak and subject to manipulation by more powerful political and economic elites in development assistance. The creation of safeguards sheds light on the workings of international institutions in this particular case. Social issues respecting the status of local and indigenous peoples' socio-economic conditions and positioning in deciding REDD projects that affect their livelihoods were constituted by justice and are established in international development assistance practices. The ethical considerations framed by justice suggest thick governance values in REDD projects and may influence the social and political governance of REDD target countries.[49]

Safeguards represent the system by which donors and REDD countries can verify that environmental and social standards are met in REDD projects. They, however, differ from the principles, criteria, and indicator (PC&I) systems that private bodies have developed in order to guarantee the quality and legitimacy of governance with particular respect to environmental and social standards.[50] When PC&I systems aim to forge comprehensive governance by controlling different phases of production using environmental, social, and economic principles, safeguards represent a compromise between national sovereignty (that is, a government's control) and international monitoring. Safeguards thus play a merely guiding role in national REDD projects. Therefore, thick governance values which consist of the participation of all stakeholders in decision-making and benefit sharing seem to drain away in the REDD project routines and represent thin governance values at best. At the regime level, thin governance values were further emphasized: in SBSTA discussions, a general understanding has prevailed regarding the content of safeguards, though disagreements persist concerning how they should be connected to the UNFCCC. Although the positions of parties regarding the safeguards' information system (SIS) were far from definite, the demarcation represents how much guidance remains for (international) UNFCCC bodies as well as national ones. Although the majority of parties accept international monitoring, Brazil refused this already in Cancun COP 16 and has strictly defended national control over safeguards.[51] Following the established

48 United Nations Framework Convention on Climate Change (UNFCCC), "Report of the Conference of the Parties on Its Sixteenth Session, Held in Cancun from 29 November to 10 December 2010," (Part Two: Action taken by the Conference of the Parties at Its Sixteenth Session FCCC/CP/2010/7/Add.1 2011), para. 72.

49 Haug and Gupta, "The Emergence of Redd on the Global Agenda," pp. 89–91.

50 See Timothy Cadman, *Quality and Legitimacy of Global Governance: Case Lessons from Forestry* (New York and Houndmills, Basingstoke, Hampshire: Palgrave Macmillan, 2011), pp. 1–24.

51 World Resources Institute, "Climate Finance," World Resources Institute, http://www.wri.org/our-work/project/climate-fina.

decision-making process of the UNFCCC based on consensus, the Brazilian interpretation clearly survived in the Paris Agreement. The further discussions in the SBSTA after Paris will deal with how REDD, including safeguards, is included in detailed rules that will be developed to support the Paris agreement.

Table 4.1 does not directly reflect the different interests and bargaining positions, as there are several items where the countries have not made submissions, although they are known to support a certain approach. The table indicates the submission that the states or state groups have presented during the development of RED to REDD+. Therefore, for example, all the countries accept the scope of the REDD+, although their original submission was more limited. Similarly, although the original initiators of RED expressed market-based mechanisms on the basis of funding, they have accepted other types of funding too. Submissions reflect concrete interests of the governments in REDD, and when the mitigation efforts threaten their particular interests, sovereignty as a basic institution in the UNFCCC process matches them together. In the REDD process this has meant that most important issues concerning funding and accounting are left to the governments. However, the emphasis of sovereignty in REDD at the UNFCCC level does not mean that the other institutions are without significance in the REDD+ governance practices. The discussions and submissions of REDD in the UNFCCC reflect back to IOs and donors and to the ways in which they define their policy line. Therefore, those aspects on which the UNFCCC has not been able to decide can appear in the policies of the IOs and other actors in different REDD+ projects.

Conclusion: International institutions and the evolution of REDD practices

When examining the experience of the REDD+ process within the new governance model confirmed in Paris in December 2015, the present chapter bases its interpretations on the relationship between fundamental institutions and REDD+ debates. Different institutions have shaped REDD's development in several ways. Despite broadly accepted objectives for climate mitigation, the basic institutions of sovereignty, the market, and justice via procedural practices of development assistance determine the borders of REDD governance. Despite UNFCCC members' broad understanding of the concept, the basic institution of sovereignty has yet to be seriously challenged in climate change governance. Through development assistance, concepts of both the market and justice have prompted unifying practices, and together with environmental norms they are apparent in REDD's adoption of safeguards from COP meetings. These established practices have quite possibly helped formulate REDD+ as well as helped highlight the need to increase context-integrity among UNFCCC members. In SBSTA submissions on REDD, this need has been emphasized by the fact that submissions have failed to follow any pure division of developed–developing countries. It furthermore suggests that climate governance has become thicker, to use the previously defined terminology.

In the Paris Agreement's finalization, very few binding rules emphasized the UNFCCC's diverse and even fragmented nature. Due to the role of the regime in reconciling different interests, different countries and country groups have congregated around certain institutions and norms: developing countries around sovereignty and justice (equity) on the one hand, and developed countries around the concepts of the market, justice, and transparency on the other hand. Although SBSTA discussions about REDD have indicated the blurring of such a division, the final preparation of the climate agreement illuminates old divisions and points to the thinness of UNFCCC governance, as well as to more reduced consistency-integrity than many SBSTA submissions have indicated. In the SBSTA REDD negotiations, such thinness and reduced integrity are apparent in the failure to align domestic conservation measures with global carbon markets in a way that is compatible with the UNFCCC and can thereby be ruled in official measures towards mitigating GHG emissions.

In investigating the international climate change regime through the framework of international institutions, several issues stand out. The UNFCCC's drafted governance architecture based on UNCED and the Kyoto Protocol follows the idea of a coherent structure for a climate change regime, which suggests that the UNFCCC has created a top-down process that is oversimplified and unrealistic. Such an approach supposes above all that an intergovernmental climate change regime regulates climate policies by way of a hierarchical governance structure. In one respect, agents such as the European Union are used to constructing climate policies on that model, though interactions in any climate change regime are far more diverse. The Paris Agreement is now a more flexible bottom-up model, quite different from the Kyoto Protocol. Whether the model leads to overall thinness of climate governance remains unclear. The SBSTA discussions about REDD+'s MRV apparently indicate this. Ongoing discussion and new proposals reflect the influence of domestic and regional practices as well as their own respective debates, all while the climate change regime expands by co-opting new elements. The regime is transformed via the constitutive role of the same international institutions of sovereignty, justice, and the market – all clearly foundational to the Kyoto Protocol – in other international practices. In the case of REDD in particular, development assistance practices are crucial to producing REDD+ governance. In that light, the climate change regime's development not only emphasizes the importance of international institutions, but also indicates how that development is a reflection of those institutions, possibly enabling new international institutions to develop in an attempt to mitigate climate change.

References

Andresen, Steinar. "International Climate Negotiations: Top-Down, Bottom-up or a Combination of Both?" *The International Spectator* 50, no. 1 (2015): 15–30.

Beymer-Farris, Betsy A., and Thomas J. Bassett. "The Redd Menace: Resurgent Protectionism in Tanzania's Mangrove Forests." *Global Environmental Change* 22, no. 2 (2012): 332–41.

Breakey, Hugh, and Timothy Cadman. "A Comprehensive Framework for Evaluating the Integrity of the Climate Regime Complex." Chap. 2 In *Ethical Values and the Integrity of the Climate Change Regime*, edited by Hugh Breakey, Vesselin Popovski and Rowena Maguire, 17–27. Surrey: Ashgate, 2015.

Breakey, Hugh, Tim Cadman, and Charles Sampford. "Conceptualizing Personal and Institutional Integrity: The Comprehensive Integrity Framework." In *The Ethical Contribution of Organizations to Society*, edited by Michael Schwartz, Howard Harris and Debra Comer, 1–40. Bingley, UK: Emerald Group Publishing Ltd, 2015.

Bull, Hedley. *The Anarchical Society: A Study of Order in World Politics*. London: Macmillan, 1977.

Buzan, Barry. *From International to World Society? English School Theory and the Social Structure of Globalisation*. Cambridge and New York: Cambridge University Press, 2004.

———. *An Introduction to the English School of International Relations: The Societal Approach*. Cambridge: Polity Press, 2014.

Cadman, Timothy. *Quality and Legitimacy of Global Governance: Case Lessons from Forestry*. New York and Houndmills, Basingstoke, Hampshire: Palgrave Macmillan, 2011.

Cass, Loren R. *The Failures of American and European Climate Policy: International Norms, Domestic Politics, and Unachievable Commitments*. Albany: State University of New York Press, 2006.

"CDM Rulebook: What Is Additionality?" <http://www.cdmrulebook.org/84.html> (accessed 6 June 2016).

Coalition of Rainforest Nations. "Submission of Views on Framework for Various Approaches." (30 August 2013). <https://unfccc.int/files/documentation/submissions_from_parties/application/pdf/fva_coalition_for_rainforest_nations_15092013.pdf> (accessed 6 June 2016).

den Besten, Jan Willem, Bas Arts, and Patrick Verkooijen. "The Evolution of REDD+: An Analysis of Discursive-Institutional Dynamics." *Environmental Science & Policy* 35, no. 1 (2014): 40–48.

Environmental Integrity Group (EIG), Liechtenstein, Mexico, Monaco, Republic of Korea, and Switzerland. "Framework for Various Approaches SBSTA 39." (2013). <https://unfccc.int/files/cooperation_support/market_and_non-market_mechanisms/application/pdf/fva_environmental_integrity_group.pdf> (accessed 6 June 2016).

European Union. *Submission by Italy and The European Commission on Behalf of The European Union and Its Member States: EU Submission 2014 on Strategies and Approaches for Scaling up Climate Finance*. Rome: Council of the European Union, 29 September 2014. <http://ec.europa.eu/clima/policies/international/finance/docs/eu_strategies_and_approaches_en.pdf> (accessed 6 June 2016).

Falkner, Robert. "Global Environmentalism and the Greening of International Society." *International Affairs* 88, no. 3 (2012): 503–22.

Foster, Caroline E. "Precaution, Scientific Development and Scientific Uncertainty under the WTO Agreement on Sanitary and Phytosanitary Measures." *Review of European Community and International Environmental Law* 18, no. 1 (2009): 50–58.

Friedner Parrat, Charlotta. "Reform of the United Nations: Between Great Power Management and Sovereign Equality." *International Studies Association Conference*. New Orleans 20 February 2015.

Geden, Oliver. "Pragmatism in Climate Policy." *Project Syndicate* (23 November 2015). <https://www.project-syndicate.org/commentary/paris-climate-change-pragmatic-approach-by-oliver-geden-2015-11> (accessed 6 June 2016).

The Global Canopy Programme. "Reference Levels." *The REDD Desk*. <http://theredddesk.org/markets-standards/design-features/reference-levels> (accessed 6 June 2016).

Haug, Constanze, and Joyeeta Gupta. "The Emergence of Redd on the Global Agenda." In *Climate Change, Forests and Redd: Lessons for Institutional Design*, edited by J. Gupta, N. van der Grijp and O. Kuik, 77–98. Abingdon and New York: Taylor & Francis, 2013.

Holsti, Kalevi Jaakko. *Taming the Sovereigns: Institutional Change in International Politics*. Cambridge and New York: Cambridge University Press, 2004.

Hurrell, Andrew. "International Society and the Study of Regimes: A Reflective Approach." In *Regime Theory and International Relations*, edited by Volker Rittberger and Peter Mayer, 49–72. Oxford and New York: Clarendon Press, 1993.

———. "Power, Institutions, and the Production of Inequality." In *Power in Global Governance*, edited by M. Barnett and R. Duvall, 33–58. Cambridge: Cambridge University Press, 2005.

Keohane, Robert, and David Victor. "The Regime Complex for Climate Change." *Perspectives on Politics* 9, no. 1 (2011): 7–23.

Krasner, Stephen D. "Structural Causes and Regime Consequences: Regimes as Intervening Variables." *International Organization* 36, no. 2 (1982): 185–205.

Maguire, Rowena. "Designing Redd+ to Be Just: Considerations for a Legally Binding Instrument." *Asian Journal of International Law* 4, no. 1 (2014): 169–96.

Makundi, Willy. "African Forests and Trees in the Global Carbon Market." In *Climate Change and African Forest and Wildlife Resources*, edited by Emmanuel Chidumayo, David Okali, Goodwin Kowero and Muhamane Larwanou, 192–213. Nairobi: African Forest Forum, 2011.

McIntyre, Owen, and Thomas Mosedale. "The Precautionary Principle as a Norm of Customary International Law." *Journal of Environmental Law* 9, no. 2 (1997): 221–40.

Menton, M., C. Ferguson, R. Leimu-Brown, S. Leonard, M. Brockhaus, A.E. Duchelle, and C. Martius. *Further Guidance for Redd+ Safeguard Information Systems?: An Analysis of Positions in the UNFCCC Negotiations*. Vol. 99: CIFOR, November 2014.

Ministry of Environment, Brazil. "Summary of Information on How the Cancun Safeguards Were Addressed and Respected by Brazil Throughout the Implementation of Actions to Reduce Emissions from Deforestation in the Amazon Biome between 2006 and 2010." (May 2015). <http://redd.unfccc.int/files/brazil_safeguards_summary_final20150508.pdf> (accessed 6 June 2016).

Navari, Cornelia. "Primary and Secondary Institutions: Quo Vadit." *Cooperation and Conflict* 51, no. 1 (2016): 121–27.

Nhamo, Godwell, and Alfred Bimha. "Carbon Pricing in Cap-and – Trade Systems: International Perspectives and Lessons for Africa." In *Green Economy and Climate Mitigation: Topics of Relevance to Africa*, edited by G. Nhamo, 97–124. Pretoria: Africa Institute of South Africa, 2011.

Okereke, Chukwumerije, and Kate Dooley. "Principles of Justice in Proposals and Policy Approaches to Avoided Deforestation: Towards a Post-Kyoto Climate Agreement." *Global Environmental Change* 20, no. 1 (2010): 82–95.

Palmujoki, Eero. "Fragmentation and Diversification of Climate Change Governance in International Society." *International Relations* 27, no. 2 (2013): 180–201.

Palmujoki, Eero, and Pekka Virtanen. "Global, National, or Market? Emerging Redd+ Governance Practices in Mozambique and Tanzania." *Global Environmental Politics* 16, no. 1 (2016): 59–78.

Peel, Jacqueline. "Interpretation and Application of the Precautionary Principle: Australia's Contribution." *Review of European Community and International Environmental Law* 18, no. 1 (2009): 11–25.

Phelps, Jacob, Edward L. Webb, and Arun Agrawal. "Does Redd+ Threaten to Recentralize Forest Governance?" *Science* 328, no. 5976 (2010): 312–13.

Pistorius, Till. "From Red to Redd+: The Evolution of a Forest-Based Mitigation Approach for Developing Countries." *Current Opinion in Environmental Sustainability* 4, no. 6 (2012): 638–45.

Reinecke, Sabine, Till Pistorius, and Michael Pregernig. "UNFCCC and the Redd+ Partnership from a Networked Governance Perspective." *Environmental Science and Policy* 35 (2014): 30–39.

Roe, Stephanie, Charlotte Streck, Luke Pritchard, and John Costenbader. *Safeguards in Redd+ and Forest Carbon Standards: A Review of Social, Environmental and Procedural Concepts and Application.* Amsterdam: Climate Focus, 2013.

Schouenborg, Laust. "The English School and Institutions: British Institutionalists?" In *Guide to the English School in International Studies*, edited by Cornelia Navari and Daniel M. Green, 77–89. Oxford: Willey Blackwell, 2014.

Spandler, Kilan. "The Political International Society: Change in Primary and Secondary Institutions." *Review of International Studies* 41, no. 3 (2015): 601–22.

Streck, Charlotte. "Financing Redd+: Matching Needs and Ends." *Current Opinion in Environmental Sustainability* 4, no. 6 (2012): 628–37.

United Nations Framework Convention on Climate Change (UNFCCC). "Background Paper for the Workshop on Reducing Emissions from Deforestation in Developing Countries." *Working Paper No. 1* (17 August 2006). <https://unfccc.int/files/land_use_and_climate_change/lulucf/application/pdf/introduction_background_paper.pdf > (accessed 6 June 2016).

———. "Conference of the Parties: COP21 Paris Agreement." FCCC/CP/2015/L.9: UNFCCC, 2015.

———. "Report of the Conference of the Parties on Its Eleventh Session, Held at Montreal from 28 November to 10 December 2005." FCCC/CP/2005/MISC.1 2005.

———. "Report of the Conference of the Parties on Its Sixteenth Session, Held in Cancun from 29 November to 10 December 2010." Part Two: Action taken by the Conference of the Parties at Its Sixteenth Session FCCC/CP/2010/7/Add.1 2011.

———. *Report of the Subsidiary Body for Scientific and Technological Advice on Its Forty-Second Session, Held in Bonn from 1 to 11 June 2015.* FCCC/SBSTA/2015/2: United Nations Office at Geneva, 2015. <http://unfccc.int/resource/docs/2015/sbsta/eng/02.pdf> (accessed 6 June 2016).

———. *Views on the Range of Topics and Other Relevant Information Relating to Reducing Emissions from Deforestation in Developing Countries. Submissions from Parties.* Bonn Climate Change Conference FCCC/SBSTA/2007/MISC.2/Add.1: United Nations Office at Geneva, 2007. <http://unfccc.int/documentation/documents/advanced_search/items/6911.php?priref=600004246> (accessed 6 June 2016).

———. *Outcome of the Work of the Ad Hoc Working Group on Long-Term Cooperative Action under the Convention: Revised Proposal by the President.* Doha Climate Change Conference – FCCC/CP/2012/L.14/Rev.1: United Nations Office at Geneva, November 2012.

United Nations Framework Convention on Climate Change (UNFCCC), and Subsidiary Body for Scientific and Technological Advice (SBSTA). *Report on the in-Session Expert Meeting on Matters Relating to Non-Market-Based Approaches to Support the Implementation of the Activities in Decision 1/CP.16, Paragraph 70. Note by the Secretariat.* Lima Climate Change Conference FCCC/SBSTA/2014/INF.13: United Nations Office at Geneva, 2014. <http://unfccc.int/documentation/documents/advanced_search/items/6911.php?priref=600008078> (accessed 6 June 2016).

United Nations Framework Convention on Climate Change (UNFCCC/CCNUCC). "Glossary CDM Terms." Version 08.0 (2 March 2012). <https://cdm.unfccc.int/Reference/Guidclarif/glos_CDM.pdf> (accessed 6 June 2016).

van Asselt, Harro. "Integrating Biodiversity in the Climate Regime's Forest Rules: Options and Tradeoffs in Greening Redd Design." *Review of European Community & International Environmental Law* 20, no. 2 (2011): 139–49.

Wendt, Alexander, and Raymond Duvall. "Institutions and International Order." In *Global Changes and Theoretical Challenges: Approaches to World Politics for the 1990s*, edited by E.O. Czempiel and J.N. Rosenau, 51–73. Lexington and Massachusetts, MA: Lexington Books, 1989.

World Resources Institute. "Climate Finance." *World Resources Institute.* <http://www.wri.org/our-work/project/climate-fina> (accessed 6 June 2016).

Zürn, Michael, and Benjamin Faude. "On Fragmentation, Differentiation, and Coordination." *Global Environmental Politics* 13, no. 3 (2013): 119–30.

5 The global governance system for climate finance

Towards greater institutional integrity?[1]

Emilie Bécault and Axel Marx

Introduction

The topic of institutional integrity lies at the core of the present volume. As aptly conceptualized by Breakey, Cadman, and Sampford,[2] the idea of institutional integrity closely resonates with that of personal integrity in that it refers to whether an institution, considered either as an agent or inanimate object, actually lives up to its publicly proclaimed values and goals – what Breakey and colleagues call its 'public institutional justification' (PIJ). Embedded within a broader conceptual framework (that is, the Comprehensive Integrity Framework), the concept of institutional integrity helps to shed light on an institution's PIJ and its substantive and normative content, as well as on an institution's various aspects of institutional integrity (for example, coherence-integrity, consistency-integrity, comprehensive-integrity, and context-integrity).

In this chapter our objective is to take a closer look at how the emerging regime for climate change financing currently fares in terms of institutional integrity. Is there, for instance, an institutional integrity deficit in the global governance of climate change finance, and if so, how can it be understood? Clearly, issues of institutional integrity are highly relevant to the global governance of climate financing, not merely because 'finance' is now one of the most pressing issues for advancing international cooperation on climate change, but also because the governance of any type of finance (for example, development, environmental, social etc.), is, by its very nature, excessively 'politicized', involving hard trade-offs between different objectives and values.

1 This work was financially supported by the Academic Research Group for Policy Support (ACROPOLIS), a programme funded by ARES-CCD and VLIR-UOS which aims to support the decision-making of the Belgian Directorate General for Development Cooperation (DGD) with evidence-based research. It was conducted within the framework of BeFinD, a consortium of four research centers at Belgian universities covering issues related to financing for development.

2 Hugh Breakey, Timothy Cadman, and Charles Sampford, "Conceptualizing Personal and Institutional Integrity: The Comprehensive Integrity Framework," *Research in Ethical Issues in Organizations* 14 (2015).

However, and in contrast to most analyses of institutional integrity, our focus here is less on a particular governing unit or agent and more on the overall system for climate finance governance. While these two approaches to studying institutional integrity (or other institutional features, for that matter) are closely related, they tend to reflect distinct modes of inquiry, with agent-oriented approaches focusing on agents' preferences and interactions in accounting for the integrity of a particular organization or collection of organizations, and structural approaches emphasizing, rather, the role that key features of the broader institutional environment might play in influencing pathways and changes at the micro-level, namely in constraining or enabling governing agents in their quest for greater institutional integrity. Our perspective in some ways links up to Breakey's and colleagues' concept of 'context-integrity', one out of three institutional types of integrity identified by the authors, which draws attention to the 'factors outside of an institution' that may contribute to it acting in accordance with its publicly proclaimed values.[3] These include, for instance, the overarching legal context: that is, the set of laws that regulate institutional activities; the values of the different communities affected (directly or indirectly) by the workings of the institution; and more importantly for our present purpose, the institution's organizational context, which refers to the larger institutional framework it forms one part of. On Breakey and colleagues' conception, however, the relationship between these contextual factors and the integrity of an institution is mainly one of direct purposeful actions or impacts. As Breakey, Cadman, and Sampford[4] explain it, indeed, '[i]n general, the institution has context-integrity when these external forces support the institution when it is working towards its PIJ, but resist it when it does not.' Our approach, somehow, is less directly causal in the sense that we are mainly interested in evaluating how some core features of the global governance system for climate finance might constrain or challenge the system's institutions overall capacity towards achieving greater institutional integrity. In line with Breakey's, Cadman's, and Sampford's distinction in this volume between *thin* and *thick* governance values, our analysis will also seek to explore how conditions of low-context integrity might make it desirable but nevertheless highly difficult to promote thicker governance values involving more elaborate and democratically demanding sets of rule-making procedures and structures.

With this having been said, the analysis proceeds as follow. We begin by laying out the conceptual basis for identifying and analyzing key features of the global governance system for climate finance. Our focus here is mainly on those macro-structural elements of broader global governance contexts that can have an impact on the context-integrity of any system of global governance. These will include the broader normative context or framework, the governance system's

3 Ibid., pp. 7–8. See also Hugh Breakey and Tim Cadman, "Governance Values and Institutional Integrity," in *Governing the Climate Change Regime*, ed. Charles Sampford, Tim Cadman, and Rowena Maguire (London: Taylor & Francis, 2016 (forthcoming)), Chapter 2 of this volume.

4 Breakey, Cadman, and Sampford, "Governance Values and Institutional Integrity," Ch. 2.

overall architecture, and its geographical scope. We then apply this framework to the emerging global governance system for climate change financing to developing countries and show how climate finance is currently characterized by a weak and contested normative framework, a highly fragmented architecture, and a restricted geographical scope. What these features suggest overall is that from a systemic perspective, the global governance system for climate financing to developing countries presently lacks context-integrity constraints. Specifically, there is at the systemic level no mechanism or normative framework strong enough to bring about sufficient policy coherence across and among the variety of institutions and actors involved in channelling and/or deploying money flows to developing countries. There is as well a persisting gap between expected and actual global policy outputs for climate-related funds. To this day they have been insufficiently targeting those developing countries that are most vulnerable to, and less capable of, dealing with climate change's physical and socio-economic impacts. A number of recommendations will follow from the perspective offered therein, each formulated with the goal of making the overarching organizational context for international climate finance more conducive to institutional integrity. In the conclusion we will have a closer look at the newly approved Paris Agreement at COP 21 in December 2015[5] and examine whether this supposedly 'historically' ambitious deal, beyond solidifying international cooperation on climate change, makes concrete and substantial progress on climate finance.

Global governance systems: A framework for analysis

Before we go into the topic of climate finance, it is necessary to provide a brief description of the notion of global governance systems and their core dimensions. Much of the early empirical literature on the phenomenon of global governance has been concerned with assessing specific institutions or organizations often operating at a single scale of governance and with a defined scope of authority. In recent years however, growing attention has been paid to broader frameworks of governance, stretching across various jurisdictional levels and geographical scales. From this more macro-structural perspective, the main units of analysis are not individual governance arrangements, but the overall structure in which these are embedded and potentially interconnected.[6]

5 United Nations Framework Convention on Climate Change (UNFCCC), "Conference of the Parties Twenty-First Session Paris, 30 November to 11 December 2015: Adoption of The Paris Agreement," (FCCC/CP/2015/L.9/Rev. 1: United Nations, 2015).

6 This strand of literature now stretches across many areas of global governance. In regard to climate change governance in particular, see especially Frank Biermann et al., *Earth System Governance People, Places, and the Planet: Science and Implementation Plan of the Earth System Governance Project Report No. 1* (Bonn: International Human Dimensions Programme on Global Environmental Change, 2009), <http://www.ihdp.unu.edu/docs/Publications/ESG/IHDP_ReportNo20_ESG_ReportNo1.pdf> (accessed 1 June 2016); Frank Biermann, Philipp Pattberg, Harro Van Asselt, and Fariborz Zelli, "The Fragmentation of Global Governance Architectures: A Framework for Analysis," *Global Environmental Politics* 9, no. 4 (2009); Robert Keohane and David Victor, "The Regime Complex for Climate Change,"

To capture such a 'meta-level' of governance',[7] the concepts of architecture, landscape, or even *system* have emerged as useful analytical tools. As advanced in the field of global environmental governance in particular, these notions essentially direct attention to 'the overarching [set] of public or private institutions, principles, norms, regulations, decision-making procedures and organizations that are valid or active in [an] issue area'.[8] In this contribution we prefer to use the notion of global governance system (hereafter referred to as GGS) to that of architecture, mainly for analytical reasons, as the latter is to be considered as a subset of the former.

As just mentioned, more systemic approaches to global governance have only recently started to emerge. In the environmental field in particular, an interest in evaluating the causes and effects of the rescaling of governance functions and activities to multiple directions and actors has stimulated new research on mapping broader governance systems, for instance, in studies on international regime complexes,[9] transnational governance complexes,[10] and polycentric governance.[11] Yet so far many of these analyses have been limited in their focus to a single aspect of these larger structures and often to issues of institutional complexity and fragmentation, making it difficult to establish a more comprehensive picture of the various ways in which these GGSs might vary across global policy domains and organizational scales. In the following, our analysis extends this literature by choosing to qualify GGSs along three core dimensions: their normative framework, overarching architecture, and geographical scope.[12]

Perspectives on Politics 9, no. 1 (2011). For a review of the literature on the fragmentation of global environmental governance architecture, see notably Marija Isailovic, Oscar Widerberg, and Philipp Pattberg, "Fragmentation of Global Environmental Governance Architectures: A Literature Review," (Online: SSRN, 2013).

7 Biermann, Pattberg, Asselt, and Zelli, "The Fragmentation of Global Governance Architectures: A Framework for Analysis," p. 32.

8 Ibid., p. 33.

9 Karen Alter and Sophie Meunier, "Nested and Overlapping Regimes in the Transatlantic Banana Trade Dispute," *Journal of European Public Policy* 13, no. 3 (2006); Keohane and Victor, "The Regime Complex for Climate Change."; Kal Raustiala and David Victor, "The Regime Complex for Plant Genetic Resources," *International Organization* 58, no. 2 (2004).

10 Kenneth Abbott, "The Transnational Regime Complex for Climate Change," *Environment & Planning C: Government & Policy* 30, no. 4 (2012).

11 Ibid.; Julia Black, "Constructing and Contesting Legitimacy and Accountability in Polycentric Regulatory Regimes," *Regulation and Governance* 2 (2008); Daniel. H. Cole, "From Global to Polycentric Climate Governance," *Climate Law* 2, no. 3 (2011); Elinor Ostrom, "Polycentric Systems for Coping with Collective Action and Global Environmental Change," *Global Environmental Change* 20, no. 4 (2010).

12 This section draws heavily on a previous work by Emilie Bécault and Colleen Carroll which distinguishes between two mutually constitutive approaches to assessing global governance (micro- and macro-dimensions) and assesses the implications of these analytical lenses for questions of democratic legitimacy. See Emilie Bécault and Colleen Carroll, "Global Governance and the Challenges of Diversity," in *Global Governance and Democracy: A Multidisciplinary Analysis*, ed. Jan Wouters, Antoon Braeckman, Matthias Lievens, and Emilie Bécault (Leuven: Edward Elgar Publishing, 2015).

Probably one of the most important dimensions of any GGSs is their normative framework. This dimension refers broadly to the set of fundamental norms or principles that more or less explicitly underpin governing institutions and activities in a particular policy area of world politics. These ideational elements of structure are a rather well-known cornerstone of structural constructivism in international relations theory (IR).[13] Scholars working in this tradition have indeed long emphasized the role that prevailing ideas or norms can play in shaping how states and other actors, and especially those with capacity for political influence, frame and respond to specific global challenges. Commonly cited guiding institutional principles or foundational norms include, for instance, equality, transparency, non-intervention, and secrecy, but also sovereignty, globalization, and environmental liberalism.

Essentially, the tracing of these broader principles can shed light on the questions of why some rules, regulations, decisional procedures, instruments, or actors have come to dominate (or not dominate) governance practices and thinking in a particular domain of global concern. The underlying logic here is that prevailing principles and discourses tend to provide legitimacy constraints for the rise as well as non-rise of some particular modes and strategies of governance at the micro-level, namely at the level of individual governance institutions or arrangements. Quite logically, when competitive or conflictive foundational norms do exist at the macro-level, this might explain why some GGSs might be characterized by a plurality of diverging, if not competing, institutional arrangements and rules.

Two types of variations in relation to these foundational norms can be considered. They can first vary along a continuum of formality, ranging from very formal (that is, firmly grounded in international law) to very informal (for example, principles, ideologies . . .). They can also vary depending on their depth or degree of institutionalization, which as Bernstein explains it, 'can be inferred primarily from norms frequency or "density" in social structures, that is, the amount and range of instruments, statements, and so on that invoke them'.[14] Highly institutionalized principles or norms can form the basis for the construction of collective identities and interests at the micro-level and are often less prone to be challenged by competing norms and values. Second, and relatedly, variations in foundational norms can be evaluated in terms of their degree of contestedness.

13 See, for instance, Martha Finnemore and Kathryn Sikkink, "International Norm Dynamics and Political Change," *International Organization* 52 (1998); Martha Finnemore and Kathryn Sikkink, "Taking Stock: The Constructivist Research Program in International Relations and Comparative Politics," *Annual Review Political Science* 4, no. 3 (2001); James G. March and Johan P. Olsen, "The Institutional Dynamics of International Political Orders," *International Organization* 52 (1998); Christian Reus-Smit, "The Constitutional Structure of International Society and the Nature of Fundamental Institutions," ibid.51, no. 4 (1997); Antje Wiener, *The Invisible Constitution of Politics* (New York: Cambridge University Press, 2008).

14 Steven Bernstein, *The Compromise of Liberal Environmentalism* (New York: Columbia University Press, 2001), p. 30.

In this regard, identifying which groups of actors come to dispute the legitimacy, legal validity, or even application of foundational norms, and on which grounds, can shed light on perceived legitimacy gaps or deficits in a specific domain of global governance.[15]

A second core dimension of GGSs is their 'architecture', a term which, as just suggested, is narrower than that of 'system' and which points to the overall structure of global governance taking place in a specific issue area of world politics. As understood, the concept of architecture essentially helps to account for variations in the ways the main governing units in an issue area of global governance are spatially dispersed and potentially interconnected with one another. It is especially in close resonance with the concept of 'fragmentation' that the notion of global governance architecture has developed in the last few years as a useful tool to evaluate the nature and effects of different types and degrees of authority dispersion at the global level.[16]

A traditional point of departure for characterizing and assessing the architecture of global governance systems is the notion that global governance (as opposed to traditional interstate cooperation) necessarily implies some kind of multi-levelled or networked structure, where authority to make and implement rules is somewhat diffused or fragmented vertically between different layers of social organizations (for example, supranational, national, regional, subnational) and horizontally between different groups of actors and sectors.[17] Based on these qualifications in turn, the architecture of global governance systems can vary depending on whether governance is dispersed across multiple levels and actors or concentrated in the hands of a few actors and/or located in a limited number of levels or jurisdictions. In both cases, however, the governance structure is still multi-levelled in the sense that the system is not anarchical and involves some measure of functional and stratificatory differentiation.

A third and final core dimension of GGSs can be inferred from analyzing their *geographical scope*. If the term 'global' is usually used to describe the rescaling of

15 This perspective is clearly indebted to Antje Wiener's recent work on contestation in international politics, which convincingly argues how contestation, as a discursive social practice, is ultimately both indicative and productive of legitimacy in global governance. See Antje Wiener, *A Theory of Contestation*, Springerbriefs in Political Science (Heidelberg, Germany: Springer, 2014).

16 For a framework analyzing different degrees or types of fragmentation of global governance architecture, see especially Biermann et al., *Earth System Governance People, Places, and the Planet: Science and Implementation Plan of the Earth System Governance Project Report No. 1.*

17 For a classic discussion of different types of multi-levelled or networked forms of governance see, most notably, Liesbet Hooghe and Gary Marks, *Multi-Level Governance and European Integration* (Lanham, MD: Rowman & Littlefield, 2001); Liesbet Hooghe and Gary Marks, "Unraveling the Central State, but How?: Types of Multi-Level Governance," *American Political Science Review* 97, no. 2 (2003); Michael Zürn, "Global Governance as Multi-Level Governance," in *Handbook on Multi-Level Governance*, ed. Henrik Enderlein, Sonja Wälti, and Michael Zürn (Cheltenham, UK and Northampton, MA: Edward Elgar Publishing, 2011).

politics and governance beyond traditional state borders, it does not say much about the actual 'spatial configuration(s)' of governance activities – whether, for instance, the governance of a host of global problems is truly global in reach or restricted to a particular geographical area of the world. In this respect the notion of geographical scope helps to capture what has thus far been a largely neglected element of the global governance phenomenon, namely the 'governed' or 'subjects of global governance', understood here as the set of people or communities for whom and over whom global governance activities are being exercised. Accordingly, a useful conceptual distinction can be established between the intended and the actual geographical scope of governance systems. In most areas of global governance, such as global environmental degradation, human rights, health, finance, and trade, the 'governed' are generally intended to encompass the globe's entire population (that is, the global citizenry), and it is in that sense, then, that governance can be qualified as being truly global in reach. But in most cases, however, in-depth empirical mapping exercises tend to unveil systems of governance that are in reality far more restricted in geographical scope, affecting the behaviour, values, or interests of a limited segment of the world's population.[18]

In a nutshell, these three core dimensions of GGSs constitute useful *analytical lenses* through which to better understand the key dynamics of broader systems of global governance. These, of course, are not the only possible ones, for additional variables could easily be taken into account depending on the focus and scale of analysis. This notwithstanding, our framework can already serve to highlight three important and related features of the emerging GGS for climate finance: a weak and contested normative framework, a highly fragmented architecture, and last but not least, a restricted geographical scope.

18 The scholarship on transnational climate/environmental governance is particularly suggestive in this regard. Despite their relative inclusive membership composition, empirical research has revealed how most transnational governance schemes tend, for the most part, to operate in Western industrialized countries, the geographical location of the most resource-rich and powerful actors in the system. In their assessment of a database of sixty transnational climate initiatives, Bulkeley and colleagues conclude that despite the increasing participation of actors from developing and least-developed countries, the vast majority of initiatives (87 percent) are actually initiated and operational in the 'Global North' (See Harriet Bulkeley, Liliana B. Andonova, Michele Betsill, Daniel Compagnon, Thomas Hale, Matthew J. Hoffmann, Peter Newell, Matthew Paterson, Charles Roger, and Stacy D. VanDeveer, "Governing Climate Change Transnationally: Assessing the Evidence from a Database of Sixty Initiatives," *Environment and Planning-Part C* 30, no. 4 (2012): p. 601). In an earlier analysis of the democratic legitimacy of WSSD partnerships, Bäckstrand (2006) reached broadly similar conclusions, pointing out that despite their deliberative and multi-stakeholder qualities, PPPs in sustainable development governance rarely involve the participation of grassroots and local people from the South. 'The broad picture', she notes, 'is that the partnerships are North-driven, sponsored primarily by international organizations and a handful of industrialized countries' (See Karin Bäckstrand, "European Journal of International Relations," *European Journal of International Relations* 12, no. 4 (2006): p. 489).

Assessing the GGS for climate finance

The weak and contested normative foundations of public climate finance mobilization

The basic principles relevant to the climate finance responsibilities of developed countries towards developing countries are mainly spelled out in the UNFCCC (that is, the Convention) and further strengthened by follow-up agreements.[19] Article (4)3 of the Convention, for instance, commits developed countries to provide new and additional financial resources to meet the 'agreed full costs' of climate change in developing country parties. In the subsequent paragraph the Convention further stresses the need to scale up support for adaptation in developing countries, particularly in Least Developed Countries (LDCs) and Small Island States (SIS). In regard to the volume of national contributions, there is an implicit consensus, at least from the perspective of recipient developing countries, that the 'respective capacities' principle should apply, meaning that the amount pledged by one country should correspond to its estimates of national contributions to global GHG emissions, but also to its level of national wealth measured in terms of its gross domestic product (GDP) or gross national income (GNI). The Convention and later the Bali Action Plan adopted at COP 13 in 2007,[20] in its paragraph 1(e), and the Copenhagen Accord[21] further state that climate financing to developing countries should follow the principle of adequacy, meaning that the volume of provided funds should ultimately be equal to the scale of the task of maintaining a global temperature rise to below 2°C, as well as the principle of predictability for which the disbursement of funds should be sustainable enough over the medium term so as to allow national investment programmes to plan, scale up, and implement priority actions for adaptation and mitigation.[22]

As repeatedly highlighted by many climate finance experts in recent years,[23] this set of publicly proclaimed principles tends, in reality, to offer a rather thin

19 United Nations Framework Convention on Climate Change (UNFCCC), "Framework Convention on Climate Change," (FCCC/INFORMAL/84: United Nations, 1992). UNFCCC 1992.

20 United Nations Framework Convention on Climate Change (UNFCCC), "Conference of the Parties Serving as the Meeting of the Parties to the Kyoto Protocol, 3–15 December 2007, Bali, Decision," (FCCC/KP/CMP/2007/9/Add.1, 1/CMP.3: UNFCCC, 2007), para. 1(e).

21 United Nations Framework Convention on Climate Change (UNFCCC), "The Conference of the Parties (COP), at Its Fifteenth Session, Took Note of the Copenhagen Accord of 18 December 2009," (Decision 2/CP.152009).

22 For a more detailed account of the climate finance principles under the UNFCCC, see Neil Bird and Jessica Brown, "International Climate Finance: Principles for European Support to Developing Countries," (Belgium: European Development Cooperation to 2020, 2010).

23 See, among many others, the studies provided by Liane Schalatek, "A Matter of Principle(s). A Normative Framework for a Global Climate Finance Compact," *The Green Political Foundation*, 19 November 2010; Liane Schalatek and Neil Bird, "The Principles and Criteria of Public Climate Finance – a Normative Framework," in *The Climate Finance Fundamentals* (The Climate Funds Update, 2015).

normative framework (that is, a minimal set of informal principles) against which to judge and monitor the quality and quantity of public climate financial flows.[24] In fact, and despite a longstanding consensus among the world's leaders about the critical importance of climate finance for preventing further irreversible climate change and building climate resilience, substantial divergences between developed and developing countries have continuously embittered international negotiations up to the 2015 Paris Climate Conference. Substantial disagreements persist, in particular, over how these UNFCCC principles should be put into practice, resulting in differing, and at times potentially conflictive, understandings of what shall constitute a fair, equitable and effective provision of climate finance to developing countries.

As noted by Pickering and colleagues,[25] a notorious point of disagreement regards the relationship between climate finance and development aid, an issue closely bound up with concerns expressed in debates around the question of financial additionality. While developed countries have long privileged a complementary approach between the two types of financial flows, emphasizing in particular the efficiency and pragmatic benefits of mainstreaming climate adaptation and mitigation finance into development assistance, developing countries, for their part, have continuously favoured a clear distinction between climate change and development financing. Of grave concern to recipient countries in particular is the issue of financial diversion: that a more integrated approach to climate finance and development aid would, in the end, divert development finance away from supporting traditional development needs of poor communities.[26] According to developing countries, the two types of finance (aid and climate) should reflect different forms of 'international' responsibility, with development aid being based primarily on the moral responsibility of the wealthiest nations to help combat world poverty, and climate finance stemming mainly from the 'causal responsibility' of developed countries for their disproportional historical contributions to climate change.[27] Then, and mainly in order to maximize the

24 For a comprehensive assessment of the difficulties involved in tracking climate financial flows, see Christa Clapp, Jane Ellis, Julia Benn, and Jan Corfee-Morlot, "Tracking Climate Finance: What and How?" (Paris: OECD and IEA Secretariats, 2012).

25 Jonathan Pickering, Jakob Skovgaard, Soyeun J. Kim, Timmons Roberts, David Rossati, Martin Stadelmann, and Hendrikje Reich, *Acting on Climate Finance Pledges: Inter-Agency Dynamics and Relationships with Aid in Contributor States: CCEP Working Paper 1306* (Canberra: Crawford School of Public Policy, The Australian National University, 2013).

26 For a more extensive discussion of potentially different forms of 'diversion', see Michael Dutschke and Axel Michaelowa, "Development Aid and the CDM – How to Interpret Financial Additionality: Hamburgisches Welt-Witschafts-Archiv (Hwwa) Discussion Paper 228," (Hamburg: Hamburg Institute of International Economics, 2003).

27 Nancy Birdsall and Michele De Nevers, "Adaptation Finance: How to Get out from between a Rock and a Hard Place: Policy Paper 001," (Washington, DC: Center for Global Development, 2012); Pickering, Skovgaard, Kim, Roberts, Rossati, Stadelmann, and Reich, *Acting on Climate Finance Pledges: Inter-Agency Dynamics and Relationships with Aid in Contributor States: CCEP Working Paper 1306.*

resources they can receive, most developing countries have long argued for climate change funds and development aid to be managed separately. On this view climate finance should not count towards official development assistance (ODA) targets (that is, they should be additional to existing aid commitments); and the mobilization and delivery of climate-related funds should operate via arrangements, channels, and procedures that purposely differ from traditional development cooperation. However, and because individual donor countries still retain significant discretion over how they can deliver their commitments, it is not surprising to find that substantial flows of climate finance are currently sourced from bilateral development cooperation agencies through the use of ODA.[28] In the absence of a common baseline against which 'new and additional' finance can be measured, developed countries can also use different methodologies to account for the exact share of development aid targeting climate change adaptation and/or mitigation, rendering a comparative evaluation of national commitments very challenging from a methodological standpoint.

Essentially, the extensive normative contestation over the relationship between aid and climate finance tends to closely resonates well with deeper ethical debates about the burdens and costs associated with the public provision of global public goods. A persisting issue of contention in the climate change negotiations has been, in fact, the distribution among developed and developing countries of the costs and benefits of addressing climate change; how country Parties conceive of distributed fairness has, in the end, significant implications for their understanding of the nature of climate financial responsibilities: whether state's financial commitments shall fall, for instance, under the UNFCCC principles like the 'common but differentiated responsibilities', vulnerabilities, capacities, and polluter-pays principles, or instead be less rigidly understood, to use Pickering's and colleagues' terms, as 'part of a bargain between developed and developing countries helping the latter in better achieving their mitigation targets'.[29]

The highly contested question of distributive fairness and equity in international climate change politics has been constitutive of many other lines of conflict between developed and developing countries. Despite an increasing consensus that the bulk of long-term finance should come from a variety of sources (public, private, and hybrid) and involve diverse financial instruments and mechanisms, there remains a considerable disagreement about the preferable mix and design of these sources. As is well known, there is as yet no internationally agreed

28 Schalatek, "A Matter of Principle(s). A Normative Framework for a Global Climate Finance Compact." Note that the UNFCCC also requires that Parties report on climate change financing for developing countries. Guidelines require Parties to indicate what 'new and additional' funds they have provided pursuant to Article 4.3 of the Convention and to explain how they have determined that such financial resources are in fact 'new and additional'.

29 Pickering, Skovgaard, Kim, Roberts, Rossati, Stadelmann, and Reich, *Acting on Climate Finance Pledges: Inter-Agency Dynamics and Relationships with Aid in Contributor States: CCEP Working Paper 1306*, p. 5.

definition of what should constitute climate finance to developing countries and whether private sources of funds should count towards national climate financial contributions. Developed and developing countries also largely differ on which type of climate finance (mitigation vs. adaptation) should be given more importance. While several developing countries, and particularly those from the least developed countries (LDCs) and Small Island States (SISs), have long called for a distribution of climate finance that privileges adaptation finance over mitigation finance, developed countries for their part have long been more interested in contributing to mitigation finance,[30] which unlike adaptation finance brings more direct benefits to developed countries' mitigation efforts.[31]

A highly fragmented architecture

These substantial disagreements over the meaning and/or applicability of UNFCCC principles for climate finance have been conducive to the development of a diversely populated architecture. As demonstrated by several mapping studies, climate financial flows to developing countries are currently mobilized, channelled, and delivered through a myriad of organizational entities. These include, for instance, *global donor funds set by UN agencies* like the UNFCCC, the World Bank (WB), the United Nations Development Programme (UNDP), the United Nations Environment Programme (UNEP), and the Food and Agriculture Organisation of the United Nations (FAO) (for example, the Global Environment Fund, the Green Climate fund); *Global donor funds managed by the EU institutions* such as the Global Energy Efficiency and Renewable Energy Fund; *regional recipient funds* managed by regional development banks, bilateral financial institutions (BFIs), and national development banks (NDBs); and *national recipient funds managed* by bilateral financial institutions (BFIs) and National Development Banks (NDBs).[32] A substantial share of public climate finance to developing countries is also flowing directly via bilateral channels such as development cooperation agencies and bilateral development banks. To facilitate the distinction between climate finance and development aid, some developed countries, and most notably the UK, Germany, and Japan, have in recent years established their own climate-specific funds. Noteworthy as well are financial flows provided by private actors who range from single actors to multinational corporations,

30 Martin Stadelmann, Jessica Brown, and Lena Hornlein, "Fast-Start Finance: Scattered Governance, Information and Programmes," in *Carbon Markets for Climate Finance? Low Carbon and Adaptation Investments Choices for the Developing World*, ed. Axel Michaelowa (London: Routledge, 2012), p. 134.

31 Dirk T.G. Rubbelke, "International Support of Climate Change Policies in Developing Countries: Strategic, Moral, and Fairness Aspects," *Ecological Economics* 70, no. 8 (2011).

32 For a more detailed description of these various channels, see especially Diane Smallridge, Barbara Buchner, Chiara Trabacchi, Maria Netto, José J.G.L. Lorenzo, and Lucila Serra, "The Role of National Development Banks in Intermediating International Climate Finance to Scale Private Sector Investments: Discussion Paper No. IDB-DB-249," (Washington: Inter-American Development Bank, Institutions for Development (IFD), 2012).

international investors initiatives, and their intermediaries. Carbon finance has also become an important mechanism of climate financing, with the proliferation in the past few years of a wide range of private, public, and hybrid carbon funds. Finally, since the mid 2000s an increasing number of innovative transnational and regional public-private partnerships, gathering stakeholders from different sectors (for example, public, private, and non-profit), have been established for the purpose of scaling up private climate-related investments in developed and developing countries.

A detailed description of all of the various organizations and actors more or less directly implicated in the mobilization and deployment of climate finance to developing countries is, understandably, beyond the scope of this chapter.[33] It suffices to say that the rapid proliferation of institutions, actors, and mechanisms involved in international climate financing ultimately maps out into a governance architecture that is highly fragmented vertically, between different layers of societal organizations (for example, national, supranational, regional) and spatial scopes (for example, bilateral, multilateral, transnational), but also, and increasingly, horizontally between different groups of actors and policy sectors. In this regard the global architecture for climate finance can be described, following Abbott's[34] useful analysis of the 'transnational complex for climate change', as loosely 'polycentric', a term pioneered by Elinor Ostrom to qualify multi-leveled and multi-actor governance systems involving a multiplicity of relatively autonomous governing units, formal or informal, operating at different levels, scales, and policy fields.[35] The attributes of autonomy and formal independence

33 The literature on this topic is indeed quite extensive. See especially the Climate Policy Initiative (CPI)' s annual overviews of the overall landscape of climate finance: Climate Policy Initiative (CPI), "The Landscape of Climate Finance 2011," *A Climate Policy Initiative Report* (2011), http://climatepolicyinitiative.org/publication/the-landscape-of-climate-finance; Climate Policy Initiative (CPI), "The Landscape of Climate Finance 2012," *A Climate Policy Initiative Report* (2012), http://climatepolicyinitiative.org/wp-content/uploads/2012/12/The-Landscape-of-Climate-Finance-2012.pdf; Climate Policy Initiative (CPI), "The Global Landscape of Climate Finance 2013," *A Climate Policy Initiative Report* (2013), http://climatepolicyinitiative.org/wp-content/uploads/2014/11/The-Global-Landscape-of-Climate-Finance-2014.pdf; "The Global Landscape of Climate Finance 2014," Climate Policy Initiative (CPI), *A Climate Policy Initiative Report* (2014), http://climatepolicyinitiative.org/wp-content/uploads/2014/11/The-Global-Landscape-of-Climate-Finance-2014.pdf; Climate Policy Initiative (CPI), "The Global Landscape of Climate Finance 2015," *A Climate Policy Initiative Report* CPI (2015), http://climatepolicyinitiative.org/wp-content/uploads/2015/11/Global-Landscape-of-Climate-Finance-2015.pdf; Smallridge, Buchner, Trabacchi, Netto, Lorenzo, and Serra, "The Role of National Development Banks in Intermediating International Climate Finance to Scale Private Sector Investments: Discussion Paper No. IDB-DB-249." For comprehensive analyses of the evolving architecture for international climate finance, see United Nations Framework Convention on Climate Change (UNFCCC), "2014 Biennial Assessment and Overview of Climate Finance Flows Report," (Bonn, Germany: UNFCCC Standing Committee on Finance, 2014).

34 Abbott, "The Transnational Regime Complex for Climate Change."

35 Ostrom, "Polycentric Systems for Coping with Collective Action and Global Environmental Change."

are often crucial to differentiate polycentric systems from mere vertically multi-levelled hierarchies, where a central organizing unit often plays a leading steering role. Determining how the climate finance architecture fares in terms of polycentricity would require, understandably, a more in-depth empirical analysis with a close focus on the degree of interactions between the various key organizational components of the system. At this phase of the discussion, however, we can say that both the lack of a strongly centralized coordinating mechanism coupled with the diversity of values and interests of key parties have been key drivers for the development of a weakly polycentric governance architecture for climate finance, a macro-structural feature which, as we will later in this chapter, presents substantial disadvantages for questions of institutional integrity.[36]

The restricted geographical deployment of climate finance

As noted earlier, a third important characteristic of global governance systems is their geographical scope, a dimension which in the context of climate finance draws attention to how much and where climate finance flows globally. As is well known, since 2009 some progress has been made, at least towards a short-term finance pathway up to 2020. At the fifteenth and sixteenth sessions of the COP in 2009 and 2010, developed countries promised 'new and additional' finance to address the needs of developing countries, with a balanced allocation between adaptation and mitigation. They agreed to provide USD 30 billion a year for the period 2010–2012 and to 'jointly mobilize USD 100 billion per year by 2020 to address the needs of developing countries (. . .) from a wide variety of sources (. . .)'.[37]

Thus far, however, relatively little information has been made available on past and current volumes of climate finance, and especially on mobilized climate finance in the context of the USD 100 billion a year goal promised by developed countries at COP 15 in 2009. The problem is that in the absence of an internationally agreed definition of what constitutes appropriate climate finance, it remains very difficult to properly track and monitor financial flows for climate actions. Organizations set up to collect data on international climate finance do exist, but tend to include different types of financing sources and flows.

Notwithstanding these difficulties, some reports taking stock of a variety of climate financial flows do exist. To date, the most comprehensive estimates of total climate finance are provided by the Climate Policy Initiative (CPI) through

36 At COP 16 in Cancun, the Green Climate Fund (GCF) was formally adopted, which, when fully operational and appropriately funded, shall become the central channel for climate financing to developing countries. However, there is still much uncertainty about the overarching role of the GCF: whether, for instance, the fund can help to bring about much-needed transparency and accountability in the mobilization, administration, and use of money flows.

37 United Nations Framework Convention on Climate Change (UNFCCC), "Report of the Conference of the Parties on Its Sixteenth Session, Held in Cancun from 29 November to 10 December 2010," (Part Two: Action Taken by the Conference of the Parties at Its Sixteenth Session FCCC/CP/2010/7/Add.1: United Nations, 2011).

its annual overviews of the overall landscape of climate finance flows;[38] by the UNFCCC Standing Committee on Finance (SCF) in its 2014 Biennial Assessment and Overview of Climate Finance;[39] and most recently by an OECD/CPI joint report which was released in advance of COP 21 and which specifically focuses on assessing developed countries' progress towards achieving their USD 100 billion a year commitments (hereafter referred to as 'mobilized climate finance').[40]

A positive trend common to all of these studies is that over the past five years or so climate finance for both mitigation and adaptation has been on an upward trajectory. According to the latest CPI's report, in 2014 the total amount of climate finance increased by 18 percent (compared to 2013 levels) to reach an estimated USD 391 billion. In that period public climate finance increased by 8 percent and accounted for about USD 148 billion, while private finance, mainly in the form of investments in renewable energies, grew by 26 percent, totalizing approximately USD 243 billion, accounting for around 62 percent of total climate financial flows. Significant progress has been made in regard to mobilized climate finance as well. According to the OECD/CPI report, the total volume of public and private mobilized climate finance increased from USD 52.2 billion in 2013 to USD 61.8 billion in 2014, the equivalent of a yearly average of USD 57.0 billion for 2013–2014, with about 71 percent of the total originating from the public sector.[41]

38 Climate Policy Initiative (CPI), "The Landscape of Climate Finance 2011"; Climate Policy Initiative (CPI), "The Landscape of Climate Finance 2012"; Climate Policy Initiative (CPI), "The Global Landscape of Climate Finance 2013"; Climate Policy Initiative (CPI), "The Global Landscape of Climate Finance 2014"; Climate Policy Initiative (CPI), "The Global Landscape of Climate Finance 2015"; United Nations Framework Convention on Climate Change (UNFCCC), "Report of the Conference of the Parties on Its Sixteenth Session, Held in Cancun from 29 November to 10 December 2010."

39 United Nations Framework Convention on Climate Change (UNFCCC), "2014 Biennial Assessment and Overview of Climate Finance Flows Report."

40 Organisation for Economic Cooperation and Development (OECD) and Climate Policy Initiative (CPI), "Climate Finance in 2013–14 and the USD 100 Billion Goal: A Report by the OECD in Collaboration with Climate Policy Initiative," (Paris: OECD-CPI, 2015). It is important to note that the OECD/CPI report has been largely criticized by developing countries and civil society organizations. Some counting and methodological issues have been raised, especially in regard to climate-related development aid, which does not have climate change as a primary or significant objective. The government of India, for instance, in a released paper discussing the report, concludes on a highly critical note by stating that '[t]he OECD report is deeply flawed and unacceptable [the authors are careful to say their report is preliminary and aggregate]. It repeats a previous experience we had of double-counting, mislabelling and misreporting when rich countries provided exaggerated claims of 'fast-start climate financing' in the period 2010–12 – which were widely criticized by independent observers.' See Climate Change Finance Unit, "Climate Change Finance, Analysis of a Recent OECD Report: Some Credible Facts Needed," (Delhi: Department of Economic Affairs Ministry of Finance, Government of India, 2015), p. 10.

41 Organisation for Economic Cooperation and Development (OECD) and Climate Policy Initiative (CPI), "Climate Finance in 2013–14 and the USD 100 Billion Goal: A Report by the OECD in Collaboration with Climate Policy Initiative." The report is careful to note, however, that a large share of the rise in funds from 2013–14 came about with a substantial increase in outflows from multilateral banks.

Despite these encouraging trends, however, there are several points of concern, especially when paying closer attention to the geographical and thematic allocation of mobilized climate funds. The CPI report, for instance, seems to confirm a long-standing pattern in the spatial deployment of climate finance, which is that the majority of total money flows (about 74 percent) has been actually deployed in the country of origins. Such a pattern is confirmed by the OECD/CPI report on mobilized climate finance, which despite reaching USD 57 billion on average for 2013/2014, actually accounts for a small share of total climate financial flows (14–16 percent). Not surprisingly, a substantial share of total climate finance (31 percent) has been directed at the East Asian and the Pacific regions, with China alone accounting for 22 percent (USD 84 billion) of total finance. The second-largest destination of climate finance was Western Europe, which received 24 percent of total finance, or USD 93 billion. The OECD/CPI report also highlights the persistent massive neglect of adaptation needs of the most vulnerable and poorest countries (LDCs, SIDs, and African states). According to the report, only 16 percent of mobilized climate finance has been actually allocated towards climate adaptation. In addition to this unequal thematic and geographical allocation of funds, it is vastly agreed that the figure of USD 100 billion per year, pledged by developed countries up 2020, will not be sufficient to meet the actual climate finance needs of developing countries and especially LDCs. While estimates of these needs vary substantially, all tend to far exceed the annual provision of USD 100 billion per year, at times calling for trillions and not billions in new additional innovative climate financing. According to a recent study conducted by the International Institute for Environment and Development (IIED),[42] the total cost of financing post-2020 climate actions in all 48 LDCs should at least amount to around USD 93.7 billion per year, a figure which far exceeds the amount of available international public climate finance that currently reaches the LDCs (less than a third).

Taking stock: The GGS for climate finance and its weak capacity for promoting institutional integrity

Taking stock

Our analysis reveals that at the systemic level, climate finance has thus far been characterized by extensive normative contestation over its publicly proclaimed principles and values and, relatedly, by a highly fragmented governance architecture characterized by a myriad of loosely coordinated institutions and mechanisms. In regard to its geographical scope, in turn, climate finance has been insufficiently targeting those countries or communities that are most in need of financial support, especially for adaptation measures. Taken together, it seems

42 International Institute for Environment and Development (IIED), "A Fair Climate Deal in Paris Means Adequate Finance to Deliver INDCs in LDCs," *IIED Briefing*, November 2015.

that these three features of the broader organizational context for climate finance tend to be indicative and conducive of the difficulty of achieving greater institutional integrity at the micro-institutional level and in regard to all sequential phases of climate finance support to developing countries (for example, mobilization, administration/governance, and delivery).[43] Three main issues stand out in this respect.

The first has to do with the lack of strong normative guidance, which in the context of a large and highly diversified institutional environment can work to hinder rather than to facilitate the move towards greater policy coherence and institutional consistency among and across the variety of institutions that compose the system. As mentioned before, the persistence of deep disagreements between developed and developing countries over the meaning and/or normative force of some key principles of climate finance has thus far prevented or, if not, significantly delayed the development of an overarching normative framework for tracking and monitoring climate financial flows, as well as for safeguarding the legitimacy and effectiveness of the use of funds for concrete mitigation and adaptation measures. Conversely, one major implication of the lack of a strong normative framework is that developed countries and other contributing actors have so far been able to exercise considerable discretion in choosing how to mobilize, count, administer, and deliver their funds. All in all, it seems that without strongly agreed and institutionalized guiding norms and principles as well as enforcement mechanisms, the climate finance system ultimately runs the risk of evolving into an incoherent governance complex involving incompatible or even conflictive, mutually harmful rules and institutions.

A second and related issue concerns the impact of fragmentation on the capacity for democratic control and oversight, a condition which remains crucial for greater context-integrity. What is clear is that the proliferation of a variety of institutions involved in the mobilization and/or delivery of climate funds since the mid 2000s makes it even more challenging today to trace and monitor the quality of climate funds, and particularly those originating from mixed sources or flowing through a variety of institutional channels. The problem essentially is that if the diffusion of governing activities to multiple levels and directions can open up considerable opportunities for the emergence of more inclusive and pluralistic modes of governance, the high degree of institutional fragmentation that this trend implies invariably raises crucial challenges of democratic control and oversight. As repeatedly highlighted in the emerging literature on 'international' and 'transnational' regime complexity, institutional fragmentation tends to reduce not only the clarity of legal obligations but also, and perhaps more fundamentally, the visibility of the exercise of political power in global governance. In other terms, institutional fragmentation, in the absence of centralized or networked mechanisms of coordination, not only leaves the terms of legal

43 For a more detailed description of these three sequential phases and the principles that shall apply to it, see especially Schalatek and Bird, "The Principles and Criteria of Public Climate Finance – a Normative Framework."

compliance unclear; it also makes it harder to identify where political authority over a particular issue ultimately resides, thereby potentially undermining transparency and accountability.[44]

A third and final issue revolves around the implications of institutional fragmentation for the geographical and/or thematic allocation of climate finance. As Keohane and Victor[45] explain in their analysis of the regime complex for climate change politics, fragmented governance architectures have for sure several advantages over the more traditional modes of governance based on centralized/integrated top-down structures, chief among them being their greater adaptability and flexibility over time. In addition, bottom-up forms of international cooperation, are, quite logically, expected to be more attuned to the interests, concerns, or values of local communities. These advantages, however, are no guarantee of effective policy outcomes, at least at the systemic level. The climate finance landscape, in its unequal geographical and thematic allocation of funds, aptly illustrates how highly fragmented structures of governance often run the risks of failing to produce meaningful results or policy outcomes. Of particular concern here are the costly inefficiencies generally associated with the presence of overlapping sets of institutions with little to no history of intra-organizational cooperation or competition. More importantly, institutional fragmentation, when poorly coordinated, tends also to facilitate negative outcomes including non-compliance, forum shifting, and exit from legally demanding institutional settings.

Towards greater institutional integrity? Key policy recommendations

In light of the perspective offered therein, three main recommendations can be offered with the purpose of enhancing the institutional integrity capacity of the overall institutional landscape for climate finance.

First, bringing about greater policy coherence in all the aspects and levels of climate change financing is crucial to promoting a more effective, efficient, and equitable global climate financing system. Achieving this goal, however, essentially depends on the implementation of a thicker normative framework for climate finance, a condition which, in turn, crucially hinges upon the capacity and willingness of both developed and developing countries to build a solid and legitimate consensus on the meaning or application of core climate change funding principles and concepts. In a recent policy brief published by Climate Funds Update in advance of COP 21 in Paris, Schalatek and Bird,[46] both renowned

44 For an overview of the potential impacts of international regime complexity, see especially Karen Alter and Sophie Meunier, "The Politics of International Regime Complexity," *Perspective on Politics, Symposium* 7, no. 1 (2009).

45 Keohane and Victor, "The Regime Complex for Climate Change."

46 Schalatek and Bird, "The Principles and Criteria of Public Climate Finance – a Normative Framework."

climate finance analysts, echo our analysis by calling for the urgent need to apply relevant principles and criteria to all sequential phases of public climate finance to developing countries (for example, mobilization, administration, and disbursement). Drawing on Schalatek's earlier work,[47] the brief also recommends the implementation of a human rights perspective to climate change financing, one which complements existing UNFCCC principles with core human rights values and obligations. In a nutshell, then, the quest for building a more policy-coherent climate change financing system depends on bringing about principles that are more widely agreed upon, as well as integrating other relevant international norms and values to the climate change regime.

This emphasis on the importance of reaching a 'global' normative consensus on key climate finance principles and values, however, should not be taken to mean the end of conflicts and contestation in climate politics, for an adequate consensus-based approach is one that should recognize both the inevitability and desirability of political contestation for the sustained institutional integrity of climate finance institutions. The main rationale for this is easy to understand. Under conditions of 'deep pluralism' and high institutional fragmentation, it is principally through practices and discourses of contestation that perceived institutional integrity gaps and deficits can be brought to the fore and potentially remedied. Accordingly, then, the building of a stronger normative consensus between developed and developing countries over the principles of climate finance should be accompanied by the promotion of thicker governance values and procedures that can help strengthen and broaden the perceived legitimacy of consensus-based decision-making and, when disagreements or conflicts are deep, provide meaningful, inclusive, and cooperative opportunities for contestation and legitimate mechanisms for conflict managements.

Second, there is an obvious need to enhance the systemic effectiveness and distributive fairness of the GGS for climate finance to developing countries. Two main issues are of particular concern here. The first has to do with the need to promote vertical as well as horizontal modes of coordination capable of enhancing synergies between institutional arrangements outside and inside the UNFCCC. Core institutions such as the board of the GCF could play an important role in in that process by closely monitoring funds channelled outside the UNFCCC and by helping to overcome some of the main obstacles that commonly inhibit a highly fragmented and diversified institutional architecture to reach its full potential (that is, institutional overlaps, low intra-organizational collaboration, the lack of political visibility/transparency, and accountability regarding the mobilization and deployment of money flows). A second set of proposals relates to the distributive fairness and effectiveness of climate finance and the need for a more balanced allocation between mitigation and adaptation finance. More attention in particular should be devoted to providing greater volumes of adaptation finance to countries most vulnerable to the adverse

47 Schalatek, "A Matter of Principle(s). A Normative Framework for a Global Climate Finance Compact."

effects of climate change. The good news in this respect is that when fully operationalized and adequately capitalized, the GCF promises to allocate funding equally to mitigation and adaptation measures, and for at least 50 percent of the adaptation finance to go to particularly vulnerable countries. Another positive trend is that over the past few years, increasing attention has been paid to the role that the public sector can play in catalyzing and leveraging private capital for both mitigation and adaptation. The Private Sector Facility (PSF) of the GCF can play a crucial role in that process, for it has been established for the purpose of improving the role of private sector actors in developing countries, in particular local actors such as small and medium enterprises and local financial intermediaries.[48]

Finally, and perhaps more importantly, enhancing the distributive fairness and effectiveness of climate finance will depend not only on the greater availability of a variety of financing resources, but also on the capacities of recipient countries, and especially the most vulnerable ones to absorb and implement money flows. One major obstacle to the fair and effective deployment of climate finance is that unlike developed countries that have the internal capacities to use and generate climate finance, developing countries and especially LDCs often lack the necessary institutional, policy, and skills systems to use climate finance effectively. While some progress has been made, especially with regard to the establishment in certain countries of climate-specific national funds, greater interest and effort on the part of both developed and developing countries should be devoted to improving recipient countries' national climate finance readiness, that is, their capacity to plan for, access, and deliver both domestic and international climate finance, as well as to monitor and report its use and results.[49]

These three sets of recommendations for enhancing the institutional-integrity capacity of the GGS for climate finance are evidently not without limitations. Chief among them is the fact that under conditions of low context-integrity, whereby systemic issues of normative contestation, institutional fragmentation, and distributive ineffectiveness weaken the capacity of international climate finance institutions to pursue greater institutional integrity, the promotion of thicker governance values entailing stronger sets of democratic procedural norms is, in the end, as necessary as it is difficult. Following Breakey's, Cadman's, and Sampford's argument regarding the relationship between low context-integrity and thick governance values (see Chapter 2 of this volume), it is clear indeed that in light of persisting, deep ethical (or substantive) disagreements over the basic meaning of climate change financing norms, achieving thicker governance values might appear impracticable and even undesirable.

48 See Green Climate Fund, "Governing Instrument for the Green Climate Fund," (Republic of Korea: Secretariat, Green Climate Fund, 2010), para. 41.

49 For an extensive analysis of the emerging concept of 'climate finance readiness', see especially United Nations Development Programme (UNDP), "Readiness for Climate Finance: A Framework for Understanding What It Means to Be Ready to Use Climate Finance," (New York: UNDP Environment and Energy Group, 2012).

Conclusion – The 2015 Paris agreement: A missed opportunity for climate finance

The goal of this contribution was to take a closer look at some of the core dimensions of the GGS for climate finance to developing countries and to evaluate their implications especially in terms of institutional integrity. In conclusion, it is worth briefly exploring what the newly approved Paris Agreement and its related decision actually do for climate finance: whether the agreement in particular can over the long term ensure more adequate, effective, transparent, and equitable deployment of climate funds to developing countries.[50]

As is well known, immediate reactions to the Paris Outcome have been strongly positive. It is clear that the agreement, which was adopted with universal agreement at the twenty-first session of the COP in Paris in December 2015, constitutes a major breakthrough in the deadlock over the negotiations of a successor agreement for the post-Kyoto period. One of the most lauded outcomes of the Paris Agreement in particular is its long-term ambition and the establishment of a clear process for progress reporting and assessments (that is, transparency, stocktaking, and compliance). Indeed, the agreement, which will enter into force after 2020, firmly commits to keep the global average temperature 'well below' 2°C above pre-industrial levels by the end of the century, and even set out to pursue a more ambitious limit of below 1.5°C, a target which gives the most vulnerable countries and especially small island states a higher chance of survival.[51] The agreement also establishes clear procedures to monitor these commitments. From 2020 onwards all countries will be legally required to update their mitigations plans every five years,[52] and from 2023 onwards to publicly monitor and report on their emissions levels and reductions using a universal accounting system (that is, global stocktaking).[53] The Paris Decision itself nevertheless acknowledges with concern how current mitigation plans resulting from the total of Intended Nationally Determined Contributions (INDCs) are currently not enough to meet the 2°C goal, let alone the 1.5°C target, and subsequently highlights the need to address this 'significant' gap of close to 15 Gigatons (Gt) or billion tons by the year 2020.[54] As such, then, much of the success of the Paris Agreement will ultimately depend on greater emission reduction efforts at least to levels compatible with the agreed temperature limit of 2°C and on the

50 It is important to note that the Paris climate text (that is, the Paris Outcome) actually involves two different documents: the legally binding agreement (that is, the Paris Agreement) and the Paris Decision, which is not legally binding and which spells out in more detail how the agreement should be carried out. See United Nations Framework Convention on Climate Change (UNFCCC), "Conference of the Parties Twenty-First Session Paris, 30 November to 11 December 2015: Adoption of The Paris Agreement," Decision 1/CP.21.

51 Ibid., Article 2 (1a), p. 21.

52 Ibid., Artcle 4 (9), p. 22

53 Ibid., Artcle 14, p. 28.

54 Ibid., Item 17.

provision of more adequate technological and financial support to the most vulnerable and poor countries.

Unfortunately, the Paris Agreement remains very vague on the area of climate finance, a topic which, as expected, turned out to be one of the most dividing issues at the Paris talks. While the Paris Decision (in paragraphs 54 and 115) strongly urges developed country Parties to pursue their existing collective mobilization goals through 2025, and to establish prior to 2025 a new collective quantified target that shall go beyond the commitment of USD 100 billion a year, the agreement itself does not include any specific quantified commitments. The main rationale for this was that with no clear monetary targets, the agreement will not require the approval of the US Senate, thereby making it easier for President Obama to ratify it by executive decree only.

Accordingly, the agreement, in rather vague terms, simply states that '[d]eveloped country Parties shall provide financial resources to assist developing country Parties with respect to both mitigation and adaptation in continuation of their existing obligations under the Convention'.[55] Somewhat more precisely, the text further indicates that the developed country parties 'should continue to take the lead in mobilizing climate finance' and that 'such mobilization (. . .) should represent a progression beyond previous efforts', meaning, probably, above the promise of USD 100 billion a year.[56] Developing countries and especially high-income countries (for example, Gulf states, Singapore, and Korea) are encouraged to provide financial support, but only on a voluntary basis. The agreement further stresses the need to achieve a balance between adaptation and mitigation finance, and that climate financing should more actively prioritize the needs of least-developed countries and small-island developing states, especially in the form of grant-based funding for adaptation.[57] Overall, though, it is clear that the agreement fails to establish specific monetary goals for after 2025, and remains equally vague with respect to what exactly should be counted towards international climate finance, whether public funds, for instance, should mainly consist of grants, loans, or grant equivalents. On a more positive note, the agreement unveils a relatively transparent and consistent monitoring, reporting, and verification (MRV) system for tracking the volumes and impacts of international climate finance contributions. Indeed, according to the agreement, developed country parties are required to report biannually on their financial commitments, and this information should from 2023 onwards be included in the global stocktaking process.[58] Finally, the agreement urges the operating entities of the Financial Mechanisms of the Convention to ensure greater climate finance readiness support for recipient countries and for the poorest and most vulnerable developing country Parties.

In sum, it is clear that the Paris Climate talks have not generated any genuinely new actions and targets on climate change financing to developing countries

55 Ibid., Article 9(1), p. 25.
56 Ibid., Article 9(3), pp. 25–26.
57 Ibid., Article 9(4), p. 26.
58 Ibid., Article 9(5–6), p. 26.

beyond 2025. There remains notably much uncertainty about several key issues, including what should be included in climate finance, what 'new and additional' means, and what roles new actors and sources can play for the long-term mobilization of climate finance. In fact, the indeterminate language of Article 9 of the agreement leaves, in the end, considerable scope for interpretative flexibility, making it difficult to foresee the short- as well as long-term evolution of the global governance system for climate finance. Hopefully, despite making little concrete progress on long-term finance, the Paris deal can be understood, as many have started to suggest, not as an end in itself, but as the beginning of a process, whereby both developing and developed countries will, in the next few years of the UNFCCC negotiations, finally succeed to bridge persisting normative divides and bring about more concrete and effective options on the table, capable of enhancing the institutional integrity and hence the efficiency of the global governance system for climate finance.

References

Abbott, Kenneth. "The Transnational Regime Complex for Climate Change." *Environment & Planning C: Government & Policy* 30, no. 4 (2012): 570–90.

Alter, Karen, and Sophie Meunier. "Nested and Overlapping Regimes in the Transatlantic Banana Trade Dispute." *Journal of European Public Policy* 13, no. 3 (2006): 363-82.

———. "The Politics of International Regime Complexity." *Perspective on Politics, Symposium* 7, no. 1 (2009): 13–24.

Bäckstrand, Karin. "European Journal of International Relations." *European Journal of International Relations* 12, no. 4 (2006): 467–98.

Bécault, Emilie, and Colleen Carroll. "Global Governance and the Challenges of Diversity." In *Global Governance and Democracy: A Multidisciplinary Analysis*, edited by Jan Wouters, Antoon Braeckman, Matthias Lievens and Emilie Bécault, 191–225. Leuven: Edward Elgar Publishing, 2015.

Bernstein, Steven. *The Compromise of Liberal Environmentalism.* New York: Columbia University Press, 2001.

Biermann, Frank, Michele Betsill, Aarti Gupta, Norichika Kanie, Louis Lebel, Diana Liverman, Heike Schroeder, et al. *Earth System Governance People, Places, and the Planet: Science and Implementation Plan of the Earth System Governance Project Report No. 1.* Bonn: International Human Dimensions Programme on Global Environmental Change, 2009. <http://www.ihdp.unu.edu/docs/Publications/ESG/IHDP_ReportNo20_ESG_ReportNo1.pdf> (accessed 26 May 2016).

Biermann, Frank, Philipp Pattberg, Harro Van Asselt, and Fariborz Zelli. "The Fragmentation of Global Governance Architectures: A Framework for Analysis." *Global Environmental Politics* 9, no. 4 (2009): 14–40.

Bird, Neil, and Jessica Brown. "International Climate Finance: Principles for European Support to Developing Countries." Belgium: European Development Cooperation to 2020, 2010.

Birdsall, Nancy, and Michele De Nevers. "Adaptation Finance: How to Get out from between a Rock and a Hard Place: Policy Paper 001." Washington, DC: Center for Global Development, 2012.

Black, Julia. "Constructing and Contesting Legitimacy and Accountability in Polycentric Regulatory Regimes." *Regulation and Governance* 2 (2008): 137–64.

Breakey, Hugh, and Tim Cadman. "Governance Values and Institutional Integrity." Chap. 2 In *Governing the Climate Change Regime*, edited by Charles Sampford, Tim Cadman and Rowena Maguire. London: Taylor & Francis, 2016 (this volume).

Breakey, Hugh, Timothy Cadman, and Charles Sampford. "Conceptualizing Personal and Institutional Integrity: The Comprehensive Integrity Framework." *Research in Ethical Issues in Organizations* 14 (2015): 1–40.

Bulkeley, Harriet, Liliana B. Andonova, Michele Betsill, Daniel Compagnon, Thomas Hale, Matthew J. Hoffmann, Peter Newell, Matthew Paterson, Charles Roger, and Stacy D. VanDeveer. "Governing Climate Change Transnationally: Assessing the Evidence from a Database of Sixty Initiatives." *Environment and Planning-Part C* 30, no. 4 (2012): 591–612.

Clapp, Christa, Jane Ellis, Julia Benn, and Jan Corfee-Morlot. "Tracking Climate Finance: What and How?" Paris: OECD and IEA Secretariats, 2012.

Climate Change Finance Unit. "Climate Change Finance, Analysis of a Recent OECD Report: Some Credible Facts Needed." 1–16. Delhi: Department of Economic Affairs Ministry of Finance, Government of India, 2015.

Climate Policy Initiative (CPI). "The Global Landscape of Climate Finance 2013." *A Climate Policy Initiative Report* (2013). <http://climatepolicyinitiative.org/wp-content/uploads/2014/11/The-Global-Landscape-of-Climate-Finance-2014.pdf> (accessed 1 June 2016).

———. "The Global Landscape of Climate Finance 2014." *A Climate Policy Initiative Report* (2014). <http://climatepolicyinitiative.org/wp-content/uploads/2014/11/The-Global-Landscape-of-Climate-Finance-2014.pdf> (accessed 1 June 2016).

———. "The Global Landscape of Climate Finance 2015." *A Climate Policy Initiative Report* CPI(2015). <http://climatepolicyinitiative.org/wp-content/uploads/2015/11/Global-Landscape-of-Climate-Finance-2015.pdf> (accessed 1 June 2016).

———. "The Landscape of Climate Finance 2011." *A Climate Policy Initiative Report* (2011). <http://climatepolicyinitiative.org/publication/the-landscape-of-climate-finance> (accessed 1 June 2016).

———. "The Landscape of Climate Finance 2012." *A Climate Policy Initiative Report* (2012). <http://climatepolicyinitiative.org/wp-content/uploads/2012/12/The-Landscape-of-Climate-Finance-2012.pdf> (accessed 1 June 2016).

Cole, Daniel H. "From Global to Polycentric Climate Governance." *Climate Law* 2, no. 3 (2011): 395–413.

Dutschke, Michael, and Axel Michaelowa. "Development Aid and the CDM – How to Interpret Financial Additionality: Hamburgisches Welt-Witschafts-Archiv (Hwwa) Discussion Paper 228." Hamburg: Hamburg Institute of International Economics, 2003.

Finnemore, Martha, and Kathryn Sikkink. "International Norm Dynamics and Political Change." *International Organization* 52 (1998): 887–917.

———. "Taking Stock: The Constructivist Research Program in International Relations and Comparative Politics." *Annual Review Political Science* 4, no. 3 (2001): 391–416.

Green Climate Fund. "Governing Instrument for the Green Climate Fund." Republic of Korea: Secretariat, Green Climate Fund, 2010.

Hooghe, Liesbet, and Gary Marks. *Multi-Level Governance and European Integration*. Lanham, MD: Rowman & Littlefield, 2001.

———. "Unraveling the Central State, but How?: Types of Multi-Level Governance." *American Political Science Review* 97, no. 2 (2003): 233–43.

International Institution for Environment and Development (IIED). "A Fair Climate Deal in Paris Means Adequate Finance to Deliver INDCs in LDCs." *IIED Briefing*, November 2015.

Isailovic, Marija, Oscar Widerberg, and Philipp Pattberg. "Fragmentation of Global Environmental Governance Architectures: A Literature Review." SSRN, 2013.

Keohane, Robert, and David Victor. "The Regime Complex for Climate Change." *Perspectives on Politics* 9, no. 1 (2011): 7–23.

March, James G., and Johan P. Olsen. "The Institutional Dynamics of International Political Orders." *International Organization* 52 (1998): 943–70.

Organisation for Economic Cooperation and Development (OECD), and Climate Policy Initiative (CPI). "Climate Finance in 2013–14 and the USD 100 Billion Goal: A Report by the OECD in Collaboration with Climate Policy Initiative." Paris: OECD-CPI, 2015.

Ostrom, Elinor. "Polycentric Systems for Coping with Collective Action and Global Environmental Change." *Global Environmental Change* 20, no. 4 (2010): 550–57.

Pickering, Jonathan, Jakob Skovgaard, Soyeun J. Kim, Timmons Roberts, David Rossati, Martin Stadelmann, and Hendrikje Reich. *Acting on Climate Finance Pledges: Inter-Agency Dynamics and Relationships with Aid in Contributor States: CCEP Working Paper 1306.* Canberra: Crawford School of Public Policy, The Australian National University, 2013.

Raustiala, Kal, and David Victor. "The Regime Complex for Plant Genetic Resources." *International Organization* 58, no. 2 (2004): 277–309.

Reus-Smit, Christian. "The Constitutional Structure of International Society and the Nature of Fundamental Institutions." *International Organization* 51, no. 4 (1997): 555–89.

Rubbelke, Dirk T.G. "International Support of Climate Change Policies in Developing Countries: Strategic, Moral, and Fairness Aspects." *Ecological Economics* 70, no. 8 (2011): 1470–80.

Schalatek, Liane. "A Matter of Principle(s): A Normative Framework for a Global Climate Finance Compact." *The Green Political Foundation*, 19 November 2010.

Schalatek, Liane, and Neil Bird. "The Principles and Criteria of Public Climate Finance – A Normative Framework." *The Climate Finance Fundamentals: The Climate Funds Update*, 2015.

Smallridge, Diane, Barbara Buchner, Chiara Trabacchi, Maria Netto, José J.G.L. Lorenzo, and Lucila Serra. "The Role of National Development Banks in Intermediating International Climate Finance to Scale Private Sector Investments: Discussion Paper No. IDB-DB-249." Washington: Inter-American Development Bank, Institutions for Development (IFD), 2012.

Stadelmann, Martin, Jessica Brown, and Lena Hornlein. "Fast-Start Finance: Scattered Governance, Information and Programmes." In *Carbon Markets for Climate Finance? Low Carbon and Adaptation Investments Choices for the Developing World*, edited by Axel Michaelowa, 117–45. London: Routledge, 2012.

United Nations Development Programme (UNDP). "Readiness for Climate Finance: A Framework for Understanding What It Means to Be Ready to Use Climate Finance." New York: UNDP Environment and Energy Group, 2012.

United Nations Framework Convention on Climate Change (UNFCCC). "2014 Biennial Assessment and Overview of Climate Finance Flows Report." Bonn, Germany: UNFCCC Standing Committee on Finance, 2014.

———. "The Conference of the Parties (COP), at Its Fifteenth Session, Took Note of the Copenhagen Accord of 18 December 2009." Decision 2/CP.15, 2009.
———. "Conference of the Parties Serving as the Meeting of the Parties to the Kyoto Protocol, 3–15 December 2007, Bali, Decision." FCCC/KP/CMP/2007/9/Add.1, 1/CMP.3: UNFCCC, 2007.
———. "Conference of the Parties Twenty-First Session Paris, 30 November to 11 December 2015: Adoption of The Paris Agreement." FCCC/CP/2015/L.9/Rev. 1: United Nations, 2015.
———. "Framework Convention on Climate Change." FCCC/INFORMAL/84: United Nations, 1992.
———. "Report of the Conference of the Parties on Its Sixteenth Session, Held in Cancun from 29 November to 10 December 2010." Part Two: Action taken by the Conference of the Parties at its sixteenth session FCCC/CP/2010/7/Add.1: United Nations, 2011.
Wiener, Antje. *The Invisible Constitution of Politics.* New York: Cambridge University Press, 2008.
———. *A Theory of Contestation.* Springerbriefs in Political Science. Heidelberg, Germany: Springer, 2014.
Zürn, Michael. "Global Governance as Multi-Level Governance." In *Handbook on Multi-Level Governance*, edited by Henrik Enderlein, Sonja Wälti and Michael Zürn, 80–102. Cheltenham, UK and Northampton, MA: Edward Elgar Publishing, 2011.

6 Governance of climate displacement within the UN climate regime

Rowena Maguire

Introduction

This chapter explores how the UN Climate regime[1] governs climate displacement. One of the gravest injustices of climate change is the distribution of negative impacts upon the poor and vulnerable in society. It is fundamentally unjust that people who are most likely to be displaced by climate change are the ones who have contributed least to global emissions. Such innocent people stand to lose their homes, their connections with land, community, language and family and face the challenge of starting life again in a foreign place with a potentially hostile reception from host communities. As such, it has been recognised that climate displacement is a multifaceted challenge which has generated debate about the most appropriate international institution to respond to this issue. Rather than revisiting the merits and pitfalls of international environmental human rights, and refugee law regimes to deal with climate displacement,[2] this chapter will instead explore how the UN Climate regime has responded to the issue. It is argued that the UN Climate regime will be play a critical role in assisting those displaced by climate change for two reasons. First, the UN Climate regime can create a moral argument on the basis of historical and current emissions contributions as to why Parties to the United Nations Framework Convention on Climate Change (UNFCCC) must assist those must vulnerable to climate change. Second, the UN Climate regime has the ability to generate financial assistance separate and additional from any other existing international funds for

1 UN Climate regime is used in this article to refer to all of the instruments, mechanisms, decisions, working groups, technical and scientific bodies used in the operationalisation of the United Nations Framework Convention on Climate Change.

2 See generally Jane McAdam, "Refusing Refuge in the Pacific: (De)Constructing Climate-Induced Displacement in International Law," in *Migration, Environment and Climate Change*, ed. Etienne Piguet, Antoine Pecoud, and Paul de Guchteneire (Cambridge: Cambridge University Press, 2011); Rowena Maguire and Louise Kruger, "The International Regulation of Persons Displaced by Climate Change," in *Protection of Refugees and Displaced Persons in the Asia Pacific Region*, ed. Angus Francis and Rowena Maguire (Ashgate: Farnham, Surrey, 2013), pp. 201–24.

displaced/refugee populations. The UN Climate regime therefore plays a key role in governing climate displacement by recognising the moral issues associated with excessive greenhouse gas emissions and by providing a governance structure to provide financial and other support to those displaced by climate change.

The UN Climate regime does not directly address climate displacement; rather it creates a number of obligations which, when read collectively, provide a legal basis for protection and assistance. In order to evaluate the protections or lack of protections within the UN Climate regime, it is necessary to interpret existing provisions within the United Nations Framework Convention on Climate Change (UNFCCC), the Cancun Agreement and the Warsaw Loss and Damage Mechanism and the Paris Agreement. This chapter adopts the same definition of governance as outlined by Breakey, Cadman and Sampford in Chapter 2, which envisages looking at the 'structures and processes used to steer and coordinate interactions within an institution or institutional complex'.[3] This chapter will thus examine the key instruments and Conference of the Parties (COP) decisions that recognise the special vulnerability of some groups to climate change and will explore how these provisions can be read in conjunction to assist those displaced by climate change. Within UN Climate instruments, the language of vulnerability is used as opposed to the language of liability. This framing has important ramifications, as the language of vulnerability is more scientific in nature and is used within the regime to classify risk to climate change, as opposed to creating legal obligations to assist. The language of vulnerability thus avoids issues of liability, which helps explain the lack of concrete recognition of the climate displacement issue. It is challenging to assess the thinness or thickness of the governing values of the UN Climate regime in respect of climate displacement, as the protections afforded are piecemeal and not at a stage where one can easily evaluate accountability, transparency, compliance etc. This chapter will, however, examine (where applicable) the governing values evident in the instruments that recognize the issue of climate displacement.

Understanding climate displacement

There are no concrete figures on the number of people who have migrated as a result of climate change, nor on projected future climate migration figures. The International Migration Organization (IOM) has collated data on environmental migration but finds that existing projections vary by a factor of 40, with estimates ranging from 25 million to 1 billion people being environmental migrants by 2050. Precise figures are difficult to generate, as the numbers will depend on the (a) climate change scenarios that play out, (b) the adaptation activities implemented and (c) the evolution of various socio-economic, political and demographic factors influencing the decision to migrate, such as economic growth and

3 Refer to Chapter 2 within this collection.

development, population growth and governance.[4] What is known is that climate is expected to affect the movement of people in at least four ways:

1 Greater frequency and, potentially, greater intensity of weather-related natural disasters – both sudden and slow onset may lead to higher risk of humanitarian emergencies and increased population movements.
2 Livelihood stress linked to climate change could in some places result in lower levels of outmigration, as migration requires resources. Thus people may want to move but may not be able to due to lack of resources, and as such become trapped populations.
3 Rising sea levels may make coastal areas and low-lying islands uninhabitable.
4 Competition over shrinking natural resources may exacerbate tensions and potentially lead to conflict and, in turn, displacement.[5]

This means that some of the most climate-vulnerable people will not move and will instead face greater socio-economic challenges. Empirical research demonstrates that climate-change-related migration will take place almost entirely within national borders (internal displacement); will involve predominately male movement (but the number of women are growing); largely involves individual movement as oppose to family movement and will largely be driven by livelihood needs (rainfall patterns have a large effect on migration decisions).[6] This means that a common profile of a climate-displaced person is a sole male moving internally as a result of poor rainfall. It is crucial to understand the profile of people migrating as a result of climate factors, and UN Climate regime climate displacement policies must be designed in order to assist the profile of climate migrants.

There is no consensus on how best to describe and define those people forced to migrate as a result of climate change.[7] Terms used in the literature vary from environmental refugees, climate change refugees, environmentally displaced persons and environmental migrants. Definitions also differ in whether they address dislocation that is temporary or permanent, forced or voluntary, national or international. A definition has important ramifications, as the coverage of protection will be limited to those people who meet the criteria outlined in the definition. This chapter will adopt the definition of the IOM, which defines a climate

4 International Organization for Migration (IOM), *IOM Outlook on Migration, Environment and Climate Change* (Geneva: International Organization for Migraiton (IOM), 2011), http://publications.iom.int/system/files/pdf/mecc_outlook.pdf. p. 38.

5 Ibid.

6 Koko Warner, Tamer Afifi, Walter Kälin, Scott Leckie, Beth Ferris, Susan F.C. Martin, and David Wrathall, *Changing Climate, Moving People: Framing Migration, Displacement and Planned Relocation* (Geneva: United Nations University Institute for Environment and Human Security, 2013), https://collections.unu.edu/eserv/UNU:1837/pdf11213.pdf. p. 20.

7 See generally Rosemary Lyster, “Protecting the Human Rights of Climate Displaced Persons: The Promise and Limits of the United Nations Framework Convention on Climate Change,” *Legal Studies Research Paper* no. 15/24 (2015), file:///Users/rowenamaguire/Downloads/SSRN-id2585266.pdf.

displaced person (CDP) as '[p]ersons or groups of persons who, for compelling reasons of sudden or progressive changes in the environment that adversely affect their lives or living conditions, are obliged to leave their homes or choose to do so, either temporarily or permanently, and who move either within their country or abroad'.[8] This definition has been selected because it is broad and does not exclude individuals or groups on the basis of a climate driver, and it also covers internal and international migration arising from climate change.

This chapter acknowledges that the drivers of migration are complex and non-linear. As Berringer wisely states:

> The driver is not specifically the environment but its effects. People do not migrate simply because of drought but because they cannot produce food anymore. They do not move because of a cyclone but because the cyclone has eradicated the industry that provided the community jobs.[9]

Climate change will overlay other drivers of migration and may be the tipping point which encourages migration, but it will remain difficult to quantify the contribution of a changing climate to migration decisions.[10] The definition of persons displaced by climate change adopted by this chapter suggests that the UN Climate regime should assist persons displaced by climate change that is caused by both slow and sudden changes in the natural environment. It is acknowledged that the demand for assistance will likely outweigh the support available, and as such there will be need to prioritise assistance to those deemed most vulnerable. It is suggested that those communities at risk of losing of their nationality, their land, their culture and their community as a result of rising sea levels are the communities that should receive priority assistance. The number of people in such circumstances is not expected to be huge in comparison to the movement of people internally, as suggested by the profiling work discussed previously, and this chapter makes some recommendations on where funding should come within the UN Climate regime to support these categories.

UNFCCC governance of persons displaced by climate change

This section explores the key instruments of the UN climate regime that deal with issues of vulnerability and/or climate displacement directly. The analysis in

8 International Organization for Migration (IOM), *IOM Outlook on Migration, Environment and Climate Change*. p. 22.

9 Andreas C. Berringer, "Migration and Climate Change: Global Governance Regimes and the Incorporation of Climate Change Displacement," in *Quality and Legitimacy of Global Governance: Case Lessons from Forestry*, ed. Tim Cadman (London, UK: Palgrave, Macmillan, 2011), p. 204.

10 Walter Kalin, "Conceptualising Climate-Induced Displacement," in *Climate Change and Displacement: Multidisplinary Perspectives*, ed. Jane McAdam (UK: Hart Publishers, 2010); Jane McAdam, *Climate Change, Forced Migration and International Law* (Oxford: Oxford University Press, 2012).

this section suggests that most of the instruments and decisions discussed accord with Breakey's, Cadman's and Sampford's description of 'thin governing values'. Thin governing values refer to the structures and processes that determine how things are done.[11] Accountability, transparency, compliance and rule of law are thin governing values. This section examines the wording of UNFCCC, the Cancun COP decision, the Warsaw Loss and Damage Mechanism and the Paris Agreement. In general the wording contained in these instruments is reflective of thin governing values only, with a focus on transparency, accountability, reporting and compliance, with only hints of thickish or thick governing values such as dispute settlement or equality in decision-making.

The United Nations Framework Convention on Climate Change

The UNFCCC[12] does not include any specific references to climate displacement/migration. As such it is necessary to look closely at the provisions dealing with vulnerability in order to gain an understanding of how the architecture of the Convention applies for those displaced by climate change. Vulnerability is discussed within Article 3(2) of the UNFCCC, which requires that:

> The specific needs and special circumstances of developing country Parties, especially those that are particularly vulnerable to the adverse effects of climate change, and of those Parties, especially developing country Parties, that would have to bear a disproportionate or abnormal burden under the Convention, should be given full consideration.

The definition of vulnerability within the Convention thus hinges on scientific criteria as opposed to considerations of vulnerability within populations (that is, women, children, indigenous or displaced populations). The UNFCCC recognises that certain geographical features make certain countries more vulnerable to climate change. Article 4(8) requires that full consideration be given to what actions related to funding, insurance and the transfer of technology are needed to meet the specific needs and concerns of developing country Parties arising from the adverse effects of climate change and/or the impact of the implementation of response measures, especially on:

a Small island countries;
b Countries with low-lying coastal areas;
c Countries with arid and semi-arid areas, forested areas and areas liable to forest decay;
d Countries with areas prone to natural disasters;
e Countries with areas liable to drought and desertification;

11 Breakey, Cadman and Sampford, Chapter 2, this volume.
12 United Nations Framework Convention on Climate Change (UNFCCC), "United Nations Framework Convention on Climate Change," (Rio: United Nations, 1992).

f Countries with areas of high urban atmospheric pollution;
g Countries with areas with fragile ecosystems, including mountainous ecosystems;
h Countries whose economies are highly dependent on income generated from the production, processing, export and/or on consumption of fossil fuels and associated energy-intensive products; and
i Landlocked and transit countries.

Persons displaced by climate change are likely to come from arid or semi-arid areas, small island countries, countries with low-lying coastal areas and areas prone to natural disasters. As such, persons displaced by climate change are likely to fit the geographical vulnerability definition provided by the UNFCCC. Article 4(8) therefore has utility in showing that the UNFCCC's definition of vulnerability is consistent with helping those displaced by climate change.

In respect to creating more specific obligations to assist those vulnerable to climate change, insight can be gained from Article 4(4), which requires that the developed country Parties and other developed Parties included in Annex II assist the developing country Parties that are particularly vulnerable to the adverse effects of climate change in meeting costs of adaptation to those adverse effects. Article 4(3) provides stronger wording regarding finance by requiring that developed country Parties 'shall provide new and additional financial resources to meet the agreed full costs incurred by developing countries, in meeting their obligations under Article 12(1)',[13] while Article 4(9) focuses on a group with known vulnerability (least-developed countries) and requires that the Parties take full account of the specific needs and special situations of the least-developed countries in their actions with regard to funding and transfer of technology.

These provisions collectively show that the UNFCCC deems the most appropriate response to vulnerability is to provide some sort of financial assistance and/or technology transfer. At this stage, other support such as providing migration rights for those needing to leave sinking islands is not envisaged and is arguably not within the mandate of the UN Climate regime (though the regime does have authority to generate finance to assist any such cross-border migration).

Cancun Adaptation Framework[14]

The Cancun Adaptation Framework broke tremendous ground in respect of international governance arrangements for climate displacement/migration.

13 Article 12(1) deals with reporting obligations and creates obligations on all Parties to create (a) a national inventory of anthropogenic emissions; (b) a general description of steps taken by Parties to implement the Convention and (c) any other relevant information.

14 United Nations Framework Convention on Climate Change (UNFCCC), "Report of the Conference of the Parties on Its Sixteenth Session, Held in Cancun from 29 November to 10 December 2010," (Part Two: Action taken by the Conference of the Parties at Its Sixteenth Session FCCC/CP/2010/7/Add.12011).

This occurred through paragraph 14(f), which invited Parties to enhance action on

> Measures to enhance understanding, coordination and cooperation with regard to climate change induced displacement, migration and planned relocation, where appropriate, at the national, regional and international levels.[15]

This simple paragraph had the effect of clarifying that the UN Climate regime recognised that climate-related migration and displacement was a live issue, and importantly gave the UN Climate regime authority to encourage work to address this issue. Paragraph 14(f) recognises that displacement may be internal or cross-border by referring to national, regional and international levels, thus providing some form of international institutional responsibility to assist for those who have to flee from sinking islands.

Warsaw International Mechanism for Loss and Damage associated with Climate Change Impacts[16]

The Warsaw International Mechanism for Loss and Damage associated with Climate Change Impacts (Warsaw Loss and Damage Mechanism) is relevant mechanism for people displaced by climate change. This mechanism seeks to address loss and damage arising from climate change, and it is argued that the loss of one's home combined with the costs of starting life again should receive priority attention under this mechanism. This mechanism is not yet compensatory in nature and does not create liability for those countries with unmitigated emissions, but rather focuses on institutional development, information gathering and capacity building to respond to loss suffered as a result of climate change.[17] The Alliance of Small Islands States (AOSIS) has been instrumental in ensuring that loss and damage issues remained on the UN Climate regime agenda.[18] Article 4(8) of the Convention creates the legal basis for establishing a loss and damage mechanism by requiring 'that Parties shall give full consideration to what actions are necessary under the Convention, including actions related to funding, insurance and the transfer of technology, to meet the specific needs and concerns of developing country Parties arising from the adverse effects of climate change'.

15 UNFCCC, Outcome of the work of the Ad Hoc Working Group on long-term Cooperative Action under the Convention, CP 16, 2010, http://unfccc.int/files/meetings/cop_16/application/pdf/cop16_lca.pdf#page=3, para 14(f).

16 United Nations Framework Convention on Climate Change (UNFCCC), "Warsaw International Mechanism for Loss and Damage Associated with Climate Change Impacts," http://unfccc.int/adaptation/workstreams/loss_and_damage/items/8134.php.

17 Meinhard Doelle, "The Birth of the Warsaw Loss and Damage Mechanism: Planting a Seed to Grow Ambition," *Carbon and Climate Law Review* 8, no. 1 (2014): p. 43.

18 Ian Fry, "The Paris Agreement: An Insider's Perspective – The Role of Small Island Developing States," *Environmental Policy and Law* 46, no. 2 (2016).

This obligation was created due to pressure from the Alliance of Small Island States (AOSIS) to include a right for developing countries to compensation and/or or insurance for harm arising from climate change.[19] AOSIS proposed that the compensation/insurance pool should be funded by contributions calculated according to a formula modelled on the *1963 Brussels Supplementary Convention on Third Party Liability in the Field of Nuclear Energy*, which required:

i 50 percent – the ratio between the gross national product (GNP) of each industrialised country contributing to the fund as against the total country contributor GNPs.

ii 50 percent – the ration of each industrialised country's CO_2 emissions as against the total CO_2 emissions of the group of contributing countries.[20]

The AOSIS proposal was not retained within the final UNFCCC text; however, the obligations created by Article 4(8) are reflective of the AOSIS submissions. At the COP 13 hosted in Bali, Indonesia, in 2007, the Parties established the Ad Hoc Working Group on Long-Term Cooperative Action (AWG-LCA), which required Parties to enhance action on adaptation, including risk reduction and risk sharing/transfer mechanisms such as insurance, disaster-reduction strategies and means to address loss and damage associated with climate change.[21] AOSIS made a submission to the AWG-LCA Multi-Window Mechanism to Address Loss and Damage from Climate Change Impacts outlining three necessary elements for the mechanism:[22]

1 An insurance component to assist particularly vulnerable developing countries in managing financial risk from increasingly frequent and severe weather events;
2 A rehabilitative/compensatory component to address loss and damage resulting from un-avoided and unavoidable climate change impacts such as rising sea levels, increasing land and sea temperatures and ocean acidification;
3 A risk management component to support risk assessment and risk management.

19 Vanuatu on behalf of AOSIS, United Nations Commission on Sustainable Development, "Intergovernmental Negotiating Committee for a Framework Convention on Climate Change: Fifth Session, Second Part New York, 30 April–9 May," (A/AC.237/18 (Part II)/Add.1 1992); Intergovernmental Committee Negotiating Committee for a Framework Convention on Climate Change, "Elements of a Framework Convention on Climate Change," (UN Doc A/AC237/Miscl/Add.3 18 June 1991), Clause 19.3.

20 Intergovernmental Negotiation Committee for a Framework Convention on Climate Change, "Consolidated Working Document," (UN Doc A/A C237/MISC.17/ADD.9 19 December 1991).

21 United Nations Framework Convention on Climate Change (UNFCCC), "Report of the Dialogue on Long-Term Cooperative Action by Enhancing Implementation of the Convention," (UN Doc FCCC/CP/2007/4/Add.1 19 November 2007), at 4.

22 Ilona Millar, Catherine Gascoigne, and Elizabeth Caldwell, "Making Good the Loss: An Assessment of the Loss and Damage Mechanism under the UNFCCC Process," in *Threatened Island Nations: Legal Implications of Rising Seas and Changing Climate*, ed. Michael Gerrad and Gregory Wannier (Cambridge: Cambridge University Press, 2013), p. 449.

The Cancun Adaptation Framework laid the foundations for developing a loss and damage mechanism by requiring Parties to enhance climate-change-related disaster risk-reduction strategies, including early warning systems, risk assessment and management and sharing and transfer mechanisms such as insurance at local, national, sub-regional and regional levels.[23] In 2013 the Warsaw Loss and Damage Mechanism was then formally created to address loss and damage associated with impacts of climate change, including extreme events and slow onset events, in developing countries that are particularly vulnerable to the adverse effects of climate change.[24] The specific mention of slow onset events means that damage associated with sea level rise such as displacement is covered by this mechanism. Paragraph 5 of this decision spells out the three key roles of the mechanism:

a Enhancing knowledge and understanding of comprehensive risk management approaches to address loss and damage associated with the adverse effects of climate change, including slow onset impacts;
b Strengthening dialogue, coordination, coherence and synergies among relevant stakeholders;
c Enhancing action and support including finance, technology and capacity building to address loss and damage associated with the adverse effects of climate change so as to enable countries to undertake actions to assess climate-related risks, identify risk management strategies and implement those strategies.

The IOM made a submission in respect of paragraph 5(a) and 5(b) at the Doha COP negotiations. This submission highlighted the IOM's operations support of national authorities and local communities in migration management through direct assistance, promoting political dialogue and researching the role of human mobility, including in the context of natural disasters and crises.[25] The IOM was the only international migration institution that partook in the Doha knowledge gathering, and this positioning submission suggest that the IOM is likely to become one of the key implementing bodies for any climate change displacement/migration programs using funding generated from the Loss and Damage Mechanism. Paragraph 5(c) of the Warsaw Loss and Damage mechanism is very important from a climate displacement perspective, as it creates authority for the regime to provide technical support and to facilitate the mobilisation and securing of expertise, including finance, technology and capacity building to address loss and damage associated with extreme weather events and slow-onset events. This decision thus gives the UN Climate regime through the Warsaw Loss

23 United Nations Framework Convention on Climate Change (UNFCCC), "Report of the Conference of the Parties on Its Sixteenth Session, Held in Cancun from 29 November to 10 December 2010."
24 United Nations Framework Convention on Climate Change (UNFCCC), "Warsaw International Mechanism for Loss and Damage Associated with Climate Change Impacts" at [1].
25 Ibid.

and Damage Mechanism authority to generate finance and to draw on technical skills to practically assist with climate displacement.

The initial two-year work plan of the Executive Committee of the Warsaw International Mechanism for Loss and Damage involves nine action areas, with Action 6 focusing on issues of migration, displacement and human mobility.[26] Action Area 6 focused on enhancing the understanding of and expertise on how the impacts of climate change are affecting patterns of migration, displacement and human mobility and the application of such understanding and expertise. This action invited relevant organisations, experts and UN bodies to submit information on projected migration and displacement, as well as lessons learned and good practices between January 2015 and June 2016.[27] The submission period has now ended, and the executive committee will be following up on actions from these submissions in the latter half of 2016. The Warsaw Loss and Damage Mechanism has thus put in place processes and structures to start exploring how to govern climate displacement. It is, however, too early to assess the thinness or thickness of the governing values given that the framework is only starting to emerge.

Paris Agreement

The Paris Agreement does not directly address climate displacement; however, the frames of human rights and vulnerability, along with the more substantial provisions on adaptation and loss and damage, are relevant from a climate-displacement perspective. The preamble requires Parties to respect, promote and consider their human rights obligations and the rights of indigenous peoples, local communities, migrants, children, persons with disabilities and people in vulnerable situations.[28] This preamble reference should be constructed to ensure that people displaced by climate change must have their human rights respected during any migration and resettlement processes (a rights-based approach to migration).[29]

Article 7 deals with adaptation and continues to discuss adaptation with vulnerability language. The Agreement requires for adaptation to be a 'country driven, gender-responsive, participatory and fully transparent approach, taking

26 United Nations Framework Convention on Climate Change (UNFCCC), "Initial Two-Year Workplan of the Executive Committee of the Warsaw International Mechanism for Loss and Damage," http://unfccc.int/adaptation/workstreams/loss_and_damage/items/8805.php.

27 The submissions are available at United Nations Framework Convention on Climate Change (UNFCCC), "Submissions on Internal and Cross-Border Migration, Displacement and Other Forms of Human Mobility," United Nations, http://unfccc.int/adaptation/groups_committees/loss_and_damage_executive_committee/items/9508.php.

28 United Nations Framework Convention on Climate Change (UNFCCC), "Conference of the Parties: COP21 Paris Agreement," (FCCC/CP/2015/L.9: UNFCCC, 2015), Preamble.

29 For more information on a rights-based approach in the displacement context, see Bridget Lewis and Rowena Maguire, "A Human-Rights Based Approach to Disaster Displacement in the Asia Pacific," *Asian Journal of International Law* Online (2015).

into consideration vulnerable groups, communities and ecosystems'.[30] This kind of language shows a move towards embracing thickish governing values and ensuring that adaptation activities, the most extreme being migration as adaptation, are carried out in a way that ensure consultation and procedural justice for vulnerable groups. These kinds of protections are essential for people adapting to climate change by migrating.

The Warsaw Loss and Damage Mechanism is recognised within Article 8, with a focus on extreme weather events and slow-onset events. The Agreement gives the COP explicit power to continue to develop modalities in this space[31] and to acknowledge the necessity of working with groups with relevant expertise.[32] The IOM is the international body with relevant expertise in dealing with migration and resettlement, and will likely be the body involved in rolling out programs on migration, displacement and human mobility under the mechanism.

How to compensate people displaced by climate change?

The World Wide Fund for Nature (WWF) commissioned a report examining international legal issues involved in designing a climate compensation fund.[33] The table below summarises some of the issues dealt with within the paper:[34]

Table 6.1 Options for loss and damage mechanism components

Component	*Options*
Eligibility for Compensation	• Vulnerable developing countries suffering greatest relative impacts • Developing countries that have contributed least to GHG emissions • All Parties with eligibility phases
Damage Covered	• Loss of life or personal injury • Loss of or damage to property • Economic loss • Environmental damage • Moral damage
Timing for Claim	• Upon proof of causation (climate-induced change) • Upon crossing of agreed thresholds or variations from agreed baselines
Determining Damage	• Actual valuation of damage • Estimated loss agreed in advance

30 United Nations Framework Convention on Climate Change (UNFCCC), "Conference of the Parties: COP21 Paris Agreement," Article 7(5).

31 Ibid., Article 8(2).

32 Ibid., Article 8(5).

33 Roda Verheyen and Peter Roderick, Beyond Adaptation: The Legal Duty to Pay Compensation for Climate Change Damage, (UK: WWF, 2008), http://assets.wwf.org.uk/downloads/beyond_adaptation_lowres.pdf.

34 Ibid., pp. 30–37.

Component	*Options*
Coverage	• All loss and damage • Loss and damage with respect to certain elements • A fixed share of loss and damage
Limits	• Financial limit: agreed ceilings, caps relevant to GDP, tiered financial limits • Time limits: damage occurring after a threshold year, restricting claims to within x years of discovery of damage
Contribution to the Fund	• Assessed contributions from Annex I Parties based on calculated level of 'excess' emissions • Assessed contributions based on agreed criteria (i.e. GDP, GHG Emissions, human development index, per capita emissions etc.) • Contributions derived from auctioning of specified percentage of Assigned Amount Units • Agreed level of contributions from national level auctioning revenues • Layers of contributions from different stakeholder groups (state, private entities) • National contributions based on state collections from regulated sectors

This work shows that there are a variety of options that could be included in a loss and damage mechanism.[35] Some of the options are far better suited to those displaced by climate change, such as limiting eligibility to those most vulnerable and least responsible for climate change; ensuring that the damage covered includes a wide definition of property, economic and moral damage; payment being contingent upon crossing of agreed thresholds as opposed to establishing causation; allowing for loss to be claimed in advance and allowing for generous coverage in respect of what can be claimed. The costs associated with permanent relocation, whether internal or cross-border, will be significant (cost of land, construction, travel etc.), and given that other sources of adaptation funding provide assistance for resilience-type activities, it is suggested that the Loss and Damage Mechanism should prioritise assistance for displaced populations as its number one priority.

Some work has been done by Oxfam suggesting how Parties' financial contributions should be calculated. Oxfam has suggested two criteria which should be used to establish a country's adaptation financial commitment: first, the country's contributions to greenhouse gas emission, which is then adjusted according to their capacity to finance climate action.[36] One of the biggest challenges in respect

35 Also see work by Rosemary Lyster, which proposes to create a fossil fuel–funded climate disaster response fund to compensate victims for residual risks of climate disasters in developing countries: Rosemary Lyster, "A Fossil Fuel-Funded Climate Disaster Response Fund under the Warsaw International Mechanism for Loss and Damage Associated with Climate Change Impacts," *Transnational Environmental Law* 4, no. 1 (2015).

36 Oxfam, "Adapting to Climate Change What's Needed in Poor Countries, and Who Should Pay," *Oxfam Briefing Paper*, no. 104 (May 2007), https://www.oxfam.org/sites/www.oxfam.org/files/adapting%20to%20climate%20change.pdf.

of climate funding is that demand for assistance will far outstrip the adaptation resources available. The World Bank has estimated that the annual cost of adapting to climate change in developing countries will be in the order of USD75–100 billion an annum.[37] This study takes into account the costs associated with internal and cross-border migration in the countries examined (Mozambique, Ethiopia, Ghana, Bangladesh, Bolivia, Vietnam and Samoa); however, one of the listed limitations of the study is limited data on rural-urban migration and climate-driven migration.[38] As such, projected future costs associated with climate displacement are difficult to calculate. It is known that relocation costs will be substantial based on estimated costing from early climate-migration programs.[39] It is, therefore, likely that there will not be adequate adaptation funding for those all displaced by climate change–related events, and as such some method for prioritising assistance will be required.

There is consensus that adaptation funding should be prioritised on the basis of vulnerability, although there is no agreed method or criteria for establishing vulnerability beyond the broad geographical vulnerabilities identified within the UNFCCC.[40] It is argued that vulnerability within nations also needs to be calculated, with a focus on the four groups identified as being vulnerable by the Office of the United Nations High Commissioner for Human Rights (OHCHR) report on climate change and human rights.[41] This report identifies groups at risk as including women, children, indigenous populations and populations at risk of displacement.

Using adaptation funding to assist people at risk of internal displacement

Additional funding for populations at risk of internal displacement could also come from other adaptation financial mechanisms. Adaptation funding streams are diverse: there are funds managed by the Global Environment Facility (GEF), funds managed by the UN Climate regime, funds from bi-lateral aid programs and funds provided by non-governmental organisations. This chapter will provide

37 World Bank, *Economics of Adaptation to Climate Change: Synthesis Report* (Washington, DC: World Bank, 2010), http://www-wds.worldbank.org/external/default/WDSContentServer/WDSP/IB/2012/06/27/000425970_20120627163039/Rendered/PDF/702670ESW0P10800EACCSynthesisReport.pdf.

38 Ibid., at xxii.

39 The US Army Corps of Engineers estimated that the costs of relocating three Alaskan native villages to be in vicinity of $95–125 million for Kivalina; $100–200 million for Shishmaret and $80–130 for Newtok. See University of Oregon, "Climate Change: Realities of Relocation for Alaska Native Villages," *Tribal Climate Change Profile: Relocation of Alaska Native Communities* (Portland: University of Oregon, 2011).

40 Jonathan Pickering, "Adaptation Finance in the Asia-Pacific Region: Strengthening Fairness, Effectiveness and Transparency in Allocation," *Workshop on Climate Change Governance in the Asia-Pacific Region: Agency and Adaptiveness* (Australian National University (ANU) Canberra 2012).

41 Human Rights Council, "Human Rights and Climate Change," (Resolution 7/23 2008).

a brief analysis of the Least Developed Country Fund (LDCF) as a case study to explore the possibility of utilising existing adaptation finance streams to assist those displaced by climate change. This suggestion is made on the basis that many of the existing funded activities are consistent with supporting those at risk of internal climate displacement. There is no information available at the time of publication suggesting that funds from these programs have been used to assist those at risk of internal displacement; however, it is thought that such funding is more likely to have been classified as coastal infrastructure/livelihood assistance or disaster risk related, which is probably more reflective of the scientific language used by the regime.

Article 4(9) of the Climate Convention created obligations to take a full account of LDC with respect to funding and technology transfer. The LDCF was created at the seventh meeting of the COP in Marrakesh in 2001 in order to operationalise Article 4(9).[42] The fund was designed to the address the special needs of 49 LDCs which are especially vulnerable to climate change. Three criteria were used to define an LDC: low income, weak human asset and high economic vulnerability. Funds are distributed to assist in the preparation of National Adaptation Programs of Action (NAPAs) for least-developed countries.[43]

COP 16 resulted in a decision providing further guidance on LDCF operations. Funding is distributed according to the principle of equitable access, which has been interpreted as requiring that NAPA implementation funding is available to all LDCs and not to those who apply first. A funding ceiling is established each year, which means that each LDC can apply for a certain amount each year. Reports of the LDCF only provide limited information regarding approved projects by region, making it difficult to determine how funds have been used across various regions.

The LDCF has ten priority areas:

1 Agriculture and food security
2 Water resource management
3 Coastal zone management
4 Infrastructure, including transport and energy
5 Disaster risk management
6 Natural resource management
7 Health

42 United Nations Framework Convention on Climate Change (UNFCCC), "Report of the Conference of the Parties on Its Seventh Session, Held at Marrakesh from 29 October to 10 November 2001," (FCCC/CP/2001/13/Add.1 2001).

43 National Adaptation Programmes of Action (NAPAs) provide a process for LDCs to identify priority activities that respond to their urgent and immediate needs to adapt to climate change – those for which further delay would increase vulnerability and/or costs at a later stage. See United Nations Framework Convention on Climate Change (UNFCCC), "Report of the Conference of the Parties on Its Sixteenth Session, Held in Cancun from 29 November to 10 December 2010," p. 32.

8 Climate information services
9 Climate resilient urban systems
10 Small-Island Developing States

A number of these priority areas are relevant when looking at the profile of a climate displacement discussed above (single male moving internally, often due to rainfall patterns). In order for displacement to be avoided in such instances, early investment into water resource management may play a role. The LDCF has been used to fund programs aimed at supporting climate-resilient water governance; watershed and catchment management; the transfer and adoption of technologies for water harvesting and enhanced water-use efficiency.[44] Once knowledge improves at the country level around the communities at risk of internal displacement due to climate change, investments made under adaptation funding streams should be targeted to assist such populations. The implementation of such programs should involve the thick governance values described by Breakey, Cadman and Sampford, particularly participation and deliberation with individuals and communities at risk of displacement.

Conclusion

Creating legal obligations within the UN Climate regime to assist those displaced by climate change ensures that the connection between the action of emitting greenhouse gases and the harm caused by such emissions is maintained. The Convention's recognition of vulnerable populations and creation of financial mechanisms provide the basis to assist such populations. This chapter has shown that while the UNFCCC defines vulnerability in respect of geography or economic factors, later instruments such as the Cancun Adaptation Framework and the Paris Agreement are starting to recognise vulnerability within segments of the populations. Recognition of vulnerability within populations is a speciality area of human rights bodies, and their expertise in assisting these communities should be drawn on by the UN Climate regime. While the UN Climate regime might be the institution that hosts the governance processes for climate displacement, cooperation and engagement with other organisations more experienced in rights, migration, displacement and development will be essential to ensure that climate-displaced persons are assisted appropriately. Any programme involving migration and resettlement must involve the characteristics of thick governance values of participation, stakeholder equality in decision-making and democratic decision-making discussed by Breakey, Cadman and Sampford in this collection. This chapter finds that the only funding mechanism to assist people

44 Global Environment Facility (GEF), GEF Programming Strategy on Adaptation to Climate Change Least Developed Countries Fund Special Climate Change Fund (GEF, 2014), https://www.thegef.org/gef/sites/thegef.org/files/publication/GEF_AdaptClimateChange_CRA.pdf. p. 25.

who need to cross borders as a result of climate displacement is the Warsaw Loss and Damage Mechanism, and it is suggested that this mechanism should prioritise assistance for people displaced by climate change and needing to cross borders. It is suggested that a number of adaptation funds could be strategically used to try to prevent internal displacement from arising. Such funding should be used to identify the populations at risk of climate displacement and then result in targeted interventions to assist these people to stay in their homeland where feasible. The governance arrangements required to prevent displacement and to assist those who have to migrate as a result of climate change will be complex; however, the injustice of losing one's home through no fault of oneself means that the world must fast-track action to help these people.

References

Berringer, Andreas C. "Migration and Climate Change: Global Governance Regimes and the Incorporation of Climate Change Displacement." In *Quality and Legitimacy of Global Governance: Case Lessons from Forestry*, edited by Tim Cadman, 202–16. London: Palgrave, Macmillan, 2011.

Doelle, Meinhard. "The Birth of the Warsaw Loss and Damage Mechanism: Planting a Seed to Grow Ambition." *Carbon and Climate Law Review* 8, no. 1 (2014): 35–45.

Fry, Ian. "The Paris Agreement: An Insider's Perspective – The Role of Small Island Developing States." *Environmental Policy and Law* 46, no. 2 (2016): 105–8.

Global Environment Facility (GEF). *GEF Programming Strategy on Adaptation to Climate Change Least Developed Countries Fund Special Climate Change Fund*. GEF, 2014. <https://www.thegef.org/gef/sites/thegef.org/files/publication/GEF_AdaptClimateChange_CRA.pdf> (accessed 6 June 2016).

Human Rights Council. "Human Rights and Climate Change." Resolution 7/23 2008.

Intergovernmental Committee Negotiating Committee for a Framework Convention on Climate Change. "Elements of a Framework Convention on Climate Change." UN Doc A/AC237/Miscl/Add.3 18 June 1991.

Intergovernmental Negotiation Committee for a Framework Convention on Climate Change. "Consolidated Working Document." UN Doc A/A C237/MISC.17/ADD.9 19 December 1991.

International Organization for Migration (IOM). *IOM Outlook on Migration, Environment and Climate Change*. Geneva: International Organization for Migration (IOM), 2011. <http://publications.iom.int/system/files/pdf/mecc_outlook.pdf> (accessed 6 June 2016).

Kalin, Walter. "Conceptualising Climate-Induced Displacement." In *Climate Change and Displacement: Multidisplinary Perspectives*, edited by Jane McAdam, 88–99. Oxford: Hart Publishers, 2010.

Lewis, Bridget, and Rowena Maguire. "A Human Rights–Based Approach to Disaster Displacement in the Asia Pacific." *Asian Journal of International Law* Online (2015): 1–27.

Lyster, Rosemary. "A Fossil Fuel–Funded Climate Disaster Response Fund under the Warsaw International Mechanism for Loss and Damage Associated with Climate Change Impacts." *Transnational Environmental Law* 4, no. 1 (2015): 125–51.

———. "Protecting the Human Rights of Climate Displaced Persons: The Promise and Limits of the United Nations Framework Convention on Climate Change." *Legal Studies Research Paper* no. 15/24 (2015). <file:///Users/rowenamaguire/Downloads/SSRN-id2585266.pdf> (accessed 25 May 2016).

Maguire, Rowena, and Louise Kruger. "The International Regulation of Persons Displaced by Climate Change." In *Protection of Refugees and Displaced Persons in the Asia Pacific Region*, edited by Angus Francis and Rowena Maguire, 201–24. Ashgate: Farnham, Surrey, 2013.

McAdam, Jane. *Climate Change, Forced Migration and International Law*. Oxford: Oxford University Press, 2012.

———. "Refusing Refuge in the Pacific: (De)Constructing Climate-Induced Displacement in International Law." In *Migration, Environment and Climate Change*, edited by Etienne Piguet, Antoine Pecoud and Paul de Guchteneire, 102–37. Cambridge: Cambridge University Press, 2011.

Millar, Ilona, Catherine Gascoigne, and Elizabeth Caldwell. "Making Good the Loss: An Assessment of the Loss and Damage Mechanism under the UNFCCC Process." In *Threatened Island Nations: Legal Implications of Rising Seas and Changing Climate*, edited by Michael Gerrad and Gregory Wannier, 433–72. Cambridge: Cambridge University Press, 2013.

Oxfam. "Adapting to Climate Change: What's Needed in Poor Countries, and Who Should Pay." *Oxfam Briefing Paper* no. 104 (May 2007). <https://www.oxfam.org/sites/www.oxfam.org/files/adapting%20to%20climate%20change.pdf> (accessed 6 June 2016).

Pickering, Jonathan. "Adaptation Finance in the Asia-Pacific Region: Strengthening Fairness, Effectiveness and Transparency in Allocation." *Workshop on Climate Change Governance in the Asia-Pacific Region: Agency and Adaptiveness*. Australian National University (ANU) Canberra, 2012.

United Nations Commission on Sustainable Development. "Intergovernmental Negotiating Committee for a Framework Convention on Climate Change: Fifth Session, Second Part New York, 30 April–9 May." A/AC.237/18 (Part II)/Add.1 1992.

United Nations Framework Convention on Climate Change (UNFCCC). "Conference of the Parties: COP21 Paris Agreement." FCCC/CP/2015/L.9: UNFCCC, 2015.

———. "Initial Two-Year Workplan of the Executive Committee of the Warsaw International Mechanism for Loss and Damage." <http://unfccc.int/adaptation/workstreams/loss_and_damage/items/8805.php>.

———. "Report of the Conference of the Parties on Its Seventh Session, Held at Marrakesh from 29 October to 10 November 2001." FCCC/CP/2001/13/Add.1 2001.

———. "Report of the Conference of the Parties on Its Sixteenth Session, Held in Cancun from 29 November to 10 December 2010." Part Two: Action Taken by the Conference of the Parties at Its Sixteenth Session FCCC/CP/2010/7/Add.1, 2011.

———. "Report of the Dialogue on Long-Term Cooperative Action by Enhancing Implementation of the Convention." UN Doc FCCC/CP/2007/4/Add.1 19 November 2007.

———. "Submissions on Internal and Cross-Border Migration, Displacement and Other Forms of Human Mobility." United Nations, <http://unfccc.int/adaptation/

groups_committees/loss_and_damage_executive_committee/items/9508.php> (accessed 6 June 2016).
———. "United Nations Framework Convention on Climate Change." Rio: United Nations, 1992.
———. "Warsaw International Mechanism for Loss and Damage Associated with Climate Change Impacts." <http://unfccc.int/adaptation/workstreams/loss_and_damage/items/8134.php> (accessed 6 June 2016).
University of Oregon. "Climate Change: Realities of Relocation for Alaska Native Villages." In *Tribal Climate Change Profile: Relocation of Alaska Native Communities*, edited by University of Oregon, 1–5. Portland: University of Oregon, 2011.
Verheyen, Roda, and Peter Roderick. *Beyond Adaptation: The Legal Duty to Pay Compensation for Climate Change Damage*. UK: WWF, 2008. <http://assets.wwf.org.uk/downloads/beyond_adaptation_lowres.pdf>.
Warner, Koko, Tamer Afifi, Walter Kälin, Scott Leckie, Beth Ferris, Susan F.C. Martin, and David Wrathall. *Changing Climate, Moving People: Framing Migration, Displacement and Planned Relocation*. Geneva: United Nations University Institute for Environment and Human Security, 2013. <https://collections.unu.edu/eserv/UNU:1837/pdf11213.pdf> (accessed 6 June 2016).
World Bank. *Economics of Adaptation to Climate Change: Synthesis Report*. Washington, DC: World Bank, 2010. <http://www-wds.worldbank.org/external/default/WDSContentServer/WDSP/IB/2012/06/27/000425970_20120627163039/Rendered/PDF/702670ESW0P10800EACCSynthesisReport.pdf> (accessed 6 June 2016).

7 Human rights and equity

Governing values for the international climate regime

Kate Dooley and Gita Parihar

Introduction

Climate change is overwhelmingly seen as a moral and ethical problem due to the fact that those who contributed least to the problem will be most impacted.[1] Starting from this premise, we argue that justice and fairness, as embodied in the principle of equity, must underpin a comprehensive integrity framework for governing the international climate regime. In this chapter we examine to what degree human rights and equity contribute to the Public Institutional Justification (PIJ) of the climate regime through looking at the governing values of the United Nations Framework Convention on Climate Change (UNFCCC).[2] Governing values are similar across many global regimes: they include sustainable development, the rule of law, sovereignty, human rights, democracy and equity.[3] In the context of this volume, these can be considered as governing values of the UNFCCC rather than governance values, as they govern the goals and priorities of the regime. Building on the analytical framework of thick and thin governance as presented in Chapter 2, we situate human rights and equity as 'thick' governance values – with thickness of governance referring to how substantial and wide-ranging the values under consideration are. 'Thick' governance values move beyond procedural concerns to incorporate substantial ethical goals, including the demands of morality, justice and legitimacy.[4] At the thickest end of the spectrum, the distinction between governance values and governing values begins to overlap. The thicker the governance values, the more they potentially become political and therefore more difficult to generate consensus around.

Equity can be considered a principle of international law that addresses fairness, and as such is of clear application in the international environmental

1 See John Broome, *Climate Matters: Ethics in a Warming World* (New York and London: Norton Global Ethics Series, 2012).

2 The UNFCCC is the international treaty body tasked with achieving agreement between all state parties on the distribution of responsibilities for addressing climate changes.

3 Charles Sampford, "Sovereignty and Intervention," in *Human Rights in Theory and Practice*, ed. Tom Campbell and Burton M. Leiser (London: Ashgate, 2001).

4 See Hugh Breakey, Tim Cadman and Charles Sampford, Chapter 2 of this volume, for a discussion of the 'thickness' of governance.

context.[5] The perception of equity in an international regime is crucial to generating agreement,[6] and many environmental treaties incorporate equity principles.[7] In the UNFCCC, equity is referred to in its own right and is also operationalised through the principle of common but differentiated responsibilities (CBDR),[8] which recognises that the special needs of developing countries must be taken into account.[9] Equity and CBDR form core principles (or governing norms) of the UNFCCC, guiding the substantive values and goals of the institution to achieve equitable burden-sharing through differentiated obligations, although how (on what basis) differentiation is to be achieved remains contested. By contrast, while there is an international legal framework for ensuring that human rights are recognised, protected and enforced, it has not played an instrumental role in the PIJ of the UNFCCC, although more recent texts, including the Paris Agreement, now reference human rights. In this chapter we examine how the governing values of equity and human rights contribute to the comprehensive integrity framework of the climate regime.

Such an examination is important because there is a risk that human rights and equity in the context of climate change could be used for different end goals. The centrality of human rights to achieving the goals of the climate convention[10] is widely commented on, both in terms of the impact a warming climate will have on human rights and in terms of the impact policies and actions to address climate change will have on human rights.[11] Concerns have been raised that human rights and equity could become competing discourses, with one concept displacing the other in the negotiating effort of states, potentially threatening the ability to achieve the widespread allegiance needed for climate-change norms to

5 See the discussion in Phillipe Sands, *Principles of International Environmental Law*, Second ed. (Cambridge, UK: Cambridge University Press, 2003), p. 152. Sands discusses the *North Sea Continental Shelf* cases where the ICJ described equity as a 'direct emanation of the idea of justice' and a 'general principle directly applicable as law', which should be applied as part of international law 'to balance up the various considerations which it regards as relevant in order to produce an equitable result'.

6 Harald Winkler and Lavanya Rajamani, "CBDR&RC in a Regime Applicable to All," *Climate Policy* 14, no. 1 (2014): pp. 120–21.

7 As well as the UNFCCC, see also, for example, Organization of American States (OAS), "Treaty for Amazonian Cooperation" (Brazil: OAS, 1978), Article 1. 'Ensure that these joint actions promote equitable and mutually beneficial results'; United Nations General Assembly (UNGA), "United Nations Convention on the Law of the Sea (10 December 1982): Article 59. 'When conflicts arise with regard to the attribution of rights and jurisdictions in the EEZ 'the conflict should be resolved on the basis of equity'.

8 Sands, *Principles of International Environmental Law*, p. 285.

9 Ibid.

10 "Framework Convention on Climate Change," (FCCC/INFORMAL/84: United Nations, 1992), Article 2. Article 2 defines its ultimate objective as 'stabilisation of greenhouse gas concentrations in the atmosphere at a level that would prevent dangerous anthropogenic interference in the climate system'.

11 See, for example, Environmental Investigation Agency and Centre for International Environmental Law, *A Rights-Based Approach to Land-Use in a Future Climate Agreement: Policy and Implementation Framework* (Washington, DC: EIA and CIEL, 2015).

drive effective action on the urgent timescales required.[12] Despite the widespread acceptance of human rights as a framework governing the duties and obligations of states towards individuals and peoples, the Paris outcome illustrates the difficulty of mobilising substantive references to human rights within UNFCCC decisions and outcomes. At the same time, norms associated with equity (included in the climate regime as differentiated responsibilities) are contested in their implementation. Hence, both human rights and equity provide a good example of how, as we move towards the thicker substantive values, we see more disputation over the content of these values.

This chapter is divided into three sections. We begin by examining human rights in the context of climate change and the application of equity in the UNFCCC. In the second section we examine in more detail the complementary aspects of human rights and equity – examining the congruence (that fulfilling the demands of one normative framework does not clash with or undermine the demands of the other); the specificity (to what degree equity is a specification in the human rights regime, and vice versa) and determinative aspects (that violating the duties of one normative regime violates the duties of the other). Finally, we assess what extent the Paris outcome embodies both human rights and equity in the long-term goals of finance, adaptation and loss and damage. We conclude that the specific differentiated obligations in the UNFCCC context could be regarded as delineating human rights obligations in the context of climate change. While the 'thick' governance values embodied in human rights and equity make agreement on the norms effecting action more difficult without tackling the larger moral and ethical questions that the climate crisis invokes, proposed solutions could ultimately be both inequitable and serve to undermine progress in human rights if climate responses are in breach of international standards and obligations. It should be noted that the focus of this chapter is on exploring the debates leading to the formulation of these principles and their contribution to the comprehensive integrity framework of the climate regime. Hence, while we refer to some of the relevant legal provisions in this area, our aim is not to examine the current status of these principles as a matter of law, but rather to examine the coherence between the two normative frameworks they have engendered.

The evolution of human rights and equity as governing values for the climate regime

Human rights and climate change

There is a long history to the human rights movement, which can be seen as a shared globally values system that respects the fundamental rights of all people.

12 Hugh Breakey, "Heating up Climate Change Norms – Lessons from Human Rights," in *Ethical Values and the Integrity of the Climate Change Regime*, ed. Hugh Breakey, Vesselin Popovski, and Rowena Maguire (Surrey: Ashgate, 2015).

In 1948 the UN Declaration of Human Rights (UNDHR) anchored the modern concept of human rights. Whilst recognised as 'the most successful normative project of all time',[13] human rights norms have taken decades, even since the Declaration, to reach this level of impact. Climate change norms, on the other hand, need to achieve widespread allegiance in a much shorter timespan,[14] while incorporating the impacts of both climate change and actions to address climate change on human rights.

The classic conception of human rights focuses on obligations of states primarily relating to the people within their territory. Human rights commentators writing in the area of climate change have pointed out that this model can often work for civil and political rights, but is less useful for harm to economic, social and cultural rights.[15] Civil and political rights, historically championed by Western democracies, can be seen as liberty oriented, focusing on the right to life, security and freedom of expression. Economic, social and cultural rights can be viewed as collective rights, focusing on rights to work, food, shelter and health. These two bodies of rights were recognised in the International Covenant on Civil and Political Rights and the International Covenant on Economic, Social and Cultural Rights respectively. It is increasingly recognised that there is value to upholding a collective dimension of human rights, as integral to the protection, promotion and fulfilment of individual rights. A newer grouping of rights, often referred to as 'third generation' rights, such as the right to a healthy environment, can be seen as providing the conditions for states, particularly those in the developing world, to uphold the rights of the individual, and can only be realised through the collective effort of all international actors.

It is progress in the human rights field relating to the collective aspect of rights, and in particular obligations existing at the extra-territorial level, that is particularly relevant to considering the implementation of equitable burden-sharing of obligations to address climate change. Rajamani has documented the increasing currency of rights-based perspectives in the international climate negotiations, observing that the human rights–centred approach to climate change, advocated by scholars and human rights bodies, places the rights of the individual centrally and draws attention to the impact that climate change could have upon a range of rights.[16] Here we go one step further to suggest that while holding states accountable to deliver on human rights obligations that prevent harm and provide security and freedom to individuals is essential, recognising a collective aspect in terms of cross-boundary obligations and responsibility for realising the global conditions that enable states to uphold human rights obligations is equally

13 Rowena Maguire, David M. Douglas, Vesselin Popovski, and Hugh Breakey, "Ethical Values and Global Carbon Integrity System," ibid.
14 Hugh Breakey, "Heating up Climate Change Norms – Lessons from Human Rights," ibid.
15 Sands, *Principles of International Environmental Law*.
16 Lavanya Rajamani, "The Increasing Currency and Relevance of Rights Based Perspectives in the International Climate Negotiations," *Journal of Environmental Law* 22, no. 3 (2010).

important. In the context of climate change, recognising this collective aspect of human rights supports the requirement for differential obligations. As recognised by the Office of the UN High Commissioner for Human Rights, 'looking at climate change vulnerability and adaptive capacity in human rights terms highlights the importance of analysing power relationships, addressing underlying causes of inequality and discrimination, and gives particular attention to marginalized and vulnerable members of society'.[17]

Human rights were not directly referenced in the treaty text of the UNFCCC, and indeed the links between the two normative frameworks do not appear to have been considered until much later. Civil society advocacy in the run-up to the 2009 Copenhagen climate summit resulted in the first explicit reference to human rights in the 2010 Cancun Agreements (a series of decisions under the UNFCCC rather than a treaty), which states that countries should fully respect human rights in all climate-related actions.[18] The lead-up to the Paris climate summit in 2015 saw an increase in advocacy efforts for human rights language to be included in a new post-2020 climate agreement. In 2014, a special procedures committee of the Human Rights Council issued an open letter to parties to the UNFCCC, calling for human rights protections in the Paris climate agreement.[19] The following year over 200 organisations made a submission to the UNFCCC with the same demand.[20] Following the increasing recognition of the impact of climate change on human rights and calls for greater attention to be paid to human rights in developing the next phase of the climate regime, initial versions of the draft of the Paris Agreement included reference to human rights obligations in the overall objective. The final outcome from Paris saw human rights language included in the new climate treaty, albeit limited to the preamble:

> Acknowledging that climate change is a common concern of humankind, Parties should, when taking action to address climate change, respect, promote and consider their respective obligations on human rights, the right to health, the rights of indigenous peoples local communities, migrants, children, persons with disabilities and people in vulnerable situations and the

17 Human Rights Council, "Report of the Office of the United Nations High Commissioner for Human Rights on the Relationship between Climate Change and Human Rights," (A/HRC/10/612009), para. 81.

18 "Report of the Conference of the Parties on Its Sixteenth Session, Held in Cancun from 29 November to 10 December 2010," (Part Two: Action Taken by the Conference of the Parties at Its Sixteenth Session FCCC/CP/2010/7/Add.1: United Nations, 2011).

19 Office of the High Commissioner for Human Rights (OHCHR), "Letter to Member States," *High Commissioner for Human Rights* (2012), http://www.ohchr.org/Documents/Issues/Development/OpenLetterHC.pdf.

20 Bridget Lewis, "The Contribution of Human Rights to the Effectiveness and Integrity of the Global Carbon Regime," in *Ethical Values and the Integrity of the Climate Change Regime*, ed. Hugh Breakey, Vesselin Popovski, and Rowena Maguire (Surrey: Ashgate, 2015).

right to development, as well as gender equality, empowerment of women and intergenerational equity.[21]

This language recognises that parties have existing human rights obligations and points to the need to take these into account when acting on climate change. It explicitly draws these existing obligations into the climate-change framework by making a direct link between the two sets of obligations, with consequences for how parties take human rights into account when developing or implementing climate policies under the future climate regime. This a significant development towards a regime that considers the impact on human rights obligations when states take action to address climate change, as well as the differentiated responsibilities of states to ensure the global conditions necessary for upholding human rights for all.

Equity in international environmental law

An important step forward for the application of equity in international environmental law was the 1992 Rio 'Earth Summit' – otherwise known as the UN Conference on Environment and Development (UNCED). Coming after the 1982 Stockholm Conference on the Environment, and against a backdrop of increasing emphasis on the environment-development nexus from developing countries, the inclusion of development in the title of the Rio conference was significant.[22] During discussions in Rio, China and the G77 built on the sustainable development concept cemented in Stockholm by stressing the need to link environment and development in order to address poverty eradication, North-South equity issues and causes and responsibilities for environmental problems.[23] This focussed discussions on the recognition of differentiated obligations among states depending on their relative contributions to the problem and differing capacities to act.

The Rio Declaration is a soft law instrument that lays down a number of general principles, many of which are now seen as a 'basic normative framework' reflecting customary international law.[24] Principle 7 of the Rio Declaration is a key principle addressing equity, outlining the basis for sharing responsibilities between countries by introducing the concept of common but differentiated responsibilities (CBDR). CBDR incorporates two distinct concepts: the common responsibility of states to protect the environment, and the need to take into

21 UNFCCC, "Conference of the Parties Twenty-First Session Paris, 30 November to 11 December 2015: Adoption of The Paris Agreement," (FCCC/CP/2015/L.9/Rev. 1: United Nations, 2015), Preamble, para. 11.

22 Lavanya Rajamani, "From Stockholm to Johannesburg: The Anatomy of Dissonance in the International Environmental Dialogue," *RECIEL* 12, no. 1 (2003): pp. 23–32.

23 Chee Yoke Ling, *The Rio Declaration on Environment and Development: An Assessment* (Penang: Third World Network, 2012).

24 Günther Handl, *Declaration of the United Nations Conference on the Human Environment (Stockholm Declaration), 1972 and the Rio Declaration on Environment and Development, 1992* (New York: United Nations Audiovisual Library of International Law, 2012), p. 2.

account historical contributions to environmental pollution when determining responsibility.[25] Principle 7 links CBDR directly to contributions to global environmental harm,[26] stating:

> In view of the different contributions to environmental degradation, States have common but differentiated responsibilities. The developed countries acknowledge the responsibility that they bear in the international pursuit of sustainable development in view of the pressures their societies place on the global environment and of the technologies and financial resources they command.[27]

CBDR in the Rio Declaration can be seen as a cornerstone of the sustainable development paradigm and the basis for international action on climate change,[28] yet disagreements remain over how differentiation is interpreted. Rajamani argues that the notion of differentiated responsibility derives both from the differing contributions of states to climate change and the difference in capacity to address the problem.[29] This interpretation is widely validated, with contribution and capacity seen as the key determinants of responsibility. However, Principle 7 was one of the most contentious principles to negotiate, with observers noting that 'the obligation of the industrialised North to bear their proportionate share of the responsibility for the environmental crisis . . . were areas of deep North-South conflict'.[30]

The specific link between equity, historical responsibility and capacity is noted by academic commentators such as Philippe Sands: 'in many respects UNCED was about equity: how to allocate future responsibilities for environmental protection between states which are at different levels of economic development, have contributed in different degrees to particular problems, and have different environmental and developmental needs and priorities'.[31] The central concern with equity is reflected in each UNCED instrument.[32] The UN Human Rights Commissioner subsequently recognised the link to human rights in the Rio Declaration as having 'integrated human rights in its approach to sustainable development' and being 'thoroughly infused with human rights considerations

25 Lavanya Rajamani, "The Principle of Common but Differentiated Responsibilities and the Balance of Commitments under the Climate Regime," *RECIEL* 9, no. 2 (2000): pp. 130–31.

26 Winkler and Rajamani, "CBDR&RC in a Regime Applicable to All."

27 UNGA, "Report of The United Nations Conference on Environment and Development: Rio De Janeiro, 3–14 June 1992," (A/CONF.151/26 (Vol. I) 1992).

28 Handl, *Declaration of the United Nations Conference on the Human Environment (Stockholm Declaration), 1972 and the Rio Declaration on Environment and Development, 1992.*

29 Rajamani, "The Principle of Common but Differentiated Responsibilities and the Balance of Commitments under the Climate Regime."

30 Yoke Ling, *The Rio Declaration on Environment and Development: An Assessment.*

31 Sands, *Principles of International Environmental Law.*

32 The four instruments that came out of the Rio Earth Summit were the Rio Declaration; the UNFCCC; the Convention on Biodiversity and the Convention on Desertification.

essential to sustainable development'.[33] More broadly, international environmental law clearly recognises that environmental problems affect human well-being and are exacerbated by a lack of development.[34]

From the previous discussion it will be evident that both equity and human rights make important, albeit differing, contributions towards the context-integrity[35] of the UNFCCC: equity (as implemented through CBDR) enables a focus on obligations between states, whereas human rights focuses primarily on the relationship between individuals and states and addresses imbalances in that context. Consideration of the collective dimension of human rights obligations in a cross-state context is less evolved at present. Upholding this collective dimension requires a differentiated approach to burden-sharing as developed under the UNFCCC, and hence we argue that both normative frameworks are preconditions to achieving a comprehensive integrity framework for the UNFCCC.

The comprehensive integrity framework of the UNFCCC

After providing a background on how human rights and equity are currently understood and recognised in the climate regime, we now turn to tracing the inclusion of equity and human rights in regime documents of the UNFCCC. We observe that equity and CBDR, as core principles of the UNFCCC, retain prominence, while specific references to human rights remain rare. Even in the Paris Agreement the core objective is to be achieved 'on the basis of equity', while human rights are referenced in the preamble (to guide interpretation of the treaty).

Equity and differential obligations in the international climate regime

The concept of equity as delineated through international environmental law provides a framework for recognising and redressing the international imbalance of environmental pollution. Environmental impacts, particularly climate change, have disproportionately affected the development of poorer countries. With regards to this, human rights commentator Dinah Shelton writes that poverty has come to be seen both as a major source of environmental degradation and as a human rights issue because it means that individuals lack an adequate standard

33 Navi Pillay, "Pillay Urges States to Inject Human Rights into Rio+20," *UN High Commissioner on Human Rights*, 18 April 2012.

34 See also UNGA, "United Nations Conference on the Human Environment, 15 December 1972," (A/RES/29941972), Principle 5 on the sharing of resources, and Principle 9 on technology transfer and financial support to developing countries to help them address environmental degradation. In the Rio Declaration, Principle 3 refers to the right to development as equitably meeting development and environment needs of future generations, and Principle 6 calls for giving special priority to the situation and needs of developing countries.

35 As described in Chapter 2 of this volume.

of living – including the food, shelter, medical care and education that are guaranteed by international human rights law.[36] As noted above, the UNFCCC recognises the unequal position of developed and developing countries in terms of contribution to and impacts of climate change within its first governing principle (Article 3.1), which recognises that Parties should act 'on the basis of equity and in accordance with their common but differentiated responsibilities and respective capabilities,' and that 'developed country Parties should take the lead'.[37]

Further recognition of the link between differentiated responsibilities and achieving equity is found throughout the Convention. Article 3.2 refers to the 'specific needs and special circumstances' of developing country parties and highlights the need to consider disproportionate or abnormal burdens.[38] Article 4.8 refers to the need to consider funding, insurance and transfer of technology that may be necessary to meet the specific needs of specially affected states.[39] Developed country parties also agree to take into account the need for 'equitable and appropriate contributions' by each of them to the global effort regarding the achievement of the objective of the Convention.[40] Equity has maintained its central status in the Paris Agreement, with Article 2.1 noting the agreement is to be implemented 'to reflect equity'; the mitigation goal described in Article 4.1 is to be achieved 'on the basis of equity' and the global stocktake is to assess implementation of the Agreement 'in light of equity'.

This language makes clear that differential treatment retains a central role in the international climate legal framework, continuing to give effect to the principle of equity. However, while Principle 7 of the Rio Declaration states that CBDR is based on contribution to environmental problems, to what extent CBDR should be based on *historical* responsibility remains contested. There are those who argue that given climate change is caused by the cumulative concentration of greenhouse gasses (GHG) in the atmosphere over centuries, responsibility should be based on historical contributions to cumulative emissions. When examining the question through the lens of a 'thick' notion of governance, there are compelling moral arguments for CBDR to be understood as including historical responsibility. For example, Shelton notes that:

> Fairness and a morally coherent response suggest that these states, which attained their current developed status through imposing non-internalized

36 Dinah Shelton, "Equitable Utilization of the Atmosphere: A Rights-Based Approach to Climate Change," in *Human Rights and Climate Change*, ed. Stephen Humphries (Cambridge, UK: Cambridge University Press, 2010).

37 UNFCCC, "Framework Convention on Climate Change," Article 3.1.

38 Ibid., Article 3.2. Highlights 'countries that are particularly vulnerable to the adverse effects of climate change and of those parties, especially developing country Parties that would have to bear a disproportionate or abnormal burden under the convention'.

39 Ibid., Article 4.8. Requires all parties to consider what actions, including funding, insurance and transfer of technology, may be necessary to meet the specific needs of specially affected states.

40 Ibid., Articles 3.1 and 4.2(a).

> costs on the environment, take the major abatement actions, rather than demanding that everyone equally mitigate the externalities, including those not responsible for initially creating the problem. Equity, in this sense, is justified as a means of corrective justice, requiring remedial conduct to correct past wrongs.[41]

For this reason, Shelton makes an ethical argument for historical responsibility with compensatory or reparative justice as a basis for differential treatment of developing countries, especially where they were colonies whose resources were exploited as the means for development by industrialised countries. Objectors to this argue that it is unfair to penalise current generations for harm caused by their forebears.[42] Henry Shue contends, however, that although present and future generations in developed countries did not request or consent to the carbon emissions of their ancestors, they benefit by living amidst national affluence produced by means of those emissions.[43] Hence arguments for historical responsibility are based on ability to pay and fairness.

While theories of justice can be effectively employed to argue for the principle of CBDR to incorporate historical responsibility, as we have demonstrated previously in this chapter, Breakey points out that human rights have achieved moral authority based on appealing to widely accepted norms, with a variety of different actors (for different reasons) endorsing the same norms. Equity as operationalised through differentiation in responsibility and obligation of states to act, was, after some political battles, accepted in the Rio Declaration as the principle of CBDR and forms a central component of the UNFCCC, including the Paris Agreement. Yet the use of historical responsibility as a facet of determining CBDR (and therefore its effectiveness in promoting equity as a governing value for the climate regime) remains limited by lack of widespread acceptance, and was not defined or specified as a basis for CBDR in the Paris Agreement. At the same time, references to human rights, as arguably the most successful international normative framework, remain limited in UNFCCC decisions, including the Paris outcome.

Differential obligations in human rights law

The previous discussion demonstrates broad congruence between human rights and equity in the context of CBDR, indicating that the CBDR principle could become more widely accepted by appealing to different actors through these different normative frameworks, despite abiding contestations over *how*

41 Shelton, "Equitable Utilization of the Atmosphere: A Rights-Based Approach to Climate Change," p. 121.

42 Maguire, Douglas, Popovski, and Breakey, "Ethical Values and Global Carbon Integrity System."

43 Henry Shue, "Historical Responsibility, Harm Prohibition and Preservation Requirement: Core Practical Convergence on Climate Change," *Moral Philosophy and Politics* 2, no. 1 (2015): pp. 7–31.

differentiation is to be implemented. In this section we examine specificity and determinism between the two regimes – is equity a specification of the human rights regime, and does violating the duties of one regime violate the duties of the other? To do this we focus on extra-territorial rights related to the social and international order – namely economic, social and cultural rights, and the relationship between non-discrimination and differentiation.

Human rights instruments recognise the important role played by the social and international order in the achievement of human rights.[44] This is something that is often overlooked in human rights discourse, even though problems such as global poverty, which violate human rights, can become entrenched through a particular international order.[45] As with the environment, the international economic order has a significant impact on the attainment of human rights. In the run-up to Rio+20, then UN Human Rights Commissioner Navi Pillay stated that 'strategies based on the narrow pursuit of economic growth without due regard for equity and related environmental, social and human rights considerations will both fail in their environmental objectives, and risk damaging the planet and the fundamental rights of the people who live here'.[46]

The attainment of such a social and international order is best expressed in the UN Declaration on the Right to Development, which has a specific entitlement – the right to participate in, contribute to and enjoy economic, social, cultural and political development. The Declaration includes the 'fair distribution of the benefits of development',[47] and Pillay also highlights the role of equity here, stating that 'equity is codified in the right to development'.[48] While the legal status of the right to development is contested by some states, other international instruments reflect similar obligations. The International Covenant on Economic, Social and Cultural Rights (ICESCR) contains specific obligations,

44 UNGA, "Universal Declaration of Human Rights," (10 December 1948, 217 A (III)1948), Article 28 states that '[e]veryone is entitled to a social and international order in which the rights and freedoms set forth in this Declaration can be fully realised'.

45 Stephen Humphries, "Conceiving Justice: Articulating Common Causes in Distinct Regimes," in *Human Rights and Climate Change*, ed. Stephen Humphries (Cambridge, UK: Cambridge University Press, 2010), p. 299: 'There is an at least plausible case . . . that the persistence and exacerbation of global poverty over recent decades is itself a human rights violation, largely sustained by international actors and systematised through international law.' See also Shelton, "Equitable Utilization of the Atmosphere: A Rights-Based Approach to Climate Change," ibid., p. 92. Shelton points out that this model can generally address violations of civil and political rights but 'harm to economic, social and cultural rights, as well as to the underlying environmental conditions necessary to the enjoyment of all rights, often originates outside the jurisdiction where the harm is felt'.

46 OHCHR, "Letter to Member States."

47 UNGA, "Declaration on the Right to Development: Resolution/Adopted by the General Assembly, 4 December 1986," (A/RES/41/128: United Nations, 1986).

48 OHCHR, "Introduction Statement by the High Commissioner," *Development is a Human Right for all* (2011), http://www.ohchr.org/EN/Issues/Development/Pages/IntroductionStatement.aspx (accessed 25 May 2016).

including requiring parties to support international assistance and cooperation 'with a view to achieving progressively the full realization of the rights recognized in the present Covenant by all appropriate means . . .' and to '. . . recognize the benefits to be derived from the encouragement and development of international contacts and cooperation in the scientific and cultural fields'.[49]

Hence we see equity as a specification in the collective dimension of human rights, which is particularly relevant to the context of climate change. Collective efforts by the international community to realise these rights will become increasingly crucial to maintain a social and international order where states can uphold their obligations pertaining to civil and political rights.

In a report to the Human Rights Council, Special Rapporteur on human rights and the environment John Knox makes direct reference to the language relating to CBDR in the UNFCCC text:

> To be clear, the duty of international cooperation does not require each state to take exactly the same actions in response to climate change . . . all states have a duty to work together to address climate change but the particular responsibilities necessary and appropriate for each state will depend in part on its situation.[50]

Knox regards the treatment of climate change as a global problem requiring a global response as an application of the duty of international cooperation. There are other potential bases for highlighting the synergies between climate change and human rights. While non-discrimination obligations are generally considered to apply to the relationship between individual and state, in reality there is also a clear international aspect to inequality. As Shelton explains:

> Non-discrimination is clearly a predominant obligation in international human rights law. Yet equal treatment is required only for those equally situated . . . Requiring all states to reduce GHGs in an identical fashion would make many developing countries, or groups in those countries, worse off . . . From the perspective of equity toward the most vulnerable or least well-off, environmental protection should not result in further deterioration of their well-being. In order to address this problem, non-equal or differential obligations can, and are, being imposed as an equitable means of fostering substantive equality in the long term.[51]

49 See OCHCHR, 'International Covenant on Economic, Social and Cultural Rights', (16 December 1966), Articles 2.1 and 15.4.

50 Human Rights Council, "Report of the Special Rapporteur on the Issue of Human Rights Obligations Relating to the Enjoyment of a Safe, Clean, Healthy and Sustainable Environment," (A/HRC/31/52, 1 February 2016), paras. 42 and 46.

51 Shelton, "Equitable Utilization of the Atmosphere: A Rights-Based Approach to Climate Change," p. 125.

Based on this discussion, we suggest that differential obligations are a determinative element of the human rights regime where inequitable development, or distribution of benefits and burdens, would breach human rights obligations. While the focus in the climate regime is on states rather than on individuals, it is clear that individuals within states that are already suffering from climate harm for which they are not responsible will be disadvantaged further through an inequitable outcome. As Shelton puts it, 'any allocation of benefits and burdens that makes vulnerable populations worse off, even if harm is felt outside the boundaries of the state, cannot be regarded as equitable or in conformity with international human rights law'.[52]

This recognition of the different positions of states, of the need for progressive realisation of human rights and of the division of responsibility among states to ensure these rights are fulfilled has strong parallels with the concept of differentiation in the international climate context. However, the duty to cooperate in the context of the ICESCR is rather vague and not as developed as, for example, the clear obligations in relation to climate finance and technology transfer in the international climate framework.[53] The specific obligations in the UNFCCC agreement could be regarded as specifying what the duty to cooperate means in the context of international action on climate change. Further work by the treaty bodies and the Human Rights Council on the substance of protecting, promoting and fulfilling economic, social and cultural rights could help advance thinking on integrating human rights considerations into climate change responses.

Addressing climate change: Applying human rights and equity in practice

Having established to varying degrees the congruence, specificity and determinism of human rights and equity as complementary normative frameworks, we now turn to what this means for implementation of duties and obligations in the context of the UNFCCC. The core role of equity is evident from its placing in Article 2 of the Paris Agreement, while inclusion of human rights language in the preamble recognises that parties have existing human rights obligations and guides interpretation of the treaty towards taking these obligations into account when acting on climate change. Here we examine the coherence of specific aspects of the Paris Agreement in terms of how the governing values of human rights and equity form part of the Public Integrity Justification of the climate regime.

Mitigation and the long-term goal

The Paris Agreement includes a long-term temperature goal which commits Parties to limit global temperature rise to 'well below' 2°C above pre-industrial

52 Ibid.

53 See, for example, UNFCCC, "Framework Convention on Climate Change," Articles 4 and 11.

levels, and to 'pursue efforts' towards 1.5°C while reflecting equity and CBDR in implementation.[54] While ambitious, there is also a risk that actions to mitigate climate change could adversely impact human rights, which would be contrary to the spirit of the treaty as articulated through the human rights language in the preamble.

The mitigation objective requires parties to achieve a 'balance between anthropogenic emissions by sources and removals by sinks . . . in the second half of this century' in order to meet the temperature limit.[55] Removing CO_2 from the atmosphere primarily relies on land via bioenergy with geological storage, or growing trees to sequester carbon. Estimates for required land area range from 300–970 million hectares (up to two-thirds of current agricultural land).[56] Land use on this scale would likely undermine food security, poverty eradication and other development objectives, suggesting that relying on land for large-scale removal of emissions would be incompatible with both human rights and an equitable approach to climate mitigation. The Paris Agreement acknowledges poverty eradication and sustainable development as the context in which both the temperature goal and the long-term mitigation goal are to be achieved.[57]

As an alternative, a number of civil society groups have proposed an international 'Fair Shares' approach, which advocates a steeper and more ambitious rate of emission reductions without relying on large-scale CO_2 removal. The Fair Shares approach suggests a burden-sharing framework based on the equity principles of historical responsibility, respective capacity and a right to development.[58] This approach requires emission reduction targets of well in excess of 100 percent for developed countries (who have already used their 'fair share' of the carbon budget). Larger developing countries will need to reduce emissions to zero while the poorest countries have a few decades before emissions must decrease.

54 UNFCCC, "Conference of the Parties Twenty-First Session Paris, 30 November to 11 December 2015: Adoption of The Paris Agreement," Article 2.

55 Ibid., Article 4.1.

56 Pete Smith, Steven J. Davis, Felix Creutzig, Sabine Fuss, Jan Minx, Benoit Gabrielle, Etsushi Kato, Robert B. Jackson, Annette Cowie, Elmar Kriegler, Detlef P. van Vuuren, Joeri Rogelj, Philippe Ciais, Jennifer Milne, Josep G. Canadell, David McCollum, Glen Peters, Robbie Andrew, Volker Krey, Gyami Shrestha, Pierre Friedlingstein, Thomas Gasser, Arnulf Grubler, Wolfgang K. Heidug, Matthias Jonas, Chris D. Jones, Florian Kraxner, Emma Littleton, Jason Lowe, Jose Roberto Moreira, Nebojsa Nakicenovic, Michael Obersteiner, Anand Patwardhan, Mathis Rogner, Ed Rubin, Ayyoob Sharifi, Asbjorn Torvanger, Yoshiki Yamagata, Jae Edmonds, and Cho Yongsung, "Biophysical and Economic Limits to Negative CO_2 Emissions," *Nature Climate Change* advance online publication, no. 1–9 (2015): pp. 42–50.

57 UNFCCC, "Conference of the Parties Twenty-First Session Paris, 30 November to 11 December 2015: Adoption of The Paris Agreement," Articles 2.1 and 4.1.

58 ActionAid International, "Fair Shares: A Civil Society Equity Review of INDCs 2015," Oxfam, https://www.oxfam.org/en/research/fair-shares-civil-society-equity-review-indcs.

Finance and technology transfer

Article 4.7 of the UNFCCC confers obligations for developed countries relating to finance and technology transfer. The Paris Agreement builds on this by requiring developed countries to provide financial resources to developing countries to assist with mitigation and adaptation.[59] While intended to apply at the national level, human rights discourse concerning the duty to take affirmative action to address discrimination could provide a relevant analogy here, reinforcing the need to provide finance and technology transfer to address the unfair position in which developing countries find themselves.[60] The Human Rights Council also recognises the need for international dialogue and cooperation concerning the adverse impacts of climate change on the enjoyment of human rights in developing countries, and calls for the implementation of practical steps to promote capacity-building, financial resources and technology transfer in accordance with the UNFCCC.[61]

As pointed out by Stephen Humphreys, 'without robust and detailed policies of technology transfer and adaptation, [the poorest countries'] development and policy options will shrink with deleterious effects on basic rights'.[62] The need for an energy transition is a global one, requiring financial support. As Shue states:

> If we are to escape from the fossil-fuel energy regime into an alternative-energy regime, the costs of the transition must be borne by someone. So we must have an allocation of the burdens (and benefits), and it ought to be fair.[63]

The Fair Shares approach introduced above suggests that the difference between developed countries' domestic emissions reductions and their international fair share can be met through the transfer of financial and technological resources to allow developing countries to undertake more than their fair share of emission reductions.[64]

Adaptation and loss and damage

Human rights and equity considerations in relation to adaptation include the provision of funding and technology transfer by developed countries to assist developing countries in dealing with the harm they are already experiencing as a result

59 UNFCCC, "Conference of the Parties Twenty-First Session Paris, 30 November to 11 December 2015: Adoption of The Paris Agreement," Article 9.1.

60 Noting that discrimination duties are states' obligations towards their citizens in the human-rights context, as mentioned earlier.

61 Human Rights Council, "Resolution 26/27: Human Rights and Climate Change," (A/HRC/26/L.33/Rev.1, 2014), para. 5.

62 Humphries, "Conceiving Justice: Articulating Common Causes in Distinct Regimes," p. 307.

63 Henry Shue, *Climate Justice: Vulnerability and Protection* (Oxford: Oxford University Press, 2014), pp. 23–24.

64 ActionAid International, "Fair Shares: A Civil Society Equity Review of INDCs 2015."

of climate change. Adaptation is necessary to protect basic human rights, which also provides a framework to ensure that climate finance is directed to the most vulnerable.

Loss and damage, which focuses on the injustice of climate impacts that cannot be adapted to in countries that bear little or no responsibility for climate harm, also brings the question of human rights and equity to the fore. A central aspect of recognising loss and damage concerns redress for communities and countries facing this injustice, including, for example, compensation and the development of a legal framework to assist climate refugees. The Paris Agreement establishes a loss and damage mechanism (Article 8) and sets out that parties should enhance understanding action and support 'on a cooperative and facilitative basis'.[65] However, compensation and liability when there is no consensus on attributing historical responsibility for the climate crisis remains controversial, and accompanying text states that Article 8 'does not involve or provide a basis for any liability or compensation'.[66] It remains to be seen how this will be interpreted over time and what discussions develop in practice. Loss and damage clearly results in a breach of human rights, and the Special Rapporteur on human rights and the environment has urged parties to incorporate a human rights perspective in identifying the types of loss and damage to be addressed.[67]

Conclusion

In this chapter we have shown that complementarity between human rights and equity principles in the context of climate change goes beyond simple congruence, specifying human rights obligations in the context of equity and vice versa, and determining that human rights cannot be upheld in the context of climate change without implementing actions on the basis of equity. In other words, it is not possible for a proposal on climate action to be compliant with human rights obligations if it does not address differential obligations in a manner that protects the most vulnerable. The differentiated obligations in the UNFCCC context could be regarded as delineating the human rights obligations required in the specific context of climate change, where poverty reduction, environmental protection and the attainment of human rights are inextricably linked. While there is general agreement that equity, as operationalised through CBDR, means developed countries should take the lead, to what extent historical use of atmospheric space should curtail future use of atmospheric space remains a central controversy, impeding international agreement on climate action.

65 UNFCCC, "Conference of the Parties Twenty-First Session Paris, 30 November to 11 December 2015: Adoption of The Paris Agreement," Article 8.3.

66 UNFCCC, Decision 1/CP.21, para. 52.

67 Human Rights Council, "Report of the Special Rapporteur on the Issue of Human Rights Obligations Relating to the Enjoyment of a Safe, Clean, Healthy and Sustainable Environment," para. 64.

As we have demonstrated, a key aspect of the principle of equity, as developed through international environmental law, concerns the differential obligations of states, with clear moral arguments for greater responsibility of developed states in responding to climate change based on greater historical contribution to the problem, as well as greater capacity to act. Hence, with fuller understanding of the shared values between the two normative frameworks of human rights and equity, we conclude that these frameworks operate as complementary approaches – two pillars with a shared ethical basis, appealing to different political actors and institutions, but serving the same end goal – an effective global climate regime that delivers justice, or fairness, in its outcomes. We note that human rights and equity bring different governing values on which to build a normative framework for climate action, thereby strengthening the acceptance of these norms by appealing to a wider audience. Further consideration is needed of the ways in which the two sets of obligations overlap to contribute to a model of institutional integrity in the global climate regime that is both equitable and compatible with protecting and promoting human rights for all.

References

ActionAid International. "Fair Shares: A Civil Society Equity Review of INDCs 2015." Oxfam, <https://www.oxfam.org/en/research/fair-shares-civil-society-equity-review-indcs> (accessed 23 May 2016).

Breakey, Hugh. "Heating up Climate Change Norms – Lessons from Human Rights." In *Ethical Values and the Integrity of the Climate Change Regime*, edited by Hugh Breakey, Vesselin Popovski and Rowena Maguire, 225–37. Surrey: Ashgate, 2015.

Broome, John. *Climate Matters: Ethics in a Warming World*. New York and London: Norton Global Ethics Series, 2012.

Environmental Investigation Agency and Centre for International Environmental Law. *A Rights-Based Approach to Land-Use in a Future Climate Agreement: Policy and Implementation Framework*. Washington, DC: EIA and CIEL, 2015.

Handl, Günther. *Declaration of the United Nations Conference on the Human Environment (Stockholm Declaration), 1972 and the Rio Declaration on Environment and Development, 1992*. New York: United Nations Audiovisual Library of International Law, 2012.

Human Rights Council. "Report of the Office of the United Nations High Commissioner for Human Rights on the Relationship between Climate Change and Human Rights." A/HRC/10/61, 2009.

———. "Report of the Special Rapporteur on the Issue of Human Rights Obligations Relating to the Enjoyment of a Safe, Clean, Healthy and Sustainable Environment." A/HRC/31/52, 1 February 2016.

———. "Resolution 26/27: Human Rights and Climate Change." A/HRC/26/L.33/Rev.1, 2014.

Humphreys, Stephen. "Conceiving Justice: Articulating Common Causes in Distinct Regimes." In *Human Rights and Climate Change*, edited by Stephen Humphries, 299–319. Cambridge, UK: Cambridge University Press, 2010.

Lewis, Bridget. "The Contribution of Human Rights to the Effectiveness and Integrity of the Global Carbon Regime." In *Ethical Values and the Integrity of the Climate*

Change Regime, edited by Hugh Breakey, Vesselin Popovski and Rowena Maguire, 179–88. Surrey: Ashgate, 2015.

Maguire, Rowena, David M. Douglas, Vesselin Popovski, and Hugh Breakey. "Ethical Values and Global Carbon Integrity System." In *Ethical Values and the Integrity of the Climate Change Regime*, edited by Hugh Breakey, Vesselin Popovski and Rowena Maguire, 3–27. Surrey: Ashgate, 2015.

Office of the High Commissioner for Human Rights (OHCHR). "International Covenant on Economic, Social and Cultural Rights: Entry into Force 3 January 1976, in Accordance with Article 27." Adopted and Opened for Signature, Ratification and Accession by General Assembly resolution 2200A (XXI) of 16 December 1966.

———. "Introduction Statement by the High Commissioner." *Development is a Human Right for all* (2011). <http://www.ohchr.org/EN/Issues/Development/Pages/IntroductionStatement.aspx> (accessed 23 May 2016).

———. "Letter to Member States." *High Commissioner for Human Rights* (2012). Published electronically 30 March 2012. <http://www.ohchr.org/Documents/Issues/Development/OpenLetterHC.pdf> (accessed 23 May 2016).

Pillay, Navi. "Pillay Urges States to Inject Human Rights into Rio+20." *UN High Commissioner on Human Rights*, 18 April 2012.

Rajamani, Lavanya. "From Stockholm to Johannesburg: The Anatomy of Dissonance in the International Environmental Dialogue." *RECIEL* 12, no. 1 (2003): 23–32.

———. "The Increasing Currency and Relevance of Rights-Based Perspectives in the International Climate Negotiations." *Journal of Environmental Law* 22, no. 3 (2010): 391–429.

———. "The Principle of Common but Differentiated Responsibilities and the Balance of Commitments under the Climate Regime." *RECIEL* 9, no. 2 (2000): 120–31.

Sampford, Charles. "Sovereignty and Intervention." In *Human Rights in Theory and Practice*, edited by Tom Campbell and Burton M. Leiser, 335–51. London: Ashgate, 2001.

Sands, Phillipe. *Principles of International Environmental Law*. Second ed. Cambridge, UK: Cambridge University Press, 2003.

Shelton, Dinah. "Equitable Utilization of the Atmosphere: A Rights-Based Approach to Climate Change." In *Human Rights and Climate Change*, edited by Stephen Humphries, 91–125. Cambridge, UK: Cambridge University Press, 2010.

Shue, Henry. *Climate Justice: Vulnerability and Protection*. Oxford: Oxford University Press, 2014.

———. "Historical Responsibility, Harm Prohibition and Preservation Requirement: Core Practical Convergence on Climate Change." *Moral Philosophy and Politics* 2, no. 1 (2015): 7–31.

Smith, Pete, Steven J. Davis, Felix Creutzig, Sabine Fuss, Jan Minx, Benoit Gabrielle, Etsushi Kato, Robert B. Jackson, Annette Cowie, Elmar Kriegler, Detlef P. van Vuuren, Joeri Rogelj, Philippe Ciais, Jennifer Milne, Josep G. Canadell, David McCollum, Glen Peters, Robbie Andrew, Volker Krey, Gyami Shrestha, Pierre Friedlingstein, Thomas Gasser, Arnulf Grubler, Wolfgang K. Heidug, Matthias Jonas, Chris D. Jones, Florian Kraxner, Emma Littleton, Jason Lowe, Jose Roberto Moreira, Nebojsa Nakicenovic, Michael Obersteiner, Anand Patwardhan, Mathis Rogner, Ed Rubin, Ayyoob Sharifi, Asbjorn Torvanger, Yoshiki Yamagata, Jae Edmonds, and Cho Yongsung. "Biophysical and Economic Limits to Negative

CO_2 Emissions." *Nature Climate Change* advance online publication, no. 1–9 (7 December 2015).

Organization of American States (OAS). "Treaty for Amazonian Cooperation: The Republics of Bolivia, Brazil, Colombia Ecuador, Guyana, Peru, Suriname and Venezuela." Brazil: OAS, 1978.

United Nations Framework Convention on Climate Change (UNFCCC). "Conference of the Parties Twenty-First Session Paris, 30 November to 11 December 2015: Adoption of The Paris Agreement." FCCC/CP/2015/L.9/Rev. 1: United Nations, 2015.

———. "Framework Convention on Climate Change." FCCC/INFORMAL/84: United Nations, 1992.

———. "Report of the Conference of the Parties on Its Sixteenth Session, Held in Cancun from 29 November to 10 December 2010." Part Two: Action taken by the Conference of the Parties at its sixteenth session FCCC/CP/2010/7/Add.1: United Nations, 2011.

United Nations General Assembly (UNGA). "Declaration on the Right to Development: Resolution / Adopted by the General Assembly, 4 December 1986." A/RES/ 41/128: United Nations, 1986.

———. "Report of The United Nations Conference on Environment and Development: Rio De Janeiro, 3–14 June 1992." A/CONF.151/26 (Vol. I), 1992.

———. "United Nations Conference on the Human Environment, 15 December 1972." A/RES/2994, 1972.

———. "United Nations Convention on the Law of the Sea (Final Act Dated 3 March 1986 and 26 July 1993)." Concluded at Montego Bay on 10 December 1982: UNGA, 1982.

———. "Universal Declaration of Human Rights." 10 December 1948, 217 A (III), 1948.

Winkler, Harald, and Lavanya Rajamani. "CBDR&RC in a Regime Applicable to All." *Climate Policy* 14, no. 1 (2014): 102–21.

Yoke Ling, Chee. *The Rio Declaration on Environment and Development: An Assessment.* Penang: Third World Network, 2012.

8 Tradable permits and emissions reduction under the Paris Agreement

Governance values and integrity challenges

Tanvi Mani

Introduction: The history of tradable permits and emission reduction schemes under the Kyoto Protocol

The scheme of tradable pollution permits is a market-driven policy that aims at self-regulation through reliance on private decision-making. It comprises an agreement of trade entered into between two or more parties to ensure that that they are both better off than they would have been had they complied with only the pollution permit awarded to them exclusively.[1] The permits are usually allocated by the government, ideally through an auction.[2] Irrespective of the initial distribution of permits, the parties attempt to arrive at an equitable bargain through a trade where everyone is better off, no one is worse off and the outcome is economically efficient.[3]

The option of emissions trading is provided for within Article 17 of the Protocol.[4] Developed countries were assigned particular emission caps that they were required to comply with during the period between 2008 and 2012. This cap had been assigned based on a percentage of their emission levels, taken according to the base year 1990 or 1995, depending on the greenhouse gas in question. If, however, a country has made an assessment (in its own capacity and without

1 John K. Stranlund, Carlos A. Chavez, and Barry C. Field, "Enforcing Emissions Trading Programs: Theory, Practice and Performance," *Policy Studies Journal* 30, no. 3 (2002).

2 Thomas H. Tietenberg, "Economic Instruments for Environmental Regulation," *Oxford Review of Economic Policy* 6, no. 1 (1990).

3 James E. Krier and Robert B. Bell, "Old Economics and New Environmental Controls," *Stanford Lawyer* 15, no. 1 (1979–80).

4 See United Nations Framework Convention on Climate Change (UNFCCC), "Kyoto Protocol to the United Nations Framework Convention on Climate Change, 10 December 1997," (U.N. Doc FCCC/CP/1997/7/Add.1, 37 I.L.M.: United Nations, 1997), Articel 17. ('The Conference of the Parties shall define the relevant principles, modalities, rules and guidelines, in particular for verification, reporting and accountability for emissions trading. The Parties included in Annex B may participate in emissions trading for the purposes of fulfilling their commitments under Article 3. Any such trading shall be supplemental to domestic actions for the purpose of meeting quantified emission limitation and reduction commitments under that Article'.)

any external review mechanism) that it would exceed the emission limit during this period of compliance, the provision of emissions trading would allow it to purchase emission reduction credits from another country.[5] The latter country would make a determination of whether it would be able to reduce its emissions beyond the required cap and then sell its credits to the purchaser country accordingly. An emissions trading scheme therefore incentivizes countries to reduce their emissions beyond the required limit and to profit from the sale of credits to other countries, where the cost of emission reduction may be higher. Under this cap and trade program, both the purchaser and the seller country benefit from the overall reduced cost of emission reduction.[6]

This mechanism has, however, been restricted within the Protocol to Annex I parties upon whom the obligations to reduce emissions are legally binding. An additional restriction that has been imposed upon parties engaging in this mechanism of carbon credit trading is that it must 'be supplemental to domestic actions for the purpose of meeting quantified emission limitations and reduction commitments'.[7] The Protocol, however, does not seem to provide any clarification on the interpretation of the term 'supplemental', and this has been interpreted in myriad different ways by different countries.

In compliance with the Kyoto Protocol, European policymakers instituted The European Emission Trading Scheme (EU ETS), which aided in the allocation and trade of emissions permits from energy-intensive industries in 25 countries. Presently the scheme has been extended to 27 countries and includes 80 percent of the world's marketable permits.[8]

In addition to the cap and trade mechanism, the Kyoto protocol has also included another scheme by which Annex I countries may accumulate additional credits that they can use to meet their emission cap requirements. The Clean Development Mechanism (CDM) is the only provision under the Protocol that has been extended, by effect, to non-Annex I countries.[9] Moreover, the Joint Implementation scheme provided for under the Protocol allows developed countries to receive Emission Reduction Units or ERUs for greenhouse gas reduction projects undertaken within another country in a collaborative manner. These ERUs can be used by the various participating countries to meet their compliance requirements beginning in 2008. The procedure laid down for the governance

5 Larry Parker, "Climate Change: Design Approaches for a Greenhouse Gas Reduction Program," (Congressional Research Service, 2008), CRS Report RL33799.

6 Ibid.

7 United Nations Framework Convention on Climate Change (UNFCCC), "Kyoto Protocol to the United Nations Framework Convention on Climate Change, 10 December 1997," Article 17.

8 A. Denny Ellerman and Barbara K. Buchner, "The European Union Emissions Trading Scheme: Origins, Allocation, and Early Results," *Review of Environmental Economics and Policy* 1, no. 1 (2007).

9 United Nations Framework Convention on Climate Change (UNFCCC), "Kyoto Protocol to the United Nations Framework Convention on Climate Change, 10 December 1997," Article 12.

of these mechanisms affords a fuller level of integrity to the process. Unlike the emissions trading schemes, these mechanisms are strengthened with significant internal safeguards to prevent any gaps in their efficient governance.

Clean Development Mechanism Projects are required to have a baseline methodology in order to calculate the expected levels of emissions in the absence of the project. The participating states may use the existing methodology recommended by the UNFCCC or, alternatively, develop their own methodology, which is required to be approved and vetted by the UNFCCC to assess the actual level of emissions recorded pursuant to the implementation of the project. All projects prior to being submitted for registration under the Joint Implementation (JI) or CDM scheme must receive a letter of approval from the host country. The participants of the project must also have a letter of authorization from a party to the Kyoto Protocol.[10] An additional layer of verification has been incorporated through the requirement of independent, third-party certifiers who would attest to the quality of the project proposals and assess the veracity of the claimed emission reductions.[11]

The CDM has been suggested as an instrument to channel climate finance to less-developed nations, thus securing a role for it in the post-Paris regime.[12] It has issued 1.6 billion CERs but is presently waning due to a significant lack of investment as a consequence of declining demand from developing countries.[13] The Green Climate Fund's recent guidelines requiring funding are in consonance with the CDM infrastructure that verifies mitigation activities against baselines, thus making the CDM an effective structural support system for climate finance.[14] The CDM has been attempting to simplify its process for developers so as to lower administrative costs and has recently launched an online platform to voluntarily cancel CERs, thereby seeking to facilitate a new source of demand while promoting a broader use of intended offsetting.[15]

Joint implementation

The Subsidiary Body for Implementation, in its fortieth session held at Bonn from 4–15 June 2014, reviewed the Joint Implementation guidelines.[16] The draft

10 United Nations Framework Convention on Climate Change (UNFCCC), "Clean Development Mechanisms: Designated National Authorities," UNFCC, <https://cdm.unfccc.int/DNA/index.html?click=dna_forum>.
11 United Nations Framework Convention on Climate Change (UNFCCC), "List of DOES," UNFCCC: CDM, <http://cdm.unfccc.int/DOE/list/index.htm>.
12 Ben Garside, "Observers Tout CDM Link to GCF as Market Seeks Future Role," *Carbon Pulse* (2015), <http://carbon-pulse.com/12747/>.
13 Ibid.
14 Ibid.
15 Ibid.
16 United Nations Framework Convention on Climate Change (UNFCCC), "Review of the Joint Implementation Guidelines: Matters Relating to the Mechanisms under the Kyoto Protocol, Bonn, 4–15 June 2014," (FCCC/SBI/2014/L.112014).

conclusions proposed by the chair reviewed the guidelines for the implementation of Article 6 of The Kyoto Protocol. Accordingly, the Conference of the Parties serving as the meeting of the parties to the Kyoto Protocol (CMP) would be able to exercise authority over the Joint Implementation Supervisory Committee (JISC) and provide it with the requisite advice and guidance.[17] The JISC is tasked with setting a minimum standard of technical requirement for the Joint Implementation Activities. This standard would be set after due consultation with the host parties, the stakeholders and the Executive Board of the Clean Development Mechanism, as appropriate.[18]

An additional facilitative measure is the requirement of the JISC to set a requisite criterion for the development of project-cycle procedures by the host parties. This would ensure a level of transparency in the decision-making process. The decisions of the JISC would be taken on the basis of a consensus wherever possible. This affords a greater level of democracy in the decision-making process, facilitating a sense of equity within the body itself. The entire text of the decisions of the JISC shall be made publicly available. The mandate of the JISC includes conducting its work in a way to ensure the transparency and impartiality of its processes and decision-making. This would enable it to be amply safeguarded against any arising conflict of interest.

The Joint Implementation (JI) mechanism in Russia, if effectively capitalized upon, could have used the country's surplus of Assigned Amount Units (AAUs) to attract foreign investment. However, the inconsistent legislation, combined with an inefficient assignment of responsibilities between the various government ministries and a lackadaisical political support structure, remained significant obstructions to the implementation of the scheme.[19] The new JISC recommendations attempt to pave the way forward for robust institutional oversight through credible verification, transparency and public participation.[20]

Market mechanisms for emissions reduction under the new agreement

December 12, 2015, will be remembered as the date the Parties to the UN Framework Convention on Climate Change sought to pave a facilitative way forward for a new climate regime. The Paris Agreement, hoped to integrate elements of a bottom-up approach and to blur the previously distinct demarcation between the obligations of developed and developing countries towards securing a common goal. However, in spite of its universal political endorsement, significant concerns

17 Ibid.

18 Ibid.

19 Igor Shishlov, "Joint Implementation in Russia: On Track to Overtake Brazil as the Third Largest Supplier of Kyoto Offsets," *Climate Brief* 8 (2011), <http://www.cdcclimat.com/Climate-Brief-no8-Joint.html>.

20 Mike Szabo, "Negotiators Learn Lessons for New Markets as Kyoto Mechanism Talks End in Paris," *Carbon Pulse* (2015), <http://carbon-pulse.com/12903/>.

have emerged about the viability of the agreement in delivering concrete results. For example, the focus of the agreement on solely the consumption of fossil fuels and not their production is problematic, as it allows countries to maximize their domestic supply while committing to restrict the demand for the same.[21] The 2°C target has been criticized as being too ambitious, and the possibility of parties reverting on their commitments, despite them having been negotiated in good faith, does spell an uncertain future. Despite these concerns, it is important to understand the pivotal role the agreement will play in defining the future of climate negotiations. It therefore forms the foundation for the articulation on addressing integrity challenges within the new regime.

The Agreement, perhaps for the first time, notes the need for 'climate justice' in ensuring the integrity of ecosystems and taking action to address climate change.[22] At the outset it emphasizes sustainable development as a key component in enabling emission reductions. This is emphasized within Article 2 of the Agreement and is further reiterated within the text.[23] The agreement also acknowledges the obligations that countries have towards the protection of human rights and the need to respect this obligation when taking action to address climate change.[24]

The binding commitments by parties to submit nationally determined contributions (NDC) progressively realized every five years in the form of a 'global stocktake' and to pursue measures to achieve these targets, as well as the call for a mechanism to enable emission reductions in one country to add to another country's NDCs through a mechanism similar to the CDM, is reflective of a form of internationalization.[25] This international approach that seeks to provide a diplomatic back door in the form of a centralized review mechanism almost effectively counterbalances the lack of legally binding provisions.

A major facet of the agreement, which is especially relevant to the discussion on governance structures for market-based mechanisms, is the recognition of the

21 George Monbiot, "Cop-Out," *Guardian (UK)*, 15 December 2015. 'For example, the Government of the U.K through the Infrastructure Act of 2015, is legally obligated to "maximize economic recovery" from oil and gas resources.'

22 See United Nations Framework Convention on Climate Change (UNFCCC), "Conference of the Parties Twenty-First Session Paris, 30 November to 11 December 2015: Adoption of The Paris Agreement," (FCCC/CP/2015/L.9/Rev. 1: United Nations, 2015), Annex. 'Noting the importance of ensuring the integrity of all ecosystems, including oceans, and the protection of biodiversity, recognized by some cultures as Mother Earth, and noting the importance for some of the concept of "climate justice", when taking action to address climate change.'

23 See ibid., Article 2. 'This Agreement, in enhancing the implementation of the Convention, including its objective, aims to strengthen the global response to the threat of climate change, in the context of sustainable development and efforts to eradicate poverty including by . . .' See also ibid., Articles 2(1), 4(1), 6(1, 4a, 9), 7(1), 10(5).

24 See ibid., Annex. 'Acknowledging that climate change is a common concern of humankind, Parties should, when taking action to address climate change, respect, promote and consider their respective obligations on human rights, the right to health, the rights of indigenous peoples, local communities, migrants, children, persons with disabilities and people in vulnerable situations and the right to development, as well as gender equality, empowerment of women and intergenerational equity . . .'

25 Ibid., Article 14.

need for a transparent and accountable process within the clearly established contours of common binding commitments. Countries are required to submit their inventories containing data regarding emissions as well as the 'information necessary to track the progress made in implementing and achieving' the NDCs.[26] With the exception of least-developed countries and small island countries, such reports must be submitted every two years.

Developed countries have also been tasked with the responsibility of reporting on the support provided, and the agreement stipulates that developing countries 'should' report on the support received and their adaptation efforts. This information reported would be made subject to an 'expert technical review', and each party is required to participate in a 'facilitative, multilateral consideration of progress' in the implementation of NDCs and the achievement of targets.[27] Developing countries have been afforded with additional support in the form of capacity-building mechanisms to assist in them meeting their transparency requirements. The scope of the review, as well as the frequency and detail with which it will be undertaken, will remain flexible.

The absence of any direct reference to market-based mechanisms is reflective of the position of the countries which oppose them and have realized their functionality as only a transitional mechanism and not one that can be relied upon in the long run. This echoes an ideological antipathy between the two blocks of nations foreshadowing the re-emergence of a neo-Marxist v. neoliberal polarity within the climate governance regime. For example, advanced developing countries and transition economies have become a primary source of carbon credits traded under The Kyoto Protocol. Over 90 percent of all CERs have been accumulated from China, India, South Korea, Brazil and Mexico.[28] During the Paris process New Zealand presented a carbon market declaration with 18 countries promulgating the need to develop standards and guidelines to inform the environmental integrity of global carbon markets, signalling the resurgence of markets in the emission reduction articulation of the post–2020 period.[29]

26 Ibid., Article 3. 'As nationally determined contributions to the global response to climate change, all Parties are to undertake and communicate ambitious efforts as defined in Articles 4, 7, 9, 10, 11 and 13 with the view to achieving the purpose of this Agreement as set out in Article 2. The efforts of all Parties will represent a progression over time, while recognizing the need to support developing country Parties for the effective implementation of this Agreement.'

27 Ibid., Artricle 13 (11, 12).

28 UNEP DTU Partnership Centre on Energy Climate and Sustainable Development, "UNEP DTU CDM/JI Pipeline Analysis and Database," Centre on Energy Climate and Sustainable Development, <http://www.cdmpipeline.org/>.

29 Tim Groser, "New Zealand Leads Declaration on Carbon Markets at Paris Climate Conference," *Beehive* (13 December 2015), <https://www.beehive.govt.nz/release/new-zealand-leads-declaration-carbon-markets-paris-climate-conference>. See also, United Nations Framework Convention on Climate Change (UNFCCC), "Report of the Conference of the Parties on Its Seventeenth Session, Held in Durban from 28 November to 11 December 2011," (Addendum. Part two: Action taken by the Conference of the Parties at its Seventeenth Session: UN, 2012).

In developing countries, however, the feasibility of instituting efficient monitoring and regulatory systems is doubtful at best. The infrastructural deficiencies do not allow for an effective system of keeping a constant check on the sources of pollution to determine whether they are within the permissible level.[30] Further, the technologies available to the enforcement directorates in developing countries are often outdated and archaic. These agencies are also subject to a number of influential factors ranging from political favours to simple cases of bribery.[31] These lapses in the system of enforcement could prove to be extremely costly in the long run, especially in a market for tradable permits, as the repercussions of these lapses would generate further social costs that would no longer be borne by the firms generating detrimental pollutants.[32]

The Paris Agreement allows parties to make use of 'internationally transferred mitigation outcomes' to implement NDCs.[33] The emphasis on the development of a new alternative to the CDM, which facilitates the offsetting of tradable emissions, reflects the collective intention to gradually phase out the emissions trading scheme. However, the market paints a different picture. The recent decade has marked an exponential rise in carbon trading within domestic and international markets. The European Emissions Trading Scheme reached an overall market value of 67 billion euros. It is predicted that in spite of the recent economic crises, the regional and domestic carbon markets in the Asia Pacific regions, as well as North America would amount to 2 trillion pounds by 2020. The European Union has expressed its intention to foster 'a robust OECD wide carbon market by 2015, which would be extended to economically advanced developing countries by 2020'.

The Agreement seems to be aligned with socially productive objectives such as equitable environmental sustainability, the mitigation of loss and damage and the adaptation of countries, targets that the Kyoto Protocol did not consider. This opens the door to possible deliberations on facilitating international linkages and perhaps creating a centralized system of review for market mechanisms in the near future. This would enable the pursuit of 'co-operative approaches' and the voluntary usage of 'internationally transferred mitigation outcomes' as envisaged under the Agreement, thus enabling the meeting of reduction targets in a transparent and accountable manner.

The new mechanism under the Agreement is to be a successor to the CDM and JI schemes and attempts to find a middle ground between the two. The key point of departure is, however, the lack of territorial restrictions and a move towards

30 Thomas H. Tietenberg, "Tradable Permit Approaches to Pollution Control: Faustian Bargain or Paradise Regained?" in *Property Rights, Economics and the Environment*, ed. M.D. Kaplowitz (Stamford, CT: JAI Press Inc., 1999).

31 Ibid.

32 Ibid.

33 United Nations Framework Convention on Climate Change (UNFCCC), "Conference of the Parties Twenty-First Session Paris, 30 November to 11 December 2015: Adoption of The Paris Agreement," Article 6.

shared responsibility within the new mechanism, thus facilitating an international platform where emission reduction projects can be instituted in developed or developing countries by any other country. Additionally, the new mechanism seeks to use the foundational structure of existing Kyoto mechanisms to bring about an overall reduction of global emissions, as opposed to an offsetting. The provisions to facilitate international transfers and resilient accounting measures reflect a collective sentiment of building this new mechanism on a stronger foundation of integrity-based governance.[34]

Governance arrangements under the new agreement: Issues and options

The success of a comparative mechanism of review and assessment requires an institutional embodiment of transparency and the credibility of the commitments that have been made. The seriousness of a commitment is an important precondition for entering into negotiations and securing agreements with parties that may espouse diverging goals.[35] This is especially true for international negotiations wherein parties act on behalf of sovereign governments who owe their legitimacy to the interest of the public and can only be subject to limited forms of coercion to take actions beyond what might be undertaken ordinarily. A uniform level of transparency on the ex ante pledges undertaken by a party and the ex post outcomes of such pledges can lend a greater level of transparency to the commitments.[36]

Accordingly, an outstanding requirement of maintaining such a level of transparency is that agreements be formulated on the basis of observable factors so that the compliance with the agreement can be objectively assessed. Ultimately, '[w]hat makes many agreements enforceable is only the recognition of future opportunities for agreement that will be eliminated if mutual trust is not created and maintained, and whose value outweighs the momentary gain from cheating in the present instance'.[37]

The importance of institutions in facilitating an unbiased information exchange cannot be overstated. Monitoring the compliance of the parties to the agreement requires the utmost level of transparency, and this can be facilitated in the most unobtrusive manner if the arrangement for the monitoring of the behaviour of other countries incorporates an element of reciprocity within its ambit.[38] The requirement of public information and securing societal confidence in the

34 Ibid., Article 6, paras. 2, 4 and 5.

35 Thomas C. Schelling, "What Makes Greenhouse Sense?" *Foreign Affairs*, no. May/June (2002), <https://www.foreignaffairs.com/articles/2002-05-01/what-makes-greenhouse-sense>.

36 Thomas C. Schelling, *The Strategy of Conflict* (Cambridge, MA: Harvard University, 1980), p. 288.

37 Ibid., pp. 301–2.

38 Scott Barrett, *Environment and Statecraft: The Strategy of Environmental Treaty-Making* (Oxford: Oxford University Press, 2003), p. 284.

processes implemented to build a relationship of trust across international institutions has been extensively validated.[39] An openly accessible information system would therefore not only facilitate a mechanism of assessing the compliance of countries to an agreement but would also instil a sense of faith in the public in the policies that have been incorporated, thus making it easier to elicit their participation and support. A variety of international and non-governmental organizations have taken the initiative to assist policymakers with the process of monitoring. In addition to this, financial assistance and capacity building to facilitate extremely technical mechanisms of data collection would prove to be beneficial in the long run.

The International Monetary Fund (IMF), under its Article VI consultations, enables member countries to consult IMF economists on matters of implementation in accordance with certain preset standards that provide a metric for assessing the performance of a particular policy and identifying the economic environments within which such policies are expected to derive tangible results.[40] Similarly, World Trade Organization (WTO) members are subject to a periodic review of their trade policies, which involves the drafting of a consolidated report by the member country and a corresponding report by a team from the WTO secretariat. The WTO Trade Policy Review provides a template against which the regular reportage from each member country can be adequately structured. The Montreal Protocol surveillance programme undertakes a similar review of national policies supported by well-defined processes and data standards that have in fact strengthened the transparency and monitoring elements of the agreement.[41] The objective of instituting a review mechanism to ensure the compliance of countries with their commitments under the stipulated agreements is to incorporate a sense transparency in the implementation of these commitments. Most often the review and surveillance responsibilities have been delegated to international organizations, which play a definitive role through their unbiased approach.[42] The efficacy of a transparent assessment and comparison depends on the objective analysis of outcomes derived from a superlative understanding of the data collected.[43]

39 Jorgen Wettestad, "Monitoring and Verification," in *The Oxford Handbook of International Environmental Law*, ed. Jutta Brunnée, Daniel Bodansky, and Ellen Hey (Oxford: Oxford University Press, 2007).

40 Armin Schäfer, "A New Form of Governance? Comparing the Open Method of Co-Ordination to Multilateral Surveillance by the IMF and the OECD," *Journal of European Public Policy* 13, no. 1 (2006).

41 Fen Osler Hampson and Michael Hart, *Multilateral Negotiations: Lessons from Arms Control, Trade, and the Environment* (Baltimore: Johns Hopkins University Press, 1995).

42 Alexander Thompson, "Screening Power: International Organizations as Informative Agents," in *Delegation and Agency in International Organizations*, ed. Darren G. Hawkins, David A. Lake, Daniel L. Nielson and Michael J. Tierney (Cambridge: Cambridge University Press, 2006).

43 Abraham Chayes, Antonia Handler Chayes, and Ronald B. Mitchell, "Managing Compliance: A Comparative Perspective," in *Engaging Countries: Strengthening Compliance with International Environmental Accords*, ed. Edith Brown Weiss and Harold Karan Jacobson (Cambridge, MA: MIT Press, 1998).

Another mechanism of facilitating a better understanding of the policy mechanisms of individual countries and their implementation is through providing a forum wherein member states can engage in a system of peer review through exchanging relevant information about design and implementation. These processes of peer review enable a candid dialogue amongst participants, and the resulting 'reciprocal multilateral scrutiny' allows for the enhancement of ambitions through a combined pledge and review approach.

Although a centrally administered carbon market would seem to be the most efficient in reducing emissions, disparate economic, political and social contextual realities within the 190 states would suggest otherwise. In addition to the divergent approaches emerging from this pluralist, multi-ethnic demographic, there is also a growing sense of scepticism in the ability of international norms, regulations and policies to guide domestic decisions within individual states.[44] In this context a decentralized approach, whose efficacy would rest on the governance of domestic institutions and their collaboration with each other, supplemented by an international system of reviewing these domestic mechanisms, would perhaps be more likely to succeed.[45]

In each scenario the common necessity that emerges is, however, the need for structured institutions built on strong foundations that would either oversee the working of a global carbon market or coordinate national efforts, and in doing so administer linkages between bilateral or multilateral networks. An emissions trading system would require an extremely strong and almost infallible regulatory infrastructure.[46] This is because carbon markets are especially sensitive to changes in regulatory frameworks, economic cycles and operational disparities between jurisdictions, as this would affect not only pricing and market operations but also the tenor of future negotiations between countries on similar lines.[47]

Laying the foundations for effective governance

The governance structure of market mechanisms is predicated on the tenor of decisions made by the institutions which oversee the market process. There are three major types of mechanism that are instituted in this respect: mechanisms that are formulated in the pursuance of international agreements and are subject to governance by a duly constituted international body, mechanisms that are

44 Jack L. Goldsmith, Eric A. Posner, and Corporation Ebooks, *The Limits of International Law* (Oxford and New York: Oxford University Press, 2005).

45 David G. Victor, Joshua C. House, and Sarah Joy, "A Madisonian Approach to Climate Policy," *Science* 309, no. 5742 (2005): pp. 1820–21.

46 Ruth Greenspan Bell, "Choosing Environmental Policy Instruments in the Real World," *OECD Global Forum on Sustainable Development: Emissions Trading* (2003), <http://www.oecd.org/environment/cc/2957706.pdf>.

47 Christian Flachsland, Robert Marschinski, and Ottmar Edenhofer, "Global Trading Versus Linking: Architectures for International Emissions Trading," *Energy Policy* 37, no. 5 (2009).

instituted domestically and made subject to a national or subnational authority through a legislation and, lastly, mechanisms that are established through non-government entities and that are voluntary in nature.[48] Market mechanisms which are subject to an international negotiation process are subject to the international agreement they operate under, and their enforcement would depend entirely upon the nature of this agreement. While the Kyoto Protocol is an example of a contractual agreement under which market flexibilities were determined within prescriptive procedures, the future instruments that may be adopted under the UNFCCC may be more facilitative.[49]

The EU-ETS is an example of a domestically administered market mechanism operating in conjunction with the internationally administered institutions under the Kyoto protocol. National regulators may develop tools and specific procedures to facilitate the international linkages, such as the institution of valid compliance units under the EU ETS.[50]

Mechanisms designed by non-governmental institutions are voluntary in nature and are usually based on pre-existing standards. For example, the Verified Carbon Standard (VCS), which makes use of international best practices, requires that all projects be independently audited by certified professionals. This is supplemented by a registry which issues and transparently tracks each Verified Carbon Unit under the oversight of the VCS.[51]

Regulation and monitoring

The viability of permits depends upon the technical prowess of enforcement institutions and their ability to effectively detect violations. Moreover, there is also a need for an efficient legal system that imposes adequate sanctions in the form of monetary penalties on firms that violate the permissible norms of discharge of pollutants.[52] An arrangement of direct and continuous monitoring of emissions is thus imperative to warrant the integrity and credibility of the entire system.

A particular study arrived at the conclusive result that inadequate penalties would result in greater pollution or an increased number of non-compliant

48 Daniel Bodansky, Elliot Diringer, Jonathan Pershing, and Xueman Wang, "Strawman Elements: Possible Approaches to Advancing International Climate Change Efforts," *Climate Dialogue at Pocantico* (Pew Center on Global Climate Change, 2004).

49 Ibid.

50 European Parliament and Council of the European Union, "Directive 2004/101/EC of the European Parliament and of the Council of 27 October 2004 Amending Directive 2003/87/EC Establishing a Scheme for Greenhouse Gas Emission Allowance Trading within the Community, in Respect of the Kyoto Protocol's Project Mechanisms," *Official Journal of the European Union* L 338/18 (2004), <http://eur-lex.europa.eu/legal-content/EN/ALL/?uri=celex:32004L0101>.

51 Verified Carbon Standard (VCS), "The VCS Program," <http//www.v-c-s.org/project/vcs-program/>.

52 Juan-Pablo Montero, "Permits, Standards, and Technology Innovation," *Journal of Environmental Economics and Management* 44, no. 1 (2002).

firms.[53] This reaffirms the fact that the efficacy of the marketable permit system relies on the competence of implementation mechanisms.

China is currently the largest offsetter of carbon units in this respect.[54] China has also instituted a system of emissions fees as a further economic measure of curbing its rapidly increasing carbon levels. It has set up a rebate system where 80 percent of the fee collected for emissions would be returned to the enterprise if it invests in pollution control mechanisms.[55]

In 2011 the Chinese government experimented with seven sector-specific carbon markets, with a view to facilitate an expansive emissions trading scheme covering billions of tonnes of carbon dioxide.[56] Each of these markets has different procedures and varied compliance strategies. However, the institution of an absolute cap within these schemes is problematic because although the emissions of each firm will reduce, under this cap the overall emissions may still increase with an increase in the number of industries.[57] Further, China's political structure has left little room for any past experience with this sort of trading, and therefore progress within this scheme has been slow.[58] Issues such as the lack of market knowledge and the transparency within the scheme have left firms apprehensive about engaging in such trade. Companies have also been unable to provide verifiable records of their past emissions, making it difficult for regulators to set a cap. More importantly, the regulators do not have adequate verification and monitoring tools to check the estimates that have been provided by the companies, making it all the more difficult to prevent instances of corruption and fraud.[59]

In spite of these issues China has announced a plan to launch a national market in 2016. It could potentially be the largest emissions trading scheme and would require adequate regulatory mechanisms to ensure its efficient functioning. Bereft of these underlying monitoring instruments and a strong legislation, the outcome of this scheme remains doubtful at best.[60]

Analysis and discussion of governance values and integrity systems in the old and new climate agreements

The original Public Institutional Justification (PIJ) of an emissions trading scheme, as outlined within the Kyoto Protocol, had been to provide parties with

53 Andrew G. Keeler, "Noncompliant Firms in Transferable Discharge Permit Markets: Some Extensions," ibid.21, no. 2 (1991).
54 Global Carbon Project, "Global Carbon Atlas" (2015), http://www.globalcarbonatlas.org/
55 Ibid.
56 Carbon Brief, "Analysis: China's Big Carbon Market Experiment," (2014), <http://www.carbonbrief.org/analysis-chinas-big-carbon-market-experiment>.
57 Ibid.
58 The Economist Intelligence Unit, "Trials & Tribulations: China Experiments with Carbon Trading," (2013), <http://www.eiu.com/public/topical_report.aspx?campaignid=ChinaCarbon>.
59 Liu Shuang and Xu Nan, "Data Gaps Hobble Carbon Trading," China Dialogue, https://<www.chinadialogue.net/article/5093-Data-gaps-hobble-carbon-tradin>.
60 Keeler, "Noncompliant Firms in Transferable Discharge Permit Markets: Some Extensions."

a degree of flexibility to ensure their optimal compliance with assigned emission caps. This PIJ has, however, morphed into an economic incentivization for countries to meet their emission reduction targets in a profitable manner while sustaining industrial growth. The Internal Organizational Arrangements (IOA) that enabled the metamorphosis of this PIJ from an optional flexibility to a profitable market incentive need to be examined further.

The IOAs include the decision-making processes of the institutions that oversee and regulate the emissions trading scheme; and the structure they use for the monitoring, review and verification of those decisions, as well as the mechanisms in place to ensure the compliance of parties with those decisions. This is predicated on its policy on transparency and accountability supplemented by a mechanism of critical feedback.[61]

The governance of an emissions trading scheme, as envisaged under the Kyoto Protocol, may be categorized as 'thin' to 'thickish'. The baseline is fixed at the achievement of its original PIJ (that is, flexibility to ensure optimal compliance) through the realization of the thinnest governance values such as democracy, rule of law, due process, transparency and accountability. The Kyoto Protocol is centred on these values and does not aspire to achieve thicker values. The progression within the text of the Kyoto Protocol is thickish at best and is characterized by its mandate to mitigate the exponential rise in the level of emissions through durable, sustainable solutions, but not necessarily through the embodiment of thicker governance values. The focus of the Protocol on thinner governance values restricted the ability of institutions to achieve fuller integrity. For example, the complete lack of mention of normative values such as climate justice, environmental equity and human rights in the text of the Kyoto Protocol resulted in the lack of a foundation from which an international system of review for market mechanisms could have otherwise drawn its legitimacy. The IOAs were therefore not strong enough to ensure that the original PIJ of the emissions trading scheme was met.

While an ideal government-endorsed initiative to facilitate emission reduction, such as carbon taxation or the Polluter Pays principle, may hold normative ideals of equity, climate justice and human rights as key objectives of its mandate, market mechanisms under the Protocol tend to prioritize thinner values of procedural efficacy and accountability in accordance with the rule of law over such normatively thick values – the reason for this hierarchization being that market mechanisms in their current form are predicated more on securing economic returns than on achieving tangible emission reductions.

The recent deliberations on climate integrity surrounding the Paris Agreement have tended towards the achievement of the thickest conceptualizations of achievable targets in the form of environmental equity, climate justice, human rights and sustainable development.[62] In this context market mechanisms in

61 Hugh Breakey and Tim Cadman, "Governance Values and Institutional Integrity," in *Governing the Climate Change Regime*, ed. Charles Sampford, Tim Cadman, and Rowena Maguire (London: Taylor & Francis, 2016 (this volume)), Chapter 2.

62 Ibid.

their current form, as under the Protocol, raise legitimate concerns with regard to their durability and effectiveness in achieving actual results in the post-Paris regime. There is, therefore, a need to strengthen the IOAs within emissions trading schemes so that they may be able meet their original PIJ and are consistent with the PIJ of the Paris Agreement.

The wording of the Paris Agreement is tempered with specific measures to increase the level of coherence integrity to an almost full level, as evidenced by the demonstrably thicker values endorsed by the Parties to the Agreement and their committed relationship to fulfilling the PIJ of the Agreement.[63] The reiteration of an equitable framework is symptomatic of commitments that are common but that take into account the disparate capabilities of countries and the individual and collective contextual obligations of governments.[64] For the first time an enhanced means of ensuring transparency has been instituted by stipulating that domestic systems of monitoring, review and verification be made subject to international review; a significant step forward in the internationalization of climate-change mitigation efforts.[65] Thus, the IOAs for emissions trading schemes within the agreement have a much stronger foundation built around thicker governance values than they do within the Kyoto Protocol.

The strengthening of the IOAs through the facilitative text of the new agreement, if appropriately capitalized upon, could enable emissions trading schemes to consistently realize their original PIJ, not as a profitable market-driven scheme but as a flexibility option to facilitate optimal compliance. The best way forward to achieve this object is therefore to use the strong IOAs within the new agreement to derive an international standard of review for emissions trading schemes.

The way forward: Towards an international standard?

The previous distinction between international and domestic systems of monitoring, review and verification (MRV) has been blurred by the facilitative approach undertaken by the Paris Agreement. This overlap in governance sets an important precedent that informs the establishment of international MRV mechanisms and

63 Andrew Prag and Gregory Briner, "Crossing the Threshold: Ambitious Baselines for the UNFCCC New Market-Based Mechanism," *OECD/IEA Climate Change Expert Group Papers* No. 2012/02 (2012), <http://www.oecd.org/env/cc/50315387.pdf>.

64 See United Nations Framework Convention on Climate Change (UNFCCC), "Conference of the Parties Twenty-First Session Paris, 30 November to 11 December 2015: Adoption of The Paris Agreement," Annex, Articles 4.1 and 14.1. '. . . acknowledging that climate change is a common concern of humankind, Parties should, when taking action to address climate change, respect, promote and consider their respective obligations on human rights, the right to health, the rights of indigenous peoples, local communities, migrants, children, persons with disabilities and people in vulnerable situations and the right to development, as well as gender equality, empowerment of women and intergenerational equity.' See also Article 2(2). 'This Agreement will be implemented to reflect equity and the principle of common but differentiated responsibilities and respective capabilities, in the light of different national circumstances'.

65 Ibid., Article 13.

their interaction with domestic and subnational systems. In this context, countries may seek an international recognition of their domestic standards with a view to meet their international commitments.[66]

A multilateral dialogue would enable the determination of these international standards, taking cues from the domestic governance structures across the EU and the US which have achieved considerable success in a short span of time in the regulation of market mechanisms.[67] The development of criterion and qualifications for international recognition under the UNFCCC process would require a consensus-based decision at the Twenty-second Conference of the Parties in Marrakesh. This would encourage the submission of domestically undertaken reviews and verifications of market mechanisms to an international permanent body of review (whose composition may be determined on a rotational basis), which would assess and verify whether these domestic audits meet the criteria for international recognition. Once these international standards have been met, the domestic or regional market mechanism would be cleared for operation. The verification process would ensure that domestically governed mechanisms meet the certification criteria of an internationally recognized body, thereby enabling the institution of a pervasive standard of accountability and transparent trading system.

In the event of no such legally binding agreement being reached, however, it would be beneficial for parties to agree upon certain disclosure and transparency requirements.[68] This would facilitate greater cooperation amongst nations and allow for the establishment of international linkages based on an equitable approach under which each country would subject its domestically governed mechanism to the scrutiny of the international community.

This disclosure would enable an assessment of the existing governance mechanisms within each jurisdiction and a rating of the same by an internationally recognized agency.[69] The international agency would rate the information provided by countries on parameters of accuracy, transparency and efficacy of the scheme in meeting emissions targets. This form of rating would induce countries to submit accurate information, better their own governance mechanisms and ensure the optimal functioning of their emissions trading schemes so as to gain political traction and international recognition. The investment of the private sector in carbon market mechanisms would also be informed through this system of rating.[70] Although there would be no regulatory or enforcement body under this system, it would still instil a sense of compliance with a view to secure international good

66 Prag and Briner, "Crossing the Threshold: Ambitious Baselines for the UNFCCC New Market-Based Mechanism."
67 European Parliament., "Commission Decision on Determining Member States' Annual Emission Allocations for the Period from 2013 to 2020 Pursuant to Decision No 406/2009/EC" (D023442/01–2012/2859(RPS)2012)).
68 Prag and Briner, "Crossing the Threshold: Ambitious Baselines for the UNFCCC New Market-Based Mechanism."
69 Ibid.
70 Ibid.

faith and attract foreign investment. This form of soft governance would perhaps lend an acceptable level of integrity to the existing domestic institutions and to enable their effective oversight of emissions trading schemes.

An essential prerequisite for a governance structure to successfully oversee the facilitation of an emissions trading mechanism is its ability to enforce compliance procedures. Especially in the case of an Emissions Trading Scheme, the institution responsible for enforcing the rules of the emissions trading mechanism needs to be empowered with the legitimacy to impose penalties for non-compliance. The only previously existing international mechanism under which such legitimacy was sought was the enforcement branch of the Compliance Committee of the Kyoto Protocol. The Committee was tasked with the responsibility of enforcing the compliance of countries with their commitments made under the Protocol and was empowered to impose penalties in the case of non-compliance. The Committee could also suspend the participation of certain countries in the flexibility mechanisms envisaged in the protocol, if required.[71] The Paris Agreement has, however, been successful in establishing a more concrete enforcement mechanism. This new mechanism comprises a committee of experts, and in order to remain facilitative in nature, it will operate in a 'non-adversarial and non-punitive manner'. The committee will submit an annual report to the COP, the details of which will be formulated after the agreement comes into force.

Conclusion

This chapter has characterized the reasons for the significant gaps in the governance of market mechanisms, and through the narrative presented possible solutions to address these shortcomings. The Kyoto Protocol, although sound in its treatment of flexibilities within the climate regime, fell short of providing a durable mechanism to regulate and oversee the facilitation of market mechanisms. This can be attributed to its focus on thinner governance values such as procedural compliance and efficacy. The Paris Agreement, by pegging the target for the new climate regime at the achievement of a fuller level of integrity, paves the way forward for the development of a more robust structure of regulation and oversight, especially with regard to market mechanisms.

It is evident that in the post-Paris regime targeting thicker notions of governance would be required in order to achieve durable solutions that aid in emission reductions. Market-driven schemes tend to work from a baseline of thinner values centred on procedural efficacy. Thus, while market mechanisms may still be used as a transitional measure to aid countries with meeting their initial commitment, it has become integral to facilitate their enhanced regulation on an international level. The proposal for the internationalization of market

71 United Nations Framework Convention on Climate Change (UNFCCC). "Rules of Procedure of the Compliance Committee of the Kyoto Protocol (informal consolidated version)" (version of 3 February 2014). Available at: https://unfccc.int/files/kyoto_protocol/compliance/application/pdf/consolidated_rop_with_cmp_4&cmp9_amend_2014feb03.pdf

mechanism regulation provides a resolute foundation upon which the new sustainable mechanism, envisaged by the Paris Agreement, can be put into effect. Collaborative multilateral dialogue between nations would be integral in establishing this system of regulatory oversight. The new mechanism would benefit significantly from the creation of this international supervisory structure, as it would facilitate the achievement of a fuller level of integrity within the contours of an inherently thinly spread scheme.

It is essential for international policy formulations and modalities within the post-Paris regime to focus on collaborative initiatives and the collective achievement of common targets. It is only through the adoption of a facilitative mechanism anchored in cooperation, trust and a commitment to the same goal that tangible results will be achieved. It is therefore imperative that countries use the framework of the Paris Agreement and its spirit of responsive alliance to achieve the outstanding goal of a fuller level of integrity in the governance of market mechanisms. This would secure a more durable future for market mechanisms and transform them into reliable options of emission reduction as opposed to simply economic incentives to ease the burden of previously made commitments.

References

Barrett, Scott. *Environment and Statecraft: The Strategy of Environmental Treaty-Making*. Oxford and New York: Oxford University Press, 2003.

Bodansky, Daniel, Elliot Diringer, Jonathan Pershing, and Xueman Wang. "Strawman Elements: Possible Approaches to Advancing International Climate Change Efforts." *Climate Dialogue at Pocantico*. Pew Center on Global Climate Change, 2004.

Breakey, Hugh, and Tim Cadman. "Governance Values and Institutional Integrity." Chap. 2 In *Governing the Climate Change Regime*, edited by Charles Sampford, Tim Cadman and Rowena Maguire. London: Taylor & Francis, 2016 (this volume).

Carbon Brief. "Analysis: China's Big Carbon Market Experiment." (2014). <http://www.carbonbrief.org/analysis-chinas-big-carbon-market-experiment>.

Chayes, Abraham, Antonia Handler Chayes, and Ronald B Mitchell. "Managing Compliance: A Comparative Perspective." In *Engaging Countries: Strengthening Compliance with International Environmental Accords*, edited by Edith Brown Weiss and Harold Karan Jacobson, 39–62. Cambridge, MA: MIT Press, 1998.

The Economist Intelligence Unit. "Trials & Tribulations: China Experiments with Carbon Trading." (2013). <http://www.eiu.com/public/topical_report.aspx?campaignid=ChinaCarbon> (accessed 22 May 2016).

Ellerman, A. Denny, and Barbara K. Buchner. "The European Union Emissions Trading Scheme: Origins, Allocation, and Early Results." *Review of Environmental Economics and Policy* 1, no. 1 (2007): 66–87.

European Parliament, and Council of the European Union. "Directive 2004/101/EC of the European Parliament and of the Council of 27 October 2004 Amending Directive 2003/87/EC Establishing a Scheme for Greenhouse Gas Emission Allowance Trading within the Community, in Respect of the Kyoto Protocol's Project Mechanisms." *Official Journal of the European Union* L 338/18(2004). <http://eur-lex.europa.eu/legal-content/EN/ALL/?uri=celex:32004L0101> (accessed 22 May 2016).

Flachsland, Christian, Robert Marschinski, and Ottmar Edenhofer. "Global Trading Versus Linking: Architectures for International Emissions Trading." *Energy Policy* 37, no. 5 (2009): 1637–47.

Garside, Ben. "Observers Tout CDM Link to GCF as Market Seeks Future Role." *Carbon Pulse* (2015). Published Electronically 1 December 2015. <http://carbon-pulse.com/12747/> (accessed 22 May 2016).

Global Carbon Project. "Global Carbon Atlas." 2015. <http://www.globalcarbonatlas.org/>

Goldsmith, Jack L., Eric A. Posner, and Corporation Ebooks. *The Limits of International Law*. Oxford: Oxford University Press, 2005.

Greenspan Bell, Ruth. "Choosing Environmental Policy Instruments in the Real World." *OECD Global Forum on Sustainable Development: Emissions Trading* (2003). <http://www.oecd.org/environment/cc/2957706.pdf> (accessed 22 May 2016).

Groser, Tim. "New Zealand Leads Declaration on Carbon Markets at Paris Climate Conference." *Beehive* (13 December 2015). <https://www.beehive.govt.nz/release/new-zealand-leads-declaration-carbon-markets-paris-climate-conference> (accessed 22 May 2016).

Hampson, Fen Osler, and Michael Hart. *Multilateral Negotiations: Lessons from Arms Control, Trade, and the Environment*. Baltimore: Johns Hopkins University Press, 1995.

Keeler, Andrew G. "Noncompliant Firms in Transferable Discharge Permit Markets: Some Extensions." *Journal of Environmental Economics and Management* 21, no. 2 (1991): 180–89.

Krier, James E., and Robert B. Bell. "Old Economics and New Environmental Controls." *Stanford Lawyer* 15, no. 1 (1979–80): 15–20.

Monbiot, George. "Cop-Out." *Guardian (UK)*, 15 December 2015.

Montero, Juan-Pablo. "Permits, Standards, and Technology Innovation." *Journal of Environmental Economics and Management* 44, no. 1 (2002): 23–44.

Parker, Larry. "Climate Change: Design Approaches for a Greenhouse Gas Reduction Program." *Congressional Research Service*, 1–24. Washington: 2008.

Parliament, European. "Commission Decision on Determining Member States' Annual Emission Allocations for the Period from 2013 to 2020 Pursuant to Decision No 406/2009/EC." D023442/01–2012/2859(RPS), 2012.

Prag, Andrew, and Gregory Briner. "Crossing the Threshold: Ambitious Baselines for the UNFCCC New Market-Based Mechanism." *OECD/IEA Climate Change Expert Group Papers* No. 2012/02(2012). <http://www.oecd.org/env/cc/50315387.pdf> (accessed 22 May 2016).

Schäfer, Armin. "A New Form of Governance? Comparing the Open Method of Co-Ordination to Multilateral Surveillance by the IMF and the OECD." *Journal of European Public Policy* 13, no. 1 (2006): 70–88.

Schelling, Thomas C. *The Strategy of Conflict*. Cambridge, MA: Harvard University, 1980.

———. "What Makes Greenhouse Sense?" *Foreign Affairs*, no. May/June (2002). <https://www.foreignaffairs.com/articles/2002–05–01/what-makes-greenhouse-sense> (accessed 22 May 2016).

Shishlov, Igor. "Joint Implementation in Russia: On Track to Overtake Brazil as the Third Largest Supplier of Kyoto Offsets" *Climate Brief* 8 (2011). <http://www.cdcclimat.com/Climate-Brief-no8-Joint.html> (accessed 22 May 2016).

Shuang, Liu, and Xu Nan. "Data Gaps Hobble Carbon Trading." *China Dialogue*. <https://www.chinadialogue.net/article/5093-Data-gaps-hobble-carbon-trading> (accessed 22 May 2016).

Stranlund, John K., Carlos A. Chavez, and Barry C. Field. "Enforcing Emissions Trading Programs: Theory, Practice and Performance." *Policy Studies Journal* 30, no. 3 (2002): 343–61.

Szabo, Mike. "Negotiators Learn Lessons for New Markets as Kyoto Mechanism Talks End in Paris." *Carbon Pulse* (2015). Published electronically 4 December. <http://carbon-pulse.com/12903/> (accessed 22 May 2016).

Thompson, Alexander. "Screening Power: International Organizations as Informative Agents." In *Delegation and Agency in International Organizations*, edited by Darren G. Hawkins, David A. Lake, Daniel L. Nielson and Michael J. Tierney, 229–54. Cambridge: Cambridge University Press, 2006.

Tietenberg, Thomas H. "Economic Instruments for Environmental Regulation." *Oxford Review of Economic Policy* 6, no. 1 (1990): 17–33.

———. "Tradable Permit Approaches to Pollution Control: Faustian Bargain or Paradise Regained?" In *Property Rights, Economics and the Environment*, edited by M.D. Kaplowitz. 167–93. Stamford, CT: JAI Press Inc., 1999.

UNEP DTU Partnership Centre on Energy Climate and Sustainable Development. "UNEP DTU CDM/Ji Pipeline Analysis and Database." Centre on Energy Climate and Sustainable Development, <http://www.cdmpipeline.org/> (accessed 22 May 2016).

United Nations Framework Convention on Climate Change (UNFCCC). "Clean Development Mechanisms: Designated National Authorities." UNFCC, <https://cdm.unfccc.int/DNA/index.html?click=dna_forum> (accessed 26 May 2016).

———. "Conference of the Parties Twenty-First Session Paris, 30 November to 11 December 2015: Adoption of The Paris Agreement." FCCC/CP/2015/L.9/Rev. 1: United Nations, 2015.

———. "Kyoto Protocol to the United Nations Framework Convention on Climate Change, 10 December 1997." U.N. Doc FCCC/CP/1997/7/Add.1, 37 I.L.M.: United Nations, 1997.

———. "List of DOES." UNFCCC: CDM. <http://cdm.unfccc.int/DOE/list/index.htm> (accessed 22 May 2016).

———. "Report of the Conference of the Parties on Its Seventeenth Session, Held in Durban from 28 November to 11 December 2011." Addendum. Part two: Action taken by the Conference of the Parties at its Seventeenth Session: UN, 2012.

———. "Review of the Joint Implementation Guidelines: Matters Relating to the Mechanisms under the Kyoto Protocol, Bonn, 4–15 June 2014." FCCC/SBI/2014/L.11, 2014.

Verified Carbon Standard (VCS). "The VCS Program." <http://www.v-c-s.org/project/vcs-program/> (accessed 22 May 2016).

Victor, David G., Joshua C. House, and Sarah Joy. "A Madisonian Approach to Climate Policy." *Science* 309, no. 5742 (2005): 1820–21.

Wettestad, Jorgen. "Monitoring and Verification." In *The Oxford Handbook of International Environmental Law*, edited by Jutta Brunnée, Daniel Bodansky and Ellen Hey, 974–94. Oxford: Oxford University Press, 2007.

Part III

Improving governance within the UNFCCC

9 Enhancing good governance within the international climate regime through human rights principles

Bridget Lewis

Introduction

In addition to being a challenging issue for the fields of science, economics and politics, climate change has in recent years been increasingly acknowledged as a human rights issue due to the wide-ranging and serious consequences it has for individuals and communities. The effects of climate change, such as rising sea levels, increasingly severe and frequent extreme weather events and expansion of vector-borne diseases and pests, threaten to affect people's fundamental human rights, including the rights to health, food, water, housing, self-determination and even life itself. Further, these impacts and the costs of addressing them have the potential to limit governments' capacity to protect and fulfil a wide range of rights to which their citizens are entitled.

Given the significance of the human impacts of climate change, human rights–based approaches to climate change have often been suggested as an appropriate means of shaping our responses to climate change.[1] One benefit of such

1 See Meinhard Doelle, "Climate Change and Human Rights: The Role of International Human Rights in Motivating States to Take Climate Change Seriously," *Macquarie Journal of International and Comparative Environmental Law* 1, no. 2 (2004); Sara C. Aminzadeh, "A Moral Imperative: The Human Rights Implications of Climate Change," *Hastings International and Comparative Law Review* 30, no. 2 (2006–2007); Sumudu Atapattu, "Global Climate Change: Can Human Rights (and Human Beings) Survive This Onslaught?," *Colorado Journal of International Environmental Law and Policy* 20, no. 1 (2009); Derek Bell, "Does Anthropogenic Climate Change Violate Human Rights," *Critical Review of International Social and Political Philosophy* 14, no. 2 (2011); Michael Depledge and Cinnamon Carlane, "Sick of the Weather: Climate Change, Human Health and International Law," *Environmental Law Review* 9, no. 4 (2007); Stephen Humphreys, ed. *Human Rights and Climate Change* (Cambridge: Cambridge University Press, 2010); John Knox, "Linking Human Rights and Climate Change at the United Nations," *Harvard Environmental Law Review* 33 (2009); John Knox, "Climate Change and Human Rights Law," *Virginia Journal of International Law* 50, no. 1 (2009–2010); Marc Limon, "Human Rights and Climate Change: Constructing a Case for Political Action," *Harvard Environmental Law Review* 33, no. 2 (2009); Ole W. Pedersen, "Climate Change and Human Rights: Amicable or Arrested Development?" *Journal of Human Rights and the Environment* 1, no. 2 (2010); Daud Hassan and Awal Khan, "Climate-Change-Related Human Rights Violations," *Environmental Law and Policy* 43, no. 2 (2013); Pamela Stephens, "Applying Human Rights Norms to Climate Change: The Elusive Remedy," *Colorado Journal of International Environmental Law and Policy* 21, no. 1 (2010).

approaches is that they focus our attention on the human impacts of climate change, and particularly on those who are most vulnerable to those impacts and most in need of international assistance.[2] Another advantage of human rights–based approaches is that they draw on a framework of widely accepted human rights laws, which can be utilised to reinforce calls for stronger state action.[3]

Human rights law is not only engaged by the consequences of climate change, however. It is also relevant to the way we try to address it. The mitigation and adaptation measures which states put in place in an effort to limit the negative impacts of climate change also have the potential to impact human rights. Changes to land use or energy production, for example, have the potential to impact people's subsistence rights, while relocation of individuals or communities can have wide-ranging negative consequences. A human rights–based approach to climate change, therefore, requires that we consider the human rights impacts not only of climate change itself, but also of our responses to it.[4]

For the global climate change regime, which consists of a framework of international agreements, mechanisms and organisations, human rights law and principles are clearly relevant to the substantive work of addressing climate change. However, there is also a need for human rights values to inform the internal processes of the regime itself. One reason for this flows from the disparities in vulnerability, capacity and power which are evident among nation states. There is an injustice inherent in the fact that the states which are most vulnerable to climate change are simultaneously least responsible for greenhouse gas emissions and least capable of dealing with their impact.[5] This injustice is aggravated by the

2 Siobhan McInerney-Lankford, Mac Darrow, and Lavanya Rajamani, *Human Rights and Climate Change: A Review of the International Legal Dimensions* (Washington, DC: World Bank, 2011), p. 27; Doelle, "Climate Change and Human Rights: The Role of International Human Rights in Motivating States to Take Climate Change Seriously," p. 166; Pedersen, "Climate Change and Human Rights: Amicable or Arrested Development?" p. 248; Daniel Bodansky, "Climate Change and Human Rights: Unpacking the Issues," *Georgia Journal of International and Comparative Law* 38, no. 3 (2010): p. 517; Jane McAdam, *Climate Change, Forced Migration and International Law* (Oxford: Oxford University Press, 2012); Ibrahim Wani, "Open-Ended Consultation on the Relationship between Climate Change and Human Rights: Summary of Discussions," (Geneva: OHCHR, 22 October 2008), p. 2; Limon, "Human Rights and Climate Change: Constructing a Case for Political Action," p. 451.

3 Pedersen, "Climate Change and Human Rights: Amicable or Arrested Development?" p. 248; Stephens, "Applying Human Rights Norms to Climate Change: The Elusive Remedy," p. 200; Dalindyebo Shabalala, "Contribution to Human Rights Council Panel Discussion on the Relationship between Climate Change and Human Rights: Summary of Discussions," (Geneva: Human Rights Council 15 June 2009), p. 4 [12].

4 Bridget Lewis, "Balancing Human Rights in Adaptation and Mitigation Policies," in *Climate Change and Human Rights, an International Law Perspective*, ed. Ottavio Quirico and Mouloud Boumghar (London: Routledge, 2015); Ole W. Pedersen, "The Janus-Head of Human Rights and Climate Change: Adaptation and Mitigation," *Nordic Journal of International Law* 80, no. 3 (2011).

5 Edward Cameron, "Human Rights and Climate Change: Moving from an Intrinsic to an Instrumental Approach," *Georgia Journal of International and Comparative Law* 38, no. 3

power imbalance present in international climate negotiations. Despite efforts by small-island developing states and other vulnerable groups to lobby collectively for greater attention and stronger action, it remains a fact that the highest-polluting and richest states possess significant power when it comes to setting the agenda for climate action.

In this context, mechanisms are required within the global climate regime to ensure that action taken to address climate change is both effective and equitable. Governance values such as accountability, transparency, democracy and compliance with the rule of law are fundamental to ensuring not only that the regime addresses the needs of all affected parties and values their perspectives, but also capacity of the regime to achieve its objectives is improved. Without adequate measures to ensure these values are upheld, the global climate regime risks losing legitimacy among international actors, which in turn threatens to further undermine its effectiveness.

This chapter argues that the legitimacy of the climate regime can be enhanced through greater consideration and implementation of human rights values, and particularly by harnessing the links between human rights and good governance. The relationship between human rights, governance, integrity and legitimacy is complex, involving numerous overlapping principles and mutually supporting objectives. There are many points at which human rights can influence an institution's structures and processes as well as its substantive work, and in some cases the links can be difficult to disentangle. This chapter argues, however, that the more an institution incorporates human rights principles into its objectives and processes, the greater its potential to employ good governance, and ultimately to achieve both integrity and legitimacy.

The values and objectives of good governance and human rights are sympathetic and mutually supportive. It has long been understood that poor governance, corruption and waste undermine the protection of human rights.[6] Such conditions limit a government's capacity to provide services to vulnerable individuals and groups by diverting both attention and resources elsewhere. Further, corruption operates to restrict equal access to the law and to deny people the right to participate meaningfully in decisions which affect them. Good governance is therefore crucial to establishing conditions conducive to the fulfilment of human rights. This is true for institutions as well as for governments. An institution which aims

(2010): p. 708. See also Limon, "Human Rights and Climate Change: Constructing a Case for Political Action," p. 451; Jon Barnett, "Human Rights and Vulnerability to Climate Change," in *Human Rights and Climate Change*, ed. Stephen Humphreys (Cambridge: Cambridge University Press, 2010), p. 257; Rowena Maguire and Bridget Lewis, "The Influence of Justice Theories on International Climate Policies and Measures," *Macquarie Journal of International and Comparative Environmental Law* 8, no. 1 (2012).

6 Office of the High Commissioner for Human Rights (OHCHR), *Good Governance Practices for the Protection of Human Rights* (Geneva: United Nations, 2007); Office of the High Commissioner for Human Rights (OHCHR), *The Human Rights Case against Corruption* (Geneva: United Nations, 2013).

to contribute to the fulfilment of human rights will be ineffective if its structures and processes are not guided by the values of good governance.

Good governance is therefore a precondition to the effective protection of human rights. At the same time, human rights principles provide a set of values which can guide a state or institution towards stronger governance, and the framework of international human rights law serves to strengthen this relationship through supervision and enforcement mechanisms. This chapter seeks to explore the relationship between human rights and good governance, and to identify in more detail the ways in which human rights law and principles can help guide the global climate regime towards better institutional governance and consequently greater legitimacy.

The chapter develops an argument presented in an earlier collection which focussed on the institutional integrity of the global climate regime.[7] There it was argued that human rights principles can be utilised to enhance the integrity of the regime, that is, to improve the extent to which the regime lives up to its own objectives and standards. If we accept that climate change has negative consequences for the enjoyment of human rights, then the objective of the regime to limit the impacts of climate change must encompass, at least implicitly, the protection of human rights. Human rights therefore form part of the 'public institutional justification' (PIJ) of the regime, to use the language of the Comprehensive Integrity Framework developed by Breakey, Cadman and Sampford, which defines an institution's PIJ as the public statement of its values, commitments, principles and goals.[8] Given that the PIJ of the climate regime includes the protection of human rights, the regime can advance that objective by adhering to human rights law and principles. This would have the effect of enhancing the regime's 'consistency-integrity' – that is, the degree to which its activities facilitate achieving its PIJ.

Further, even where an institution does not include human rights within its PIJ, human rights principles can operate to improve its integrity by forming part of the organisation's 'governing values'. These are the substantive values and priorities which shape an institution's decision-making, and are distinguished from its 'governance values.'[9] Principles like non-discrimination, equality, democracy

7 Bridget Lewis, "The Contribution of Human Rights to the Effectiveness and Integrity of the Global Carbon Regime," in *Ethical Values and the Integrity of the Climate Change Regime*, ed. Hugh Breakey, Vesselin Popovski, and Rowena Maguire (Surrey: Ashgate, 2015), p. 179. A companion piece to the current chapter.

8 Hugh Breakey and Timothy Cadman, "A Comprehensive Framework for Evaluating the Integrity of the Climate Regime Complex," in *Ethical Values and the Integrity of the Climate Change Regime*, ed. Hugh Breakey, Vesselin Popovski, and Rowena Maguire (Surrey: Ashgate, 2015), p. 17; Hugh Breakey, Timothy Cadman, and Charles Sampford, "Conceptualizing Personal and Institutional Integrity: The Comprehensive Integrity Framework," *Research in Ethical Issues in Organizations* 14 (2015): p. 1.

9 See Chapter 2, this volume: Hugh Breakey, Timothy Cadman, and Charles Sampford, "Governance Values and Institutional Integrity," in *Governing the Climate Change Regime*, ed. Charles Sampford, Tim Cadman, and Rowena Maguire (London: Taylor & Francis, 2016 (this volume)).

and due process are all fundamental to human rights, and when these values are adopted by an institution they all contribute decision-making which is more equitable and accountable. In this way, adherence to human rights principles can advance the institution's 'coherence-integrity' – the degree to which its internal structures and processes allow it to achieve its PIJ, even where that PIJ is not directly related to the protection of human rights.[10]

While this earlier work focussed on the ways in which human rights principles contribute to the integrity of an institution in working towards fulfilling its PIJ, this chapter explores in more detail the contribution of human rights to achieving legitimacy by assisting the institution to employ good governance in its structures and processes. The chapter employs the framework for analyzing governance developed by Breakey, Cadman and Sampford[11] to outline the key ways in which human rights principles can contribute to good governance within institutions by helping to embed stronger (or 'thicker') governance values. The thick values identified by Breakey, Cadman and Sampford include equality, democracy and inclusiveness.[12] They go beyond the basic (thin) governance principles of accountability and transparency to encompass richer values relating to meaningful participation, democratic decision-making and dispute resolution. When fully implemented, these thick governance values can lead to effective problem-solving and durable behavioural change within an institution.[13]

This chapter identifies that human rights principles include many of these same values, or values which have similar content and purpose. Such principles include non-discrimination, the rights to information and participation, equality before the law, due process and democracy, and they underpin the promotion and protection of the full range of rights articulated in international human rights law. The chapter argues that an institution which commits to respecting these human rights principles in its activities has greater potential to employ thick governance in its internal structures and processes by virtue of the overlap between the two sets of principles and their mutually supportive relationship. Further, where an institution explicitly includes human rights values in its PIJ or its governing values, it is more likely to seek to employ thick governance measures in its operations, as doing so will help it to avoid undermining its legitimacy while making it more effective in achieving its PIJ.

After explaining these general connections between human rights and good governance, the chapter then looks to the climate regime in particular. It argues that there are two ways in which human rights values are already part of the climate regime's PIJ: first, the need to respect human rights in the implementation of climate change mitigation and adaptation measures is explicitly referred to in a number of climate instruments; and second, human rights can be implicitly incorporated into our understanding of the notion of 'harmful emissions'.

10 Ibid.
11 Ibid.
12 Ibid.
13 Ibid.

Given that human rights are part of the PIJ of the climate regime, Breakey, Cadman and Sampford's framework can be applied to argue that thick governance is required within the regime not only to ensure that those objectives are achieved, but also to maintain the regime's overall legitimacy. The chapter concludes by identifying a number of ways in which the use of human rights principles in dealing with climate change could be enhanced. It argues that future development of the climate regime would benefit from more explicit inclusion of human rights principles into both the governing values of the regime and its substantive output, and that this would ultimately improve its legitimacy and effectiveness.

How can human rights principles help achieve good governance?

The principles of human rights and good governance are mutually supportive and in many instances overlap in content and purpose. The protection of human rights requires a favourable environment, which includes establishing political and administrative systems that are conducive to identifying and responding to the needs of all people.[14] Poor governance and corruption are significant barriers to the protection and promotion of human rights, as they undermine the capacity of governments and institutions to address equitably the human rights needs of all individuals and communities and lead to discriminatory access to services which disproportionately affect economically and politically disadvantaged groups.[15] Good governance is therefore a prerequisite for the protection and promotion of human rights.

The values of accountability, transparency and non-discrimination, together with a commitment to the rule of law and democracy, are fundamental to human rights and are crucial to their effective protection.[16] These same principles also form the foundation of good governance. Breakey, Cadman and Sampford identify the values of 'thick' governance to include meaningful participation, equality in decision-making and dispute settlement strategies, in addition to the 'thinner' values of accountability, transparency and rule of law.[17] The similarity between these two sets of values means that human rights law and principles therefore provide a set of values which are not only essential for achieving full protection of human rights, but which can also be employed to guide governments and institutions towards better governance.[18] The mutually supportive relationship

14 Office of the High Commissioner for Human Rights (OHCHR), *Good Governance Practices for the Protection of Human Rights*, p. 2.
15 Office of the High Commissioner for Human Rights (OHCHR), *The Human Rights Case against Corruption*, p. 4.
16 Ibid., p. 5.
17 Breakey, Cadman, and Sampford, "Governance Values and Institutional Integrity."
18 Office of the High Commissioner for Human Rights (OHCHR), *Good Governance Practices for the Protection of Human Rights*, p. 1.

between good governance and human rights provides one way in which human rights principles can enhance the good governance of the global climate regime. This chapter further argues that a second pathway to good governance can be identified through the incorporation of human rights values into the PIJ of the regime. This section will address these two methods in turn.

Human rights principles which support good governance values

As noted above, the values which underpin good (thick) governance include many principles which are also integral to human rights, such as equality, non-discrimination, accountability, transparency, democratic processes and fair treatment.[19] This is not coincidental, but follows from the fact that both good governance and human rights are driven by similar fundamental objectives of equity and respect for morality and dignity. In practice, there is a mutually supportive relationship between the fulfilment of human rights and the pursuit of transparent and accountable governance. This is particularly so when speaking of the workings of government, where poor governance or corruption is contrary to respect for civil and political rights but also undermines effective promotion of economic, social and cultural rights. Even in relation to non-state entities, this mutually supportive relationship between human rights and good governance exists. Good governance within institutions tends to support respect for human rights and positive contributions towards their fulfilment, while poor governance can create risk of exploitation, corruption or unfair dealings leading to substantive human rights abuses. Good governance can therefore be understood as a precondition to the protection and promotion of human rights.

When examining this relationship more closely, however, it is possible to see that it also works in the opposite direction: awareness of and compliance with human rights can lead to better governance procedures. Most notably, this occurs in relation to the group of rights sometimes called 'procedural rights', but it is also true of some of the more 'substantive' human rights. This section will outline some of the key principles of human rights which can contribute to embedding the values of good governance.

Procedural rights

The set of rights known as 'procedural' rights has been developed to ensure that individuals and groups who are likely to be affected by the actions of a government or institution are provided with an opportunity to contribute to the making of those decisions. Following a decision, such persons are further entitled to a remedy when their substantive rights are negatively impacted. The rationale for these rights is that respect for dignity of the individual demands that his or her interests and preferences are given due respect in decisions likely to affect him or

19 Office of the High Commissioner for Human Rights (OHCHR), *The Human Rights Case against Corruption*, p. 5.

her, and that what matters in such decision-making is not only the final outcome but also the processes through which the decision is made. These rights also serve an important instrumental purpose, in that by ensuring that proper processes are followed the negative human rights impacts of actions can more easily be identified and avoided, or at least mitigated and compensated for.

Procedural rights are most commonly associated with environmental decision-making,[20] but there are strong arguments that they should be more broadly applied.[21] They are found in numerous international, regional and domestic laws and are underpinned by the fundamental human rights principles of respect for human dignity, equality and liberty.

a The right to information

This right entails a right to information about any proposed decision or action which is likely to impact the interests of a particular individual or group.[22] In relation to the principles of governance, the right to information facilitates transparency, accountability and meaningful participation. In so doing, it helps to combat corruption by ensuring that all interests, influences and objectives relating to the decision are made public. It is therefore a fundamental right which operates to achieve good governance.

b The right to participate

This right refers to the right of individuals and groups to participate in any decisions which might have the potential to affect them.[23] It recognises the dignity and agency of individuals by allowing them to take part in decisions which will impact their lives. It has clear similarities to the good governance values of democratic decision-making and inclusion of all stakeholders, and ensures diversity of voices to achieve democratic and deliberative decision-making. It works in tandem with the right to information to ensure that

20 See for example, United Nations, "Convention on Access to Information, Public Participation in Decision-Making and Access to Justice in Environmental Matters (Aarhus Convention)," (2161 UNTS 447 25 June 1998); United Nations, "United Nations Declaration on the Rights of Indigenous Peoples," (UN Doc A/RES/61/295 (UNDRIP) 13 September 2007).

21 Naomi Roht-Arriaza, "Human Rights in the Climate Change Regime," *Journal of Human Rights and the Environment* 3, no. 1 (2010): pp. 216–17; Dinah Shelton, "Human Rights, Environmental Rights, and the Right to Environment," *Stanford Journal of International Law* 28 (1991).

22 United Nations, "International Covenant on Civil and Political Rights," (999 UNTS 171, (ICCPR), art 19(2) 16 December 1966).

23 United Nations, "Universal Declaration of Human Rights," (UN Doc A/810 (UDHR), art 21 10 December 1979); United Nations, "International Covenant on Economic, Social and Cultural Rights," (993 UNTS 3 (ICESCR), art 1 16 December 1966); United Nations, "Convention on Access to Information, Public Participation in Decision-Making and Access to Justice in Environmental Matters (Aarhus Convention)."

participation is supported by accurate and timely information. This meaningful participation of stakeholders further contributes to transparency, accountability and compliance by enabling a wider range of interests to be represented.

c The right to access to justice

The right to access to justice applies in any situation where an individual or group has suffered a violation of another substantive human right or a negative impact on their interests. It incorporates the ability to challenge decisions through legal processes and/or to receive an effective remedy where a negative impact flows from decisions made or action taken.[24] Adherence to the right to access justice facilitates the good governance values of accountability, compliance and the rule of law. It also provides an impetus for complying with the thick governance requirement of providing strategies for effective dispute resolution. Further, the right to access to justice also works to combat corruption be requiring that all people affected have an opportunity to seek justice regardless of their means, status, location or background.

Substantive rights

The procedural rights outlined above are essential to the effective protection of the full range of substantive rights recognised under international human rights law. A number of these substantive rights also have specific application in an institutional setting, and can operate to reinforce thick governance. Human rights operate as minimum standards of treatment against which governmental or institutional conduct can be evaluated. They require states to respect human rights in all state action, but also demand that states protect against interferences by non-state actors, and can therefore be extended to the operation of private as well as public institutions. Human rights may form part of an institution's governing values, helping to guide its decision-making towards particular priorities and standards. As the following discussion shows, when certain rights are upheld by institutions they operate to reinforce good governance and not only strengthen the protection of other human rights, but also improve the effectiveness and legitimacy of institutions themselves.

a The right to freedom of opinion and expression

Freedom of opinion and expression incorporates the right to send and receive information and extends to freedom of the press.[25] It is essential to strong

24 United Nations, "Convention on Access to Information, Public Participation in Decision-Making and Access to Justice in Environmental Matters (Aarhus Convention)," Article 1; United Nations, "International Covenant on Civil and Political Rights," Article 2.

25 United Nations, "International Covenant on Civil and Political Rights," Article 18; United Nations, "Universal Declaration of Human Rights," Article 19.

democratic processes, as it enables individuals to make informed decisions and to exercise their civil and political rights in a meaningful way.[26] It also promotes economic, social and cultural rights by ensuring that information about the state of enjoyment of these rights is publicly available. While it is usually understood to apply to information held by public entities, it is also appropriately extended to any private or non-state actor whose activities have the potential to affect the lives of individuals, groups or the community more broadly.[27] It is closely linked to the procedural right of access to information, although the procedural right is typically applied to stakeholders in specific decision-making contexts who require information in order to participate meaningfully in those decisions.

The right to freedom of opinion and expression is essential to anti-corruption measures and good governance. Most obviously, it aligns to the value of transparency and contributes significantly to achieving the objectives of accountability, meaningful participation and equal inclusion of stakeholders. For these reasons, an institution which includes a commitment to upholding the right to freedom of opinion and expression among its governing values will be more likely to secure thick governance in its internal operations.

b The right to equality before the law and to a fair trial

All people have the right to be treated equally under the legal system within which they live.[28] This notion of equality before the law incorporates the human right to a fair trial, which possesses both substantive and procedural elements, including the right to the presumption of innocence, the right to have access to legal representation and the prohibition against double jeopardy.[29] These rights are closely linked with the right of all persons to have access to justice and to an effective remedy described above, and a strong and independent justice system is essential for the realisation of a wide range of human rights.[30]

Translated to a governance context, these rights require that mechanisms be put in place for the just and equitable resolution of disputes in matters where people's substantive rights and interests are in jeopardy. This may

26 United Nations, "Human Rights Committee: General Comment 34: Article 19 (Freedom of Opinion and Expression)," (UN Doc CCPR/C/GC/34 12 September 2011), 1.

27 Ibid.

28 United Nations, "International Covenant on Civil and Political Rights," Article 14; United Nations, "Universal Declaration of Human Rights," Article 7.

29 United Nations, "International Covenant on Civil and Political Rights," Article 14; United Nations, "Human Rights Committee, General Comment 32: Art 14 (Right to Equality before Courts and Tribunals and to a Fair Trial)," (UN Doc CCPR/C/GC/32 23 August 2007).

30 Office of the High Commissioner for Human Rights (OHCHR), *Seminar on Good Governance Practices for the Promotion of Human Rights* (Seoul: United Nations, 15–16 September 2004), Background note, 4.

involve referring disputes to external mechanisms within the legal system, or it may involve internal dispute resolution processes within an institution. A commitment to the right to a fair trial and equality before the law therefore contributes to the achievement of good governance by ensuring that the values of accountability, compliance, the rule of law and fair dispute settlement are implemented effectively.

c Right to self-determination

The right to self-determination is a right possessed by all peoples.[31] It includes the right to make decisions about the political and economic future of the group and the right to take advantage of natural resources.[32] A key component of the right to self-determination is the right to free, prior and informed consent for any activities which are likely to have a negative impact on the group's self-determination.[33] In this sense it builds on the rights to information and participation which are outlined previously.

The right to self-determination has an application to governance, particularly in cases where the rights of indigenous peoples or other minorities are at risk from the actions of a government or institution, which would in turn be required to consult fully with the group prior to making any decisions. Action taken which does not comply with the fulfilment of self-determination would be in breach not only with the applicable human rights law but also with governance values of including all stakeholders and affording them an equal role in decision-making. A failure to consult with affected peoples would also undermine the value of democratic decision-making and put accountability and transparency at risk. The right to self-determination can therefore guide institutions and governments towards an approach to decision-making which complies with a wide range of thick governance values, and ought also to be among the governing values of all institutions which operate within relevant fields.

Fundamental principles of human rights

The preceding discussion outlines a number of procedural and substantive human rights which have an application in a governance context and which can guide institutions and governments towards better governance. Underlying these rights are a number of fundamental principles which can also contribute to good governance, even in the absence of a connection with a specific right. These principles

31 United Nations, "International Covenant on Economic, Social and Cultural Rights," Article 1; United Nations, "International Covenant on Civil and Political Rights," Article 1.

32 United Nations, "Human Rights Committee: General Comment 12: Art 1 (Right to Self-Determination)," (UN Doc CCPR/C/GC/12 13 March 1994).

33 United Nations, "United Nations Declaration on the Rights of Indigenous Peoples," Articles 10, 11, 19, 28, 29.

include non-discrimination and equality, accountability, transparency, democratic processes and fair treatment,[34] and are clearly consistent with several of the values of thick governance. By including these fundamental principles within its governing values, an institution not only will demonstrate a respect for human rights, but also will align itself with thick governance values and take steps towards enhancing its integrity and overall legitimacy.

The contribution of human rights principles to achieving a Public Institutional Justification (PIJ)

As can be seen, the values of thick governance correlate with many of the principles of human rights and are supported by a number of procedural and substantive rights. As a consequence of the similarity between these two sets of norms, human rights principles can be seen to guide institutions towards more effective, 'thicker' governance. However, in some cases human rights might play a more explicit and direct role in achieving thick governance. This is particularly the case where an institution's PIJ incorporates the promotion of human rights (here called its PIJ-HR).

If an institution declares itself to have a PIJ-HR, its implementation of thick governance values would demonstrate a commitment to this PIJ-HR. This is due to the overlapping and mutually supportive relationship between human rights and good governance discussed above, which has the effect that efforts to promote good governance will be equally supportive of human rights. This in turn reinforces the legitimacy of the institution as a whole. As Breakey, Cadman and Sampford have explained, 'institutions genuinely pursuing a laudable PIJ have good reason to aim for Thick Governance Values' and have nothing to fear from committing themselves to transparency, accountability and democratic approaches to governance.[35] If, on the other hand, an institution with a PIJ-HR adopted a governance approach at the thinner end of the spectrum, then this would likely be seen as undermining its PIJ. Such an approach would weaken the institution's consistency-integrity and its legitimacy as a whole.

Furthermore, many of the human rights noted before can operate to make an institution more effective in achieving its PIJ. The rights to information, participation, equality before the law and access to justice all serve to combat corruption, waste and inefficiency. They therefore can be useful in ensuring that an institution is achieving its objectives to its maximum capacity. This is the case even where the PIJ does not encompass human rights objectives, but is even more pronounced where the institution possesses a PIJ-HR. Good governance can therefore be understood to be a precondition to maximising protection and

34 Office of the High Commissioner for Human Rights (OHCHR), *Good Governance Practices for the Protection of Human Rights*, p. 5.

35 Breakey, Cadman, and Sampford, "Governance Values and Institutional Integrity," Chapter 2, this volume.

promotion of human rights and consequently to the fulfilment of a PIJ-HR, where one has been articulated.

So, where the promotion of human rights is incorporated by an institution into its PIJ, adherence to human rights law and principles not only will enhance the institution's consistency-integrity and effectiveness by promoting its PIJ, but also will reinforce good governance values which contribute to its legitimacy. The following section will examine how human rights can contribute specifically to the good governance of the climate regime, most notably through recognition that human rights values fall within the regime's PIJ.

Human rights and good governance within the climate regime

There is clearly a mutually supportive relationship between human rights and good governance, and, as argued previously, an institution which commits itself to respecting human rights is likely to strive also for thick governance, leading to a greater degree of legitimacy for the institution and positive outcomes for human rights. This section will examine the international climate regime and identify a number of ways in which this relationship between human rights and good governance can be used to enhance the legitimacy of the regime.

Human rights within the PIJ of the climate regime

Human rights values can already be identified as being part of the PIJ of the global climate regime, both explicitly and implicitly. While explicit references to human rights are limited within the framework of international climate change instruments, there are statements which demonstrate the intention of states to include human rights within the overall objective of the regime, particularly when considering more recently adopted instruments.

The Cancun Agreements, decided by the Conference of Parties (COP) to the UNFCCC in 2010, followed the lead of both the United Nations Human Rights Council and the Office of the High Commissioner for Human Rights, which had previously recognised the fact that climate change has a range of direct and indirect impacts on the enjoyment of human rights.[36]

36 Human Rights Council, "Human Rights Council, Human Rights and Climate Change," (UN Doc A/HRC/Res/7/23 28 March 2008); Human Rights Council, "Human Rights and Climate Change," (UN Doc A/HRC/Res10/4 25 March 2009); Office of the High Commissioner for Human Rights (OHCHR), "Report of the Office of the United Nations High Commissioner for Human Rights on the Relationship between Climate Change and Human Rights," (UN Doc A/HRC/10/61 15 January 2009); United Nations Framework Convention on Climate Change (UNFCCC), "Report of the Conference of the Parties on Its Sixteenth Session, Held in Cancun from 29 November to 10 December 2010," (Part Two: Action taken by the Conference of the Parties at Its Sixteenth Session FCCC/CP/2010/7/Add.1 2011).

The Cancun Agreements emphasised for the first time in a climate change instrument that all state parties 'should, in all climate change related actions, fully respect human rights.'[37] A similar, though less explicit, statement can be found in the Kyoto Protocol, which provides that developed (Annex 1) states 'shall strive to implement policies and measures . . . in such a way as to minimise adverse effects, including . . . social, environmental and economic impacts on other Parties.'[38]

The requirement to ensure that human rights are respected in taking action to address climate change is reiterated in the newly adopted Paris Agreement, which states in its preamble:

> Acknowledging that climate change is a common concern of human kind, Parties should, when taking action to address climate change, respect, promote and consider their respective obligations on human rights, the right to health, the rights of indigenous peoples, local communities, migrants, children, persons with disabilities and people in vulnerable situations and the right to development, as well as gender equality, empowerment of women and intergenerational equity.[39]

At the very least these statements indicate that, along with the regime's key objective of reducing harmful emissions and keeping the warming of the planet 'well-below 2°C', to use the language of the Paris Agreement,[40] an associated purpose of the regime is to ensure that states do not further harm human rights in the action they take in pursing their ultimate goal.

Picking up on the recognition in the Cancun Agreements that climate change is an issue of concern for human rights, the protection of human rights can also be implied into the objective of the regime by incorporating human rights principles into our understanding of the harm caused by climate change. The negative

37 United Nations Framework Convention on Climate Change (UNFCCC), "Report of the Conference of the Parties on Its Sixteenth Session, Held in Cancun from 29 November to 10 December 2010," 4.

38 United Nations Framework Convention on Climate Change (UNFCCC), "Kyoto Protocol to the United Nations Framework Convention on Climate Change, 10 December 1997," (UN Doc FCCC/CP/1997/7/Add.1, 1997), Article 2(3).

39 United Nations Framework Convention on Climate Change (UNFCCC), "Conference of the Parties: COP21 Paris Agreement," (FCCC/CP/2015/L.9: UNFCCC, 2015), Preamble.

40 United Nations Framework Convention on Climate Change (UNFCCC), "Framework Convention on Climate Change," (1771 UNTS 107 1992), Article 2. 'The ultimate objective of this Convention . . . is to achieve . . . stabilization of greenhouse gas concentrations in the atmosphere at a level that would prevent dangerous anthropogenic interference with the climate system.'

United Nations Framework Convention on Climate Change (UNFCCC), "Conference of the Parties: COP21 Paris Agreement," Article 2. Purpose is to 'strengthen the global response to the threat of climate change . . . including by (a) holding the increase in the global average temperature to well below 2°C and to pursue efforts to limit the temperature increase to 1.5°C above pre-industrial levels'.

human rights impacts of climate change are by now well accepted, and they have been acknowledged by the Intergovernmental Panel on Climate Change (IPCC) as well as the United Nations human rights bodies (referred to earlier) and numerous scholars.[41]

The objective of the UNFCCC to achieve 'stabilisation of greenhouse gas concentrations in the atmosphere at a level that would prevent dangerous anthropogenic interference with the climate system' should therefore incorporate human rights considerations into our understanding of what amounts to a 'dangerous interference'.[42] The notion that 'dangerous interference' includes threats to both the physical environment and human beings is reinforced by references in later instruments to the fact that 'climate change represents an urgent and potentially irreversible threat to human societies'.[43]

A comprehensive concept of 'dangerous interference' would therefore incorporate harm caused to the enjoyment of human rights, such as the rights to life, health, food, water, housing and subsistence by environmental changes such as rising sea levels, desertification of agricultural land, salinization of water supplies, impacts on crops and livestock and the spread of vector-borne diseases.

The PIJ of the international climate regime should therefore be understood as being broad enough to encompass the addressing of the full range of human rights impacts due to climate change – the climate regime should be considered to possess a PIJ-HR. With this in mind, and given the overlap between thick governance and human rights principles outlined before, it is argued that thick governance values become relevant not just for their instrumental value in helping

41 Josef Schmidhuber and Francesco N. Tubiello, "Global Food Security under Climate Change," *Proceedings of the National Academy of Sciences* 104, no. 50 (2007); Martin Parry, Alex Evans, Mark W. Rosegrant, and Tim Wheeler, *Climate Change and Hunger: Responding to the Challenge* (Rome: World Food Programme, 2009), p. 14; Inter Governmental Panel on Climate Change (IPCC), "Summary for Policymakers," in *Climate Change 2014: Impacts, Adaptation, and Vulnerability: Part A: Global and Sectoral Aspects. Contribution of Working Group II to the Fifth Assessment Report of the Intergovernmental Panel on Climate Change*, ed. C.B. Field et al. (Cambridge, UK and New York, NY: Cambridge University Press, 2014); Knox, "Climate Change and Human Rights Law"; Doelle, "Climate Change and Human Rights: The Role of International Human Rights in Motivating States to Take Climate Change Seriously"; Raquel Rolnik (Special Rapporteur on Adequate Housing), "Contribution to Human Rights Council Panel Discussion on the Relationship between Climate Change and Human Rights: Summary of Discussions," (United Nations, 15 June 2009), p. 5 [13]; Human Rights Council, "Human Rights and Climate Change," (Resolution 7/23 and 10/4 2008).

42 United Nations Framework Convention on Climate Change (UNFCCC), "Framework Convention on Climate Change," Article 2.

43 United Nations Framework Convention on Climate Change (UNFCCC), "Report of the Conference of the Parties on Its Sixteenth Session, Held in Cancun from 29 November to 10 December 2010," Preamble; United Nations Framework Convention on Climate Change (UNFCCC), "Report of the Conference of the Parties on Its Nineteenth Session, Held in Warsaw from 11 to 23 November 2013: Part One: Proceedings," (FCCC/CP/2013/10 2013), Preamble.

the regime achieve its PIJ but also for ensuring that it does not undermine the human rights–related aspects of its PIJ by operating in a way which is incompatible with those same human rights values.

How can human rights values enhance good governance in the climate regime?

Having established that human rights have a place within the PIJ of the global climate regime, this section offers some suggestions for ways in which human rights principles and laws might support 'thicker' governance in relation to key aspects of the regime's work. A key element of this argument is that human rights principles and values need to be more explicitly incorporated into the governing values of the regime. This, it is argued, will create greater impetus and opportunity for the development of thick governance values. Some specific rights and principles are considered to have particular relevance in this respect.

The human rights of self-determination, participation, equality before the law and access to justice all operate to provide greater opportunities for individuals and communities who are vulnerable to climate change to have their concerns taken into account and addressed. Human rights–based approaches to climate change can broadly help to emphasise the human impacts of climate change, leading to a better appreciation of the needs of vulnerable individuals, communities and states.

At the international level, a stronger commitment to human rights would encourage greater inclusion of vulnerable states in order to ensure that their needs are met. Not only would this contribute to better outcomes in terms of addressing the effects of climate change on the world's most vulnerable, but also it could help to promote the thick governance values of meaningful participation and democratic decision-making.

Human rights can also help to promote the thick governance principle that all stakeholders should be given an equal role in decision-making. This principle is particularly significant in the climate context, as it helps to address the power imbalance between polluting/developed states and vulnerable/developing states. This can be achieved through upholding the human rights principle of non-discrimination and the right of participation. Within the climate regime these rights would also require that states implement national climate policies without discrimination, thereby enhancing the effectiveness and legitimacy of the regime at both international and domestic levels.

A commitment to the human rights principles of access to justice, due process and procedural rights can enhance the accountability of the global climate regime by encouraging the implementation of dispute-resolution mechanisms and avenues for hearing grievances. A greater commitment from states to human rights generally can also help to enhance the accountability of the regime. As states become more accustomed to upholding human rights within their own jurisdictions, the climate regime will be better placed to insist upon compliance with those same standards in its own work. This will result in overall better outcomes, not only in terms of combatting climate change but also in increasing the legitimacy of the climate regime itself.

Conclusion

As has been demonstrated, the mutually supportive and overlapping relationship between the values of human rights and good governance presents a number of ways in which greater adherence to human rights law and principles could enhance the governance of the global climate regime. Currently there are only limited references to human rights within the framework instruments. More explicit references to human rights and strategies for their implementation are required to improve the effectiveness of the regime, especially in addressing the needs of those who are most vulnerable to the effects of climate change and in ensuring that action taken to combat climate change does not cause further interference with human rights.

Such an expanded engagement with human rights principles would also serve to guide the regime towards better governance. It would embed human rights more substantially within the governing values of the regime and help to articulate the place of human rights within its PIJ. In order to fulfil this institutional commitment to human rights, the regime would require thicker governance values. This is because good governance is a precondition to creating an environment where human rights can be effectively promoted and protected. It is also due to the fact that the significant overlap between the values of good governance and human rights means that effort to uphold one set of values contributes to greater realisation of the other. While human rights principles can to some extent already be implied into the objectives and governing values of the regime, more specific inclusion is required to optimise the benefits for both the internal operation of the regime itself and its substantive work in addressing climate change.

References

Aminzadeh, Sara C. "A Moral Imperative: The Human Rights Implications of Climate Change." *Hastings International and Comparative Law Review* 30, no. 2 (2006–2007): 231–65.

Atapattu, Sumudu. "Global Climate Change: Can Human Rights (and Human Beings) Survive This Onslaught?" *Colorado Journal of International Environmental Law and Policy* 20, no. 1 (2009): 35–47.

Barnett, Jon. "Human Rights and Vulnerability to Climate Change." In *Human Rights and Climate Change*, edited by Stephen Humphreys, 257–71. Cambridge: Cambridge University Press, 2010.

Bell, Derek. "Does Anthropogenic Climate Change Violate Human Rights." *Critical Review of International Social and Political Philosophy* 14, no. 2 (2011): 99–124.

Bodansky, Daniel. "Climate Change and Human Rights: Unpacking the Issues." *Georgia Journal of International and Comparative Law* 38, no. 3 (2010): 511–24.

Breakey, Hugh, Tim Cadman, and Charles Sampford. "Governance Values and Institutional Integrity." Chap. 2. In *Governing the Climate Change Regime*, edited by

Charles Sampford, Tim Cadman and Rowena Maguire. London: Taylor & Francis, 2016 (this volume).
Breakey, Hugh, and Timothy Cadman. "A Comprehensive Framework for Evaluating the Integrity of the Climate Regime Complex." Chap. 2 In *Ethical Values and the Integrity of the Climate Change Regime*, edited by Hugh Breakey, Vesselin Popovski and Rowena Maguire, 17–27. Surrey: Ashgate, 2015.
Breakey, Hugh, Timothy Cadman, and Charles Sampford. "Conceptualizing Personal and Institutional Integrity: The Comprehensive Integrity Framework." *Research in Ethical Issues in Organizations* 14 (2015): 1–40.
Cameron, Edward. "Human Rights and Climate Change: Moving from an Intrinsic to an Instrumental Approach." *Georgia Journal of International and Comparative Law* 38, no. 3 (2010): 673–708.
Depledge, Michael, and Cinnamon Carlane. "Sick of the Weather: Climate Change, Human Health and International Law." *Environmental Law Review* 9, no. 4 (2007): 231–40.
Doelle, Meinhard. "Climate Change and Human Rights: The Role of International Human Rights in Motivating States to Take Climate Change Seriously." *Macquarie Journal of International and Comparative Environmental Law* 1, no. 2 (2004): 179–216.
Hassan, Daud, and Awal Khan. "Climate-Change-Related Human Rights Violations." *Environmental Law and Policy* 43, no. 2 (2013): 80–86.
Human Rights Council. "Human Rights and Climate Change." Resolution 7/23 and 10/4 2008.
———. "Human Rights and Climate Change." UN Doc A/HRC/Res10/4 25 March 2009.
———. "Human Rights Council, Human Rights and Climate Change." UN Doc A/HRC/Res/7/23 28 March 2008.
Humphreys, Stephen, ed. *Human Rights and Climate Change*. Cambridge: Cambridge University Press, 2010.
Inter Governmental Panel on Climate Change (IPCC). "Summary for Policymakers." In *Climate Change 2014: Impacts, Adaptation, and Vulnerability: Part A: Global and Sectoral Aspects. Contribution of Working Group II to the Fifth Assessment Report of the Intergovernmental Panel on Climate Change*, edited by C.B. Field, V.R. Barros, D.J. Dokken, K.J. Mach, M.D. Mastrandrea, T.E. Bilir, M. Chatterjee, T.E. Bilir, M. Chatterjee, K.L. Ebi, Y.O. Estrada, R.C. Genova, B. Girma, E.S. Kissel, A.N. Levy, S. MacCracken, P.R. Mastrandrea, and L.L. White, 1–32. Cambridge, UK and New York, NY: Cambridge University Press, 2014.
Knox, John. "Climate Change and Human Rights Law." *Virginia Journal of International Law* 50, no. 1 (2009–2010): 163–218.
———. "Linking Human Rights and Climate Change at the United Nations." *Harvard Environmental Law Review* 33 (2009): 477–98.
Lewis, Bridget. "Balancing Human Rights in Adaptation and Mitigation Policies." In *Climate Change and Human Rights, an International Law Perspective*, edited by Ottavio Quirico and Mouloud Boumghar, 673–716. London: Routledge, 2015.
———. "The Contribution of Human Rights to the Effectiveness and Integrity of the Global Carbon Regime." In *Ethical Values and the Integrity of the Climate Change Regime*, edited by Hugh Breakey, Vesselin Popovski and Rowena Maguire, 179–88. Surrey: Ashgate, 2015.
Limon, Marc. "Human Rights and Climate Change: Constructing a Case for Political Action." *Harvard Environmental Law Review* 33, no. 2 (2009): 439–63.

Maguire, Rowena, and Bridget Lewis. "The Influence of Justice Theories on International Climate Policies and Measures." *Macquarie Journal of International and Comparative Environmental Law* 8, no. 1 (2012): 16–35.

McAdam, Jane. *Climate Change, Forced Migration and International Law*. Oxford: Oxford University Press, 2012.

McInerney-Lankford, Siobhan, Mac Darrow, and Lavanya Rajamani. *Human Rights and Climate Change: A Review of the International Legal Dimensions*. Washington, DC: World Bank, 2011.

Office of the High Commissioner for Human Rights (OHCHR). *Good Governance Practices for the Protection of Human Rights*. Geneva: United Nations, 2007.

———. *The Human Rights Case against Corruption*. Geneva: United Nations, 2013.

———. "Report of the Office of the United Nations High Commissioner for Human Rights on the Relationship between Climate Change and Human Rights." UN Doc A/HRC/10/61 15 January 2009.

———. *Seminar on Good Governance Practices for the Promotion of Human Rights*. Seoul: United Nations, 15–16 September 2004.

Parry, Martin, Alex Evans, Mark W. Rosegrant, and Tim Wheeler. *Climate Change and Hunger: Responding to the Challenge*. Rome: World Food Programme, 2009.

Pedersen, Ole W. "Climate Change and Human Rights: Amicable or Arrested Development?" *Journal of Human Rights and the Environment* 1, no. 2 (2010): 236–51.

———. "The Janus-Head of Human Rights and Climate Change: Adaptation and Mitigation." *Nordic Journal of International Law* 80, no. 3 (2011): 403–23.

Roht-Arriaza, Naomi. "Human Rights in the Climate Change Regime." *Journal of Human Rights and the Environment* 3, no. 1 (2010): 211–20.

Rolnik (Special Rapporteur on Adequate Housing), Raquel. "Contribution to Human Rights Council Panel Discussion on the Relationship between Climate Change and Human Rights: Summary of Discussions." United Nations, 15 June 2009.

Schmidhuber, Josef, and Francesco N. Tubiello. "Global Food Security under Climate Change." *Proceedings of the National Academy of Sciences* 104, no. 50 (2007): 19703–08.

Shabalala, Dalindyebo. "Contribution to Human Rights Council Panel Discussion on the Relationship between Climate Change and Human Rights: Summary of Discussions." Geneva: Human Rights Council, 15 June 2009.

Shelton, Dinah. "Human Rights, Environmental Rights, and the Right to Environment." *Stanford Journal of International Law* 28 (1991): 103–38.

Stephens, Pamela. "Applying Human Rights Norms to Climate Change: The Elusive Remedy." *Colorado Journal of International Environmental Law and Policy* 21, no. 1 (2010): 49–83.

United Nations. "Convention on Access to Information, Public Participation in Decision-Making and Access to Justice in Environmental Matters (Aarhus Convention)." 2161 UNTS 447 25 June 1998.

———. "Human Rights Committee: General Comment 12: Art 1 (Right to Self-Determination)." UN Doc CCPR/C/GC/12 13 March 1994.

———. "Human Rights Committee, General Comment 32: Art 14 (Right to Equality before Courts and Tribunals and to a Fair Trial)." UN Doc CCPR/C/GC/32 23 August 2007.

———. "Human Rights Committee: General Comment 34: Article 19 (Freedom of Opinion and Expression)." UN Doc CCPR/C/GC/34 12 September 2011.

———. "International Covenant on Civil and Political Rights." 999 UNTS 171, (ICCPR), art 19(2) 16 December 1966.

———. "International Covenant on Economic, Social and Cultural Rights." 993 UNTS 3 (ICESCR), art 1 16 December 1966.
———. "United Nations Declaration on the Rights of Indigenous Peoples." UN Doc A/RES/61/295 (UNDRIP) 13 September 2007.
———. "Universal Declaration of Human Rights." UN Doc A/810 (UDHR), art 21 10 December 1979.
United Nations Framework Convention on Climate Change (UNFCCC). "Conference of the Parties: COP21 Paris Agreement." FCCC/CP/2015/L.9: UNFCCC 2015.
———. "Framework Convention on Climate Change." 1771 UNTS 107 1992.
———. "Kyoto Protocol to the United Nations Framework Convention on Climate Change, 10 December 1997." UN Doc FCCC/CP/1997/7/Add.1 1997.
———. "Report of the Conference of the Parties on Its Nineteenth Session, Held in Warsaw from 11 to 23 November 2013: Part One: Proceedings." FCCC/CP/2013/10 2013.
———. "Report of the Conference of the Parties on Its Sixteenth Session, Held in Cancun from 29 November to 10 December 2010." Part Two: Action Taken by the Conference of the Parties at Its Sixteenth Session FCCC/CP/2010/7/Add.1 2011.
Wani, Ibrahim. "Open-Ended Consultation on the Relationship between Climate Change and Human Rights: Summary of Discussions." Geneva: OHCHR, 22 October 2008.

10 Climate governance accountability challenges

Lessons from multilateral climate finance

Lisa Elges

Introduction

The threat and challenge of climate change demands urgent adaptation and mitigation actions across the globe. Governments and private-sector actors are being called on to contribute efforts to ensure against current and future serious climatic damages while ensuring sustainable economic growth. The December 2015 Paris Agreement concluded under the United Nations Framework Convention on Climate Change marks a critical step in guiding individual and collective actions of States to lead those efforts.[1] The commitments to sustainable development and averting global warming by ensuring that global average temperature increase is limited to 1.5°C above pre-industrial levels should shape national viable and transformational low-carbon growth strategies.[2]

Realising those strategies requires, among other things, money, power and integrity. The challenge of transforming economies to avoid global warming comes with a heavy price tag. So do the costs of lessening the impact of and coping with the environmental and human damages caused by increased flooding, drought, temperature and sea levels and extreme weather events.

The Paris Agreement sets out an initial framework for sequestering the financial and political capital needed to achieve this. All countries should – to varying degrees – control their long-term greenhouse gas emissions so as to ensure that global emissions are below a threshold necessary to maintain the global warming target (limit).[3] In addition to their own national efforts, developed country governments are strongly urged to provide USD 100 billion by 2020 through 2025 to support developing countries in adapting to and mitigating climate change. Subject to a further agreement, that sum may be increased after 2025.[4] On a voluntary basis, developing countries may contribute to mobilising the

1 United Nations Framework Convention on Climate Change (UNFCCC), "Conference of the Parties Twenty-First Session Paris, 30 November to 11 December 2015: Adoption of the Paris Agreement," (FCCC/CP/2015/L.9/Rev. 1: United Nations, 2015).

2 Ibid., Draft decision-/CP.21, paras. 17, 21 Article 2(1)(a).

3 Ibid., Article 4. at pp. 22–23.

4 Ibid., Draft decision-/CP, paras. 54, 115.

USD 100 billion. Both public and private sources may be accounted towards that financial goal.[5]

Given the short- and long-term environmental risks at stake, as well as the volume of money involved, the accountability and ultimately the integrity of States in fulfilling these commitments is critical. This is not a novel sentiment but rather a demand which most stakeholders have claimed for nearly a decade. Yet the transparency, accountability and integrity of governments and various global and national institutions established to deliver on environmental and financial goals equally has been a consistent challenge. This chapter attempts to address some of the institutional and policy 'constraints' which have characterised climate finance regimes as they have evolved at the international level. It discusses the dynamics of multilateral processes aimed at pooling and disbursing public resources to address global warming. In doing so, it suggests that governance policies of those processes or 'institutions' have ranged between thin and thicker but not thickest due to their organisational set-up, their mandate or public institutional justification and their politics. However, recent developments reveal a reprioritisation and movement to strengthen accountability, transparency and integrity rules amongst global climate funds. One reason for this appears to be the need to achieve sustainable and transformational governance impacts as a functional part of climate finance packages.

Thin ↔ thicker governance of global climate change processes

The UN Framework Convention on Climate Change has been central to the evolution of the global climate regime. The Convention offers its States Parties a basis to gather, negotiate and agree on their efforts and obligations to stabilise 'greenhouse gas concentrations in the atmosphere at a level that would prevent dangerous anthropogenic interference with the climate system',[6] whereby '[s]uch a level should be achieved within a time-frame sufficient to allow ecosystems to adapt naturally to climate change, to ensure that food production is not threatened and to enable economic development to proceed in a sustainable manner'.[7] This is the primary objective of the Convention and the key driver of subsequent agreements. The Convention itself obliges all States, inter alia, to account for GHG emissions and sinks and to pursue mitigation and adaptation actions.[8] Developed countries are committed specifically to developing, implementing and reporting on national climate change mitigation policies (with a view to returning emissions to pre-1990 levels).[9]

5 Ibid., Article 9.

6 United Nations Framework Convention on Climate Change (UNFCCC), "Framework Convention on Climate Change," (1771 UNTS 107 / [1994] ATS 2 / 31 ILM 849 [1992] 1992), Article 2.

7 Ibid.

8 Ibid., Article 4(1).

9 Ibid., Article 4(2).

As a legal framework, the Convention does not make such obligations time bound or quantifiable, nor is it accompanied by an institutional compliance policy. These outcomes were achieved in 2005 when the 1997 Kyoto Protocol entered into force. Targeted at developed countries, the Protocol primarily requires that States reduce national emissions by 5 percent based on 1990 levels by 2012[10] and that they set up a 'national system for the estimation of anthropogenic emissions by sources and removals by sinks'.[11] The Protocol was extended partially by the 2012 Doha Amendment,[12] whereby State Party signatories agree to cut GHG emissions 18 percent by 2020. Progressively, the Paris Agreement[13] clarifies a common environmental goal (1.5°C), draws on the participation of all States to take action towards achieving it and compels better transparency on progress achieved by States through mature reporting frameworks.

A secondary aim is to ensure that developing countries are sufficiently supported by developed countries in achieving their primary obligations. That is, given the global nature of the problem, the actions of all countries are needed to avoid catastrophe, and those countries which are least able to afford such measures should receive adequate resources. The Climate Convention enlisted developed countries, initially in the 1990s, to help developing countries in communicating their mitigation and adaptation progress.[14] Under the Kyoto Protocol, financial pledges expanded to include support for building similar accounting systems, technology transfer, capacity building, scientific research and progress reporting.[15] In 2009, more concrete financial guarantees were made in the Copenhagen Accord. Therein, developed countries committed 'to provide new and additional resources, including forestry and investments through international institutions, approaching USD 30 billion for the period 2010–2012 with balanced allocation between adaptation and mitigation'.[16] Further, '[i]n the context of meaningful mitigation actions and transparency on implementation, developed countries commit to a goal of mobilizing jointly USD 100 billion dollars a year by 2020 to

10 United Nations Framework Convention on Climate Change (UNFCCC), "Kyoto Protocol to the United Nations Framework Convention on Climate Change, 10 December 1997," (UN Doc FCCC/CP/1997/7/Add.1, 37 I.L.M.: United Nations, 1997), Article 3.

11 Ibid., Article 5.

12 United Nations Framework Convention on Climate Change (UNFCCC), "Report of the Conference of the Parties on Its Seventeenth Session, Held in Durban from 28 November to 11 December 2011," (Addendum. Part Two: Action Taken by the Conference of the Parties at Its Seventeenth Session: UN, 2012).

13 United Nations Framework Convention on Climate Change (UNFCCC), "Conference of the Parties Twenty-First Session Paris, 30 November to 11 December 2015: Adoption of the Paris Agreement."

14 United Nations Framework Convention on Climate Change (UNFCCC), "Kyoto Protocol to the United Nations Framework Convention on Climate Change, 10 December 1997," Article 4(2).

15 Ibid., Article 11.

16 United Nations Framework Convention on Climate Change (UNFCCC), "The Conference of the Parties (COP), at Its Fifteenth Session, Took Note of the Copenhagen Accord of 18 December 2009," (FCCC/CP/2009/11/Add.12009), para. 8.

address the needs of developing countries financial commitments'.[17] A year later, States Parties agreed to establish the Green Climate Fund to be the Financial Mechanism of the Convention to aid in the channelling of climate finance.[18] And, as mentioned before, the Paris Agreement promises that the scale of support to developing countries should increase to USD 100 billion by 2020.

In assessing these primary and secondary goals of the climate regime, it appears that the governance of environmental goals may be waning from a thicker to thin state, while financial accountability and transparency polices are thickening. Regarding the former, under the Kyoto Protocol an elaborate compliance system was put in place.[19] In its policy formulation (excluding operational review), that system, inclusive of the Protocol's market mechanism design, can be categorised under the thicker governance umbrella. That thicker level of governance is observable when it is contrasted with the thin approach adopted under the 2015 Paris Agreement.

The Paris Agreement aims to achieve compliance through national efforts to guarantee transparency. The flexible scheme envisions that State Parties report honestly on the progress they make in curbing greenhouse gas emissions and adapting to climate change impacts.[20] However, the present Agreement is fairly clear that the global approach to ensure compliance and accountability will not involve punitive measures or sanctions.[21] The more relaxed scheme is appreciable given that the Paris Agreement encompasses all States Parties as opposed to only developed countries, as under the Kyoto Protocol. It may be reasonable that at least in an initial phase, strict compliance rules would be burdensome for a number of countries lacking sufficient capacities to deliver. Developing countries were also hesitant to sign up without enough assurance that developed countries would help foot their mitigation and adaptation bills.

It is likely that a thicker governance framework at the global level would be operationally feasible for various legal and political reasons. The thin approach would thus seem to be more appropriate and realistic. What the scheme does evoke, however, is the need for governments to be accountable domestically in formulating and meeting their national mitigation and adaptation targets. The level of achievable accountability will relate to the fitness and functionality of each country's national integrity system. Ultimately, this corresponds with the degree to which citizens and other stakeholders can effectively hold their governments accountable on matters affecting public interests. The Paris Agreement

17 Ibid.
18 United Nations Framework Convention on Climate Change (UNFCCC), "Report of the Conference of the Parties on Its Sixteenth Session, Held in Cancun from 29 November to 10 December 2010," (Part Two: Action Taken by the Conference of the Parties at Its Sixteenth Session FCCC/CP/2010/7/Add.12011), para. 102.
19 United Nations Framework Convention on Climate Change (UNFCCC), "An Introduction to the Kyoto Protocol Compliance Mechanism," UNFCCC, <http://unfccc.int/kyoto_protocol/compliance/items/3024.php>.
20 Ibid., Article 13, paras. 5, 7, 8.
21 Ibid., Article 13 (3), Article 15 (2).

recognises this in principle by expecting that States enhance public participation and access to information in succeeding in their agreed commitments.[22] On this level, and as the terms of the Agreement are developed over the next year(s), expanding State obligations to strengthen national processes to ensure accountability and integrity will be critical to the ultimate success of the pact.

At this juncture, it is worth reflecting on the distinction between accountability and integrity and why these factors are elementary to a would-be sustainable future. According to the Transparency Accountability Initiative, the concept of accountability 'means ensuring that officials in public, private and voluntary sector organisations are answerable for their actions and that there is redress when duties and commitments are not met'.[23] This connotes an objective measure of a set of defined, agreed terms of action and an existing system of oversight, review and remediation. In contrast, integrity is a subjective concept. As discussed elsewhere in this book, integrity relates to the behaviour of an individual or institution in relation to the values which it claims to have or the values it is expected to uphold by society.[24]

On one level, the two concepts can be, but are often not, mutually exclusive. An accountability framework can function purely to monitor and guide the effectiveness and efficiency of environmental planning and project implementation processes and results which are devoid of integrity questions. Take the example of a government minister who promises to limit national GHG emissions by 50 percent by 2020. As a part of that commitment, he decides to invest in a national solar energy programme because he genuinely believes that that investment will secure a long-term, carbon-neutral energy supply in his country. However, due to unforeseeable causes such as increased cloud cover, the energy source is limited. Given the energy needs of his country, the minister may need to rely on alternative energy sources, including coal. In such a scenario, the minister should be held to account for his decisions and resulting social, economic and environmental impacts. While he may not be at fault per se, national accountability rules should allow that project results are reported and monitored regularly so that deviations can be corrected without delay. Essentially, accountability functions as an early warning system that would enable ethically correct but poor decisions to be remedied to avoid unnecessary inefficiencies and damage.

Integrity does come into play within an accountability system when decisions are unreasonable and/or corrupt. Following from the previous example, if the minister had foreseen that solar energy would not be a viable investment in the long run and knew that other renewable energy sources would have been feasible, but regardless chose the solar investment, both his integrity and accountability would be called into question. The integrity issue would be more acute if the solar investment had been driven by the minister's personal interest to receive

22 Ibid., Draft decision – /CP.21, para. 84, Article 12.

23 Open Society Foundation, "Transparency Accountability Initiative," Transparency International, <http://www.transparency-initiative.org/about/definitions>.

24 See Chapter 2, this volume.

a handsome kickback from the solar plant developer. Here the issues of integrity and accountability run together clearly. Critically, it is the accountability system which enables that integrity issues are exposed and addressed.

On another level, integrity may surpass an immediate accountability framework and beg a more robust legal accountability based on the precautionary principle. This is true when moral or ethical decision-making is demanded beyond what current accountability terms entail. Taking the previous example, the same minister may decide to introduce *some* emission controls but delay robust, comprehensive regulations. His decision is based on a reliable cost benefit analysis which shows that the short-term costs of investing in renewable energy are much higher than the expense of continued reliance on fossil fuels for the next decade. Further, his decision is accountable in that it complies with his country's international treaty obligations and his government's energy and environmental policies. However, the minister's integrity could be at issue if he was aware of (certain and potential) repercussions of his decision, that is, if sea level rising and glacier melting in his country alone could result in losses of coastal property, fresh water supply and, ultimately, economic and social welfare. Here again, the integrity question would only become a subject of accountability if the accountability regime allowed for future harms to be adequately assessed and ensured a robust application of the precautionary principle. While it may be politically difficult for States Parties to the Paris Agreement to compel each other to achieve that level of national accountability, arguably their success in averting serious catastrophe may depend on their resolve to do so.

Thin ↔ thicker governance of multilateral climate funds

In terms of finance, however, the climate regime's governance appears to be thickening on multiple tiers. Climate finance is generally understood as public and private money which is channelled from developed to developing countries through a variety of multilateral, regional and bilateral mechanisms to support developing countries' actions to mitigate and adapt to climate change.[25] The 'new' category of international financial flows best dates to 2009, when industrialised nations pledged support to the tune of USD 30 billion between 2010 and 2012.[26] It is estimated that climate finance reached USD 52.2 billion in 2013 and USD 62.8 billion in 2014.[27] This included bilateral finance totalling USD 22.5 billion

25 United Nations Framework Convention on Climate Change (UNFCCC), "2014 Biennial Assessment and Overview of Climate Finance Flows Report," (Bonn, Germany: UNFCCC Standing Committee on Finance, 2014).

26 Smita Nakhooda, Charlene Watson, Alice Caravani, Liane Schalatek, Sam Barnard, Nella Canales Trujillo, and Nick Scott, "Ten Things to Know about Climate Finance in 2012," (Berlin and Washington: ODI – Heinrich Böll Stiftung (HBF), 2012).

27 Organisation for Economic Cooperation and Development (OECD) and Climate Policy Initiative (CPI), "Climate Finance in 2013–14 and the USD 100 Billion Goal: A Report by the OECD in Collaboration with Climate Policy Initiative," (Paris: OECD-CPI, 2015).

and USD 23.2 billion in 2013 and 2014, respectively. Multilateral finance was estimated at USD 15.4 billion in 2013 and USD 20.4 billion in 2014. That money mobilised private sector co-financing to the tune of USD 12.8 billion in 2013 and USD 16.7 billion in 2014.[28] These sums do not include the amounts of public and private resources mobilised in developed or developing countries for national climate change actions. They also do not include ostensibly any globally or nationally pooled fund to assist countries in dealing with loss and damage caused by climatic changes such as extreme weather events.[29] Still, as mentioned previously, the scale of support to developing countries should increase to USD 100 billion by 2020. While this amount may still be less than what is required to meet mitigation and adaptation needs around the world,[30] it still represents a significant amount of resources which need to achieve impacts urgently.

At the UNFCCC level, the Paris Agreement marks heightened commitments for greater transparency of such climate finance flows. Under previous arrangements, transparency was limited to the developed countries disclosing how much money they had or would contribute in relation to their treaty obligations. The Paris Agreement requires such transparent reporting, but also enrols developing countries to unveil how much support they receive.[31] While the amounts of support for developing countries is to be assessed according to national needs, details on how this should be administered, and potentially how impacts can or will be assessed are to be ironed out over time.[32] These advances would arguably be important to help strengthen accountability and trust amongst States and stakeholders, particularly in countries where integrity systems are still maturing.

The next tier of climate finance accountability rests with the *sources* of public and private climate funds, as indicated above. The UNFCCC's financial mechanisms include the Green Climate Fund, the Adaptation Fund and the Global Environment Facility. These funds have been established by States Parties as conduits of climate finance to developing countries. However, these funds are not exclusive sources of climate finance. Outside the UNFCCC process, a number of other multilateral funds, as well as bilateral and private sources, exist to support climate change actions. Developed countries can fulfil their treaty obligations by contributing through these mechanisms.

28 Ibid., p. 10.

29 However, a global mechanism under the UNFCCC was established in 2013 to enable that such resources be sequestered and used for such purposes. United Nations Framework Convention on Climate Change (UNFCCC), "Warsaw International Mechanism for Loss and Damage Associated with Climate Change Impacts," <http://unfccc.int/adaptation/workstreams/loss_and_damage/items/8134.php>.

30 World Resources Institute, "Climate Finance," World Resources Institute, <http://www.wri.org/our-work/project/climate-fina>.

31 United Nations Framework Convention on Climate Change (UNFCCC), "Conference of the Parties Twenty-First Session Paris, 30 November to 11 December 2015: Adoption of the Paris Agreement," Article, 13, paras. 6, 9, 10.

32 Ibid., Article 13, para. 13.

While bilateral aid accounts for the largest percentage of climate finance contributions, the volume of public money deposited in and remitted from multilateral funds is on the rise, coming in at a close second source. At present, seven major funds exist to support and catalyse public and private enterprises to address global warming causes and effects. These include the Least Developed Countries Fund, the Special Climate Change Fund (both established by the Global Environment Facility), the Climate Investment Funds, the Forest Carbon Partnership Facility (both organised under the World Bank), the UN REDD Programme (managed by UNDP, UNEP and the FAO), the Adaptation Fund and, most recently, the Green Climate Fund. Given their mandates, the governance of these funds is becoming an issue of public scrutiny.[33] A growing number of concerned stakeholders are asking how the accountability of the use and impact of climate funds are being and will be measured and monitored, and how performance and results which fall short of what is expected are being and will be corrected. Posed differently, how will the integrity of decisions and the integrity of results to avoid serious climate change and its impacts be achieved?

From 2012–2013, Transparency International conducted anti-corruption assessments of the then six major climate funds introduced above: LDCF, SCCF,[34] CIFs,[35] FCPF,[36] UN REDD Programme[37] and the Adaptation Fund.[38] The assessments reviewed the policies of these funds under three governance parameters: transparency, accountability and integrity. Under these headings, twelve indicators were taken into account, including the scope, procedures and timeliness of information disclosure, the modalities for public participation and consultation, the existence of clear anti-corruption rules concerning conflicts of interest, bribery, embezzlement, money laundering and others, the existence of grievance mechanisms and anti-corruption hotlines, whistle-blower protection

33 In fact, numerous civil society groups, research organisation and others have attempted to map and assess the climate finance flows and how it is governed, taking into account publicly available information. See, e.g., Climate Funds Update ("CDM Rulebook: What Is Additionality?" <http://www.cdmrulebook.org/84.html>; Climate Funds Update, ODI-Heinrich Boell Stiftung, <http://www.climatefundsupdate.org/>; United Nations Framework Convention on Climate Change (UNFCCC), "Finance Portal for Climate Change," <http://www3.unfccc.int/pls/apex/f?p=116:1:3835711846125210>; United Nations Development Programme (UNDP) and World Bank, "The Climate Finance Options Platform," <http://www.climatefinanceoptions.org/cfo/node/189>.

34 Lisa Elges and C. Martin, *Protecting Climate Finance: An Anti-Corruption Assessment of the GEF LDCF and SCCF* (Berlin: Transparency International, 2013).

35 Lisa Elges and C. Martin, *Protecting Climate Finance: An Anti-Corruption Assessment of the Climate Investment Funds* (Berlin: Transparency International, 2013).

36 Lisa Elges and C. Martin, *Protecting Climate Finance: An Anti-Corruption Assessment of the Forest Carbon Partnership Facility* (Berlin: Transparency International, 2013).

37 Lisa Elges and C. Martin, *Protecting Climate Finance: An Anti-Corruption Assessment of the UN-Redd Programme* (Berlin: Transparency International, 2013).

38 Lisa Elges and C. Martin, *Protecting Climate Finance: An Anti-Corruption Assessment of the Adaptation Fund* (Berlin: Transparency International, 2013).

and investigative functions and the availability of procedures to conduct financial and integrity audits and trainings. In addition, regarding the overall governance of the funds the assessments asked, if corruption happens or, more generally, when things go wrong, what happens – that is, who is accountable to whom and under what terms? The results of the assessments observed that across all funds, their governance – based on these criteria – exhibited different levels of thickness on the policy side but relative thinness in terms of demonstrating such governance attributes in practice. In contrast, the Green Climate Fund appears to have surpassed the other funds having issued thicker governance polices across the board. Still, how thick the GCF's governance will be depends on how nascent policies (regarding independent investigations, anti-corruption and information disclosure) evolve, are implemented, and the degree to which the fund will demonstrate how effectively those policies are implemented.

One of the factors influencing the governance thickness of these funds is their complex governance structures. All of the funds are essentially intergovernmental trust funds which were established by governments through an intergovernmental process (UNFCCC) or by international organisations such as the World Bank, UNDP, UNEP, FAO and others which again are governed by States. The Climate Investment Funds, which include the Clean Technology Trust Fund and the Special Climate Change Trust Fund, were established by the World Bank in 2008.[39] The Forest Carbon Partnership Facility, also established by the Bank that same year, revolves around the administration of a Readiness Trust Fund and a Carbon Trust Fund.[40] The UN REDD Programme is a collaborative arrangement between UNDP, UNEP and the FAO and which was established in 2008. That fund is located in UNDP's Multi-Partner Trust Fund.[41] The Least Developed Countries Fund, the Special Climate Change Fund, the Adaptation Fund and the Green Climate Fund are all trust funds established by the UNFCCC Conference of States Parties.

How each of these trust funds are governed – and their overall governance fitness – is closely linked with their mandates, institutional set-up and legal personality, and the due diligence rigor of their administrative and governing bodies. It is also heavily reliant on the choice and performance of the direct recipients of the fund finance. Analysing these factors across all trust funds reveals a high level of complexity from which, with the exception of the Green Climate Fund, it has made it difficult to pinpoint where accountability rests. For example, the Global Environment Facility (GEF), although an independent institution, is a partnership among the World Bank, United Nations Development Programme (UNDP) and United Nations Environment Programme (UNEP). As such,

39 Climate Investment Funds (2015) at <http://www-cif.climateinvestmentfunds.org/about/governance>.

40 Forest Carbon Partnership Facility, "Charter and Governance Documents," World Bank Group, <https://www.forestcarbonpartnership.org/about-fcpf-0>.

41 Multi-Partner Trust Fund, "UN Redd Programme Fund," UNDP, <http://mptf.undp.org/factsheet/fund/CCF00>.

the facility is representative of all three institutions. The Facility is governed by a Council and an Assembly that comprises government officials representing countries and constituencies. Among other things, the Council directs how, for what and in what countries trust fund resources should be spent. The Facility's Secretariat comprises staff which are employed on World Bank contracts. The Secretariat can enter into legally binding Memorandums of Understanding with Global Environment Facility (GEF) Partners or Agencies which the Council had agreed to provide finance from the Least Developed Countries Fund (LDCF) or Special Climate Change Fund (SCCF) trust funds.[42] In some, but not all, cases the GEF Agencies are chosen through an accreditation process which assesses the fiduciary and other competencies based on minimum standards which the GEF has determined. Those Partners which were not accredited should adhere to those standards. The GEF Council appointed the World Bank as its trustee, which is empowered to conclude financial arrangements with GEF Partners and to enforce the term of those contracts. This includes the suspension of funding in the event the Partner misuses the funds, which the trustee may detect through audit reports.[43]

However, while the governance arrangement is progressive, its complexity and the GEF's lack of legal personality dilutes its overall governance rating. As a partnership, and not an independent legal entity, the GEF itself does not have the power to independently investigate misconduct or other corrupt and unethical behaviour of its Council members or of its Partners. It does maintain a conflict resolution function which can assist in addressing wrongdoings, but the weight of this function appears less than having a bespoke investigatory body. The Facility is also limited in the sanctions it can apply to remediate such problems. For example, the GEF's policy does not allow for the suspension of accreditation or dis-accreditation of its Partner Agencies. As explained previously, the trustee can suspend funds in cases of wrongdoing, but it is limited in reclaiming funds lost to corruption. While the trustee can verify financial discrepancies through audit reports, it cannot investigate specific cases and is therefore limited in its capacity to verify unethical conduct. Further, while the GEF's accreditation-related due diligence checks are employed to safeguard against such problems from arising, the GEF has not sufficiently communicated evidence of the fiduciary/integrity performance of its Partner Agencies.[44] Likewise, there is no procedure for appealing GEF Council decisions: a government whose funding proposal was rejected by the GEF Council has no formal way of appealing that decision. While

42 See, for example, Global Environment Facility, "Memorandum of Understanding: African Development Bank, the African Development Fund and the Secretariat of the Global Environment Facility on Direct Access to GEF Resources," (Global Environment Facility, 2005).

43 Global Environment Facility, *Instrument for the Establishment of the Restructured Global Environment Facility* (Washington: Global Environment Facility, 2015).

44 Transparency International, "Strengthening Whistleblower Protection: An International Imperative," (Berlin: Transparency International, 2013).

the Council's decision-making-by-consensus procedure helps to ensure against the need for misgivings to arise, the possibility exists for such problems to arise, particularly as Council members are not bound by a conflict-of-interest policy.

For similar reasons, the Adaptation Fund's governance could improve its thickness. The trust fund is governed by a Board that ensures balanced representation between developed and developing countries. It is administered by the Adaptation Fund Secretariat, which is functionally the GEF Secretariat staffed by World Bank employees. The Fund's trustee is also the World Bank. However, the trustee's role is concentrated on financial transactions and audits instructed by the Board. As a legal entity, the Fund Board has the power to conclude direct grant agreements with national, multilateral and regional public and non-governmental organisations, which it accredits through a comprehensive procedure. The terms of those agreements and the Fund's governing instruments allow the Fund's Board to suspend and terminate both funding and accreditation. However, the Fund's governance structure does not provide for an independent investigative function – although, like the GEF, the Secretariat has an ombudsman role which allows it to address complaints. The Fund's Board has also adopted additional zero tolerance and related policies to mitigate corruption risks and the need for investigatory powers. Still, the limitation poses functional problems. For example, at some point in or prior to 2014, a material complaint regarding an 'integrity violation' by one of the Fund's accredited entities was brought to the Secretariat's attention. Although due to confidentiality reasons the name of the entity, as well as the chronology of actions taken to address that entity's behaviour, are not disclosed, Board documents refer to an ongoing process of the complaints review, recommendations that the entity's accreditation should be suspended and efforts to allow the entity to defend itself and correct the problem.[45] Still, at its October 2015 meeting, the Board decided to defer a decision on the matter until March 2016,[46] at which point a decision was further deferred until July.[47] At that, all told, the entire process will have taken over two years – a time period which must be deemed unreasonable by acceptable governance standards.

In contrast to these experiences, the Green Climate Fund, being established as an independent legal entity, has from the outset been designed with an Independent Integrity Unit (corruption and fraud investigative function) and an Independent Review Mechanism (IRM).[48] The latter is tasked with addressing

45 See Adaptation Fund Board, "Decisions of the Twenty-Sixth Meeting of the Adaptation Fund Board," (AFB/B.26/6: World Bank, 2015); Adaptation Fund Board, "Report of the Seventeenth Meeting of the Ethics and Finance Committee for Board Report: Ethics and Finance Committee," (AFB/EFC.17/92015).

46 Adaptation Fund Board, "Decisions of the Twenty-Sixth Meeting of the Adaptation Fund Board"; Adaptation Fund Board, "Report of the Seventeenth Meeting of the Ethics and Finance Committee for Board Report: Ethics and Finance Committee."

47 Adaptation Fund Board, "Decisions of the Twenty-Sixth Meeting of the Adaptation Fund Board."

48 Green Climate Fund, "Report of the Sixth Meeting of the Board: Terms of Reference Independent Review Mechanism (IRM)," (GCF/B.06/19, Anex V2014).

grievances arising from non-compliance with the Fund's environmental and social safeguard and hearing appeals of governments not satisfied with Board decisions, such a decisions to reject project-funding proposals. In addition, the Fund has comprehensive anti-corruption and ethics policies in place for its Board members, their Advisors, the Secretariat and Consultants. They employ a rigorous accreditation process whereby actors wishing to receive GCF funding must demonstrate compliance to high levels of fiduciary standards. These include (unlike other multilateral funds) anti-money laundering and counter-terrorism financing rules. Further, if accredited actors fail to ensure these standards in financing or implementing GCF projects, the Fund may apply a wide set of remedial sanctions including debarment, finance suspension and dis-accreditation. If an accredited entity would fail to prevent corruption or fraud which amounts to GCF funds being lost or otherwise misused, the Fund can demand that the entity reimburse the Fund accordingly.[49] It appears that these conditions apply not only to accredited entities but also to readiness grant recipients.

As such, amongst most multilateral climate funds, the GCF exhibits the thickest policy-level governance model around. However, because many of its policies are not yet operationalised, a verdict on its governance strength remains premature. One current concern is that while the Fund is meant to have independent complaints review and investigatory functions, the chiefs of these bodies were not yet hired at the time of this writing. Only after these heads are in place can the unit bodies further grow in terms of rules, resources and staff. The delay in inaugurating these functions will likely foster some glitches. For instance, at the Funds' eleventh Board meeting, a number of Board members questions the ethics and independence of the GCF staff in selecting projects for approval.[50] Out of eighty some proposals, the Secretariat selected eight for approval by the Board without explanation as to why other proposals were not put forward. Quite fairly, some Board members questioned the Secretariat's rationale for doing so. Inside and outside the Boardroom the issue was discussed, consuming most of the three-day Board meeting. As a result, other agenda items, including the Fund's anti-corruption policy, had to be delayed to a future Board meeting decision.

Another stark concern across all funds is the issue of whistle-blower protection. For any accountability mechanism to work it must provide a secure forum for anyone to submit a complaint or to report wrongdoing which they witness. A reporting person should be protected against any form of retaliation.[51] All assessed climate funds require that their intermediaries provide such protection. However, the scope and extent to which both grievance and whistle-blower protection policies apply across all intermediaries are not consistent and, in a number

49 Green Climate Fund, "Twelfth Meeting of the Board: 8–10 March 2016," (GCF/B.12/322016), at 145.

50 Green Climate Fund, "Decisions of the Board – Eleventh Meeting of the Board, 2–5 November 2015," (GCF/B.11/242015).

51 Transparency International, "Strengthening Whistleblower Protection: An International Imperative," (Berlin: Transparency International, 2013).

of cases, lack sufficient transparency. Among all funds, only the Climate Investment Funds provide a bespoke web page on transparency and accountability where the accountability mechanisms of each of its implementing entities are listed with hyperlinks.[52] This makes such information easily accessible for anyone who wants to understand how to make a report and what conditions will apply. The other funds do not yet provide this level of transparency. In many cases, key accountability information is buried on the fund's website or on the website of its accredited actors, or even elsewhere. For example, one national financial institution which was recently accredited to both the Global Environment Facility and the Green Climate Fund, provides a link to report a case of fraud on its website. However, by clicking on this button, one is connected to a Deloittes website[53] for reporting such cases, without any explanation or warning, which is disconcerting and likely ineffective.[54]

Conclusion

The governance of global climate change processes and multilateral funds is thickish but demands additional reforms and evidence to demonstrate that policy-level agreements are achieving results. Some reasons why governance challenges remain within the climate policy and finance regimes include complex accountability structures with multiple actors having different legal and non-legal responsibilities, making it difficult to discern who is in charge and what rules apply. The Green Climate Fund is, so far, the best example of a denser policy governance model, but even it needs to gain weight through implementation. Such governance reforms are likely to provide an important impetus to governance changes in international, regional and national financial institutions, thereby leading to transformational advances critical to sustainable development and the prevention of serious climate change.

References

Adaptation Fund Board. 'Decisions of the Twenty-Sixth Meeting of the Adaptation Fund Board.' AFB/B.26/6: World Bank, 2015.

———. 'Report of the Seventeenth Meeting of the Ethics and Finance Committee for Board Report: Ethics and Finance Committee.' AFB/EFC.17/9, 2015.

'CDM Rulebook: What Is Additionality?' <http://www.cdmrulebook.org/84.html> (accessed 6 June 2016).

Climate Funds Update. 'ODI-Heinrich Boell Stiftung.' <http://www.climatefundsupdate.org/>.

52 See Climate Investment Funds, "Transparency & Accountability." Available at https://www-cif.climateinvestmentfunds.org/about/transparency-accountability (accessed 31 August 2016).

53 Deloittes, "Welcome to Deloitte Tip-Offs Anonymous." <https://www.tip-offs.com/>.

54 While Transparency International alerted an officer of the Development Bank of South Africa (DBSA) to this concern in November 2015, at the time of this writing the DBSA had still not corrected this issue.

Climate Investment Funds. 'Transparency & Accountability.' Available at https://www-cif.climateinvestmentfunds.org/about/transparency-accountability (accessed 31 August 2016).

Deloittes. 'Welcome to Deloitte Tip-Offs Anonymous.' <https://www.tip-offs.com/> (accessed 6 June 2016).

Elges, Lisa, and C. Martin. *Protecting Climate Finance: An Anti-Corruption Assessment of the Adaptation Fund.* Berlin: Transparency International, 2013.

———. *Protecting Climate Finance: An Anti-Corruption Assessment of the Climate Investment Funds.* Berlin: Transparency International, 2013.

———. *Protecting Climate Finance: An Anti-Corruption Assessment of the Forest Carbon Partnership Facility.* Berlin: Transparency International, 2013.

———. *Protecting Climate Finance: An Anti-Corruption Assessment of the GEF LDCF and SCCF.* Berlin: Transparency International, 2013.

———. *Protecting Climate Finance: An Anti-Corruption Assessment of the UN-Redd Programme.* Berlin: Transparency International, 2013.

Forest Carbon Partnership Facility. 'Charter and Governance Documents.' World Bank Group, <https://www.forestcarbonpartnership.org/about-fcpf-0> (accessed 6 June 2016).

Global Environment Facility. *Instrument for the Establishment of the Restructured Global Environment Facility.* Washington: Global Environment Facility, 2015.

———. 'Memorandum of Understanding: African Development Bank, the African Development Fund and the Secretariat of the Global Environment Facility on Direct Access to GEF Resources.' Global Environment Facility, 2005.

Green Climate Fund. 'Decisions of the Board – Eleventh Meeting of the Board, 2–5 November 2015.' GCF/B.11/24, 2015.

———. 'Report of the Sixth Meeting of the Board: Terms of Reference Independent Review Mechanism (IRM).' 54–57. GCF/B.06/19, Anex V, 2014.

———. 'Twelfth Meeting of the Board: 8-10 March 2016.' Agenda Item 30. GCF/B.12/32, 2016.

Multi-Partner Trust Fund. 'UN Redd Programme Fund.' UNDP, <http://mptf.undp.org/factsheet/fund/CCF00> (accessed 6 June 2016).

Nakhooda, Smita, Charlene Watson, Alice Caravani, Liane Schalatek, Sam Barnard, Nella Canales Trujillo, and Nick Scott. 'Ten things to Know about Climate Finance in 2012.' Berlin and Washington: ODI – Heinrich Böll Stiftung (HBF), 2012.

Open Society Foundation. 'Transparency Accountability Initiative.' Transparency International, <http://www.transparency-initiative.org/about/definitions>.

Organisation for Economic Cooperation and Development (OECD), and Climate Policy Initiative (CPI). 'Climate Finance in 2013–14 and the USD 100 Billion Goal: A Report by the OECD in Collaboration with Climate Policy Initiative.' Paris: OECD-CPI, 2015.

Transparency International. *Strengthening Whistleblower Protection: An International Imperative.* Berlin: Transparency International, 2013.

United Nations Development Programme (UNDP), and World Bank. 'The Climate Finance Options Platform.' <http://www.climatefinanceoptions.org/cfo/node/189> (accessed 6 June 2016).

United Nations Framework Convention on Climate Change (UNFCCC). '2014 Biennial Assessment and Overview of Climate Finance Flows Report.' Bonn, Germany: UNFCCC Standing Committee on Finance, 2014.

———. 'The Conference of the Parties (COP), at Its Fifteenth Session, Took Note of the Copenhagen Accord of 18 December 2009.' FCCC/CP/2009/11/Add.1, 2009.

———. 'Conference of the Parties Twenty-First Session Paris, 30 November to 11 December 2015: Adoption of the Paris Agreement.' FCCC/CP/2015/L.9/Rev. 1: United Nations, 2015.
———. 'Finance Portal for Climate Change.' <http://www3.unfccc.int/pls/apex/f?p=116:1:3835711846125210> (accessed 6 June 2016).
———. 'Framework Convention on Climate Change.' 1771 UNTS 107 / [1994] ATS 2 / 31 ILM 849 [1992] 1992.
———. 'An Introduction to the Kyoto Protocol Compliance Mechanism.' UNFCCC, <http://unfccc.int/kyoto_protocol/compliance/items/3024.php> (accessed 6 June 2016).
———. 'Kyoto Protocol to the United Nations Framework Convention on Climate Change, 10 December 1997.' U.N. Doc FCCC/CP/1997/7/Add.1, 37 I.L.M.: United Nations, 1997.
———. 'Report of the Conference of the Parties on Its Seventeenth Session, Held in Durban from 28 November to 11 December 2011.' Addendum. Part Two: Action taken by the Conference of the Parties at Its Seventeenth Session: United Nations, 2012.
———. 'Report of the Conference of the Parties on Its Sixteenth Session, Held in Cancun from 29 November to 10 December 2010.' Part Two: Action Taken by the Conference of the Parties at Its Sixteenth Session FCCC/CP/2010/7/Add.1, 2011.
———. 'Warsaw International Mechanism for Loss and Damage Associated with Climate Change Impacts.' <http://unfccc.int/adaptation/workstreams/loss_and_damage/items/8134.php> (accessed 6 June 2016).
World Resources Institute. 'Climate Finance.' World Resources Institute, <http://www.wri.org/our-work/project/climate-fina> (accessed 6 June 2016).

11 Co-producing climate-smart agriculture knowledge through social networks

Future directions for climate governance

Joseph Macharia and Hope Johnson

Introduction

Stakeholders commonly position agriculture as having a neglected status in climate change negotiations when compared to, for instance, improving forest management and cutting carbon emissions from other industrial sectors.[1] Yet the Public Institutional Justification of the global climate regime, found in Article 2 of the Convention, requires the:

> [s]tabilization of greenhouse gas concentrations in the atmosphere at a level that would prevent dangerous anthropogenic interference with the climate system. Such a level should be achieved within a time frame sufficient to allow ecosystems to adapt naturally to climate change, *to ensure that food production is not threatened* and to enable economic development to proceed in a sustainable manner.[2]

Thus, the rationale for the creation of the UNFCCC is closely connected with the protection of food production. This focus on food production could perhaps strengthen the public support for the UNFCCC, given the fundamental place of agriculture in economies and societies, as well as the need to meet basic human needs such as food. In fact, the Food and Agriculture Organization of the United Nations reported '[i]t was food security and hence agriculture, which was the primary concern when the world agreed to meet in Rio de Janeiro in 1992 to negotiate the first framework convention on climate change (UNFCCC)'.[3]

1 Asian Development Bank, "Making Grasslands Sustainable in Mongolia: Assessment of Key Elements in the Nationally Appropriate Mitigation Actions for Grassland and Livestock Management Development," in *Strengthening Carbon Financing for Grassland Management in the PRC and Mongolia* (Manila: Asian Development Bank, 2014), p. 1; Laura Angela Bagnetto, "Agriculture Neglected in COP21 Negotiation Says AU Expert: Interview with Esterine Fotabong," *Corporate Africa*, 15 April 2016.

2 United Nations Framework Convention on Climate Change (UNDCCC), "The Kyoto Protocol," (37 ILM 22, 1998), Article 2.

3 Food and Agriculture Organization of the United Nations (FAO), "Climate-Smart Agriculture: A Call for Action," (Paper Presented at the Asia-Pacific Regional Workshop, Bangkok, Thailand, 18 to 20 June 2015).

Article 2 applies not only to the Convention, but also to any related legal instrument that the Conference of the Parties (COP) adopts. Accordingly, ensuring food production despite changing climatic conditions is a purpose that permeates the activities of the UNFCCC. In particular, the Paris Agreement reaffirms the central objective of ensuring food production by seeking to increase 'the ability to adapt to the adverse impacts of climate change and foster climate resilience and low greenhouse gas emissions development, in a manner that does not threaten food production'.[4] Unlike the UNFCCC, however, the Paris Agreement connects the emphasis on ensuring food production with the broader objective of food security in its preamble, which recognises 'the fundamental priority of safeguarding food security and ending hunger, and the particular vulnerabilities of food production systems to the adverse impacts of climate change'.

Climate change is expected to affect agriculture and food security by resulting in more frequent, longer running and more extreme weather events, new pest invasions and diseases and changes in temperatures and rainfall.[5] Small-scale farmers in low-income countries are particularly vulnerable to the impacts of climate change on agriculture, as fluctuating world food prices or farm conditions could significantly affect both their income levels and the amount of food that they can consume.[6] Given that most of the world's hungry comprise small-scale farmers, a key way to address food security is to increase the agricultural productivity and the incomes of rural farming communities despite the impacts of climate change.[7]

At the same time as being threatened by climate change, agriculture is a major emitting sector and one of the key avenues to mitigate and adapt to the effects of climate change.[8] Subsequently, it is widely agreed that 'business as usual' is no

4 United Nations Framework Convention on Climate Change (UNFCCC), "Conference of the Parties: COP21 Paris Agreement," (FCCC/CP/2015/L.9: UNFCCC, 2015), Article 2(b).

5 For overviews see, for example, Alvaro Calzadilla, Katrin Rehdanz, Richard Betts, Pete Falloon, Andy Wiltshire, and Richard S.J. Tol, "Climate Change Impacts on Global Agriculture," *Climatic Change* 120, no. 1–2 (2013); Food and Agriculture Organization of the United Nations (FAO), "Climate Change and Food Security: A Framework Document," (Rome: FAO, 2008); Tim Wheeler and Joachim von Braun, "Climate Change Impacts on Global Food Security," *Science* 341, no. 6145 (2013); Cheryl A. Palma, Sean M. Smuklera, Clare C. Sullivana, Patrick K. Mutuoa, Gerson I. Nyadzia, and Markus G. Walsha, "Identifying Potential Synergies and Trade-Offs for Meeting Food Security and Climate Change Objectives in Sub-Saharan Africa," *Proceedings of the National Academy of Sciences of the United States of America* 107, no. 46 (2010); Henrik Urdal, "Demographic Aspects of Climate Change, Environmental Degradation and Armed Conflict," (New York: United Nations, 2008); Heidi Sulda, John Coveney, and Michael Bentley, "An Investigation of the Ways in Which Public Health Nutrition Policy and Practices Can Address Climate Change," *Public Health Nutrition* 13, no. 3 (2010).

6 Food and Agriculture Organization of the United Nations (FAO), "Climate Change and Food Security: A Framework Document."

7 World Bank, "World Development Report 2008: Agriculture for Development," (2007), p. 109, Available at http://siteresources.worldbank.org/INTWDR2008/Resources/WDR_00_book.pdf (accessed 30 August 2016), pp. 1–365; Pedro Sanchez, Philip Dobie, and Nolan Yuksel, *Halving Hunger: It Can Be Done* (London: United National Development Programme; UN Millennium Project, 2005), p. 6.

8 See e.g. Sonja J. Vermeulen, Bruce M. Campbell, and John S.I. Ingram, "Climate Change and Food Systems," *Annual Review of Environment and Resources* 37, no. 1 (2012).

longer an option for agriculture.[9] Instead, more sustainable forms of agriculture are required to ensure food production and security in changing climatic conditions than the currently widespread, industrial approaches. As a result, different models of agriculture have been put forward as alternatives to 'business-as-usual'. These models encompass different practices, philosophies and objectives and include 'organic', 'conservation' and 'sustainably intensive' agriculture.

One of the most prominent concepts is 'climate-smart agriculture'. This concept acknowledges that agriculture can be carried out in ways that contribute to food security while mitigating and adapting to climate change.[10] Climate-smart agriculture (CSA) is defined as 'agriculture that sustainably increases productivity, resilience (adaption), reduces/removes GHGs (mitigation), and enhances achievement of national food security and development goals'.[11] Accordingly, farming practices and regulatory interventions that are consistent with the goals of improving food security, adapting farms to changing conditions and mitigating climate change are considered 'climate-smart'.[12]

A key barrier to the adoption of CSA is a lack of knowledge or awareness regarding appropriate farming practices.[13] The Paris Agreement recognises the need to build capacity within developing countries to mitigate climate change,[14] and also will allow parties to obtain mitigation co-benefits resulting from adaptation actions/economic diversification plans that can contribute to mitigation outcomes.[15] This means that the UN Climate regime encourages the implementation of CSA practices by allowing for such practices to be counted as both mitigation and adaptation activities and also creates obligations for developed countries to assist developing countries to implement such initiatives.

Breakey, Cadman and Sampford argue that governance can be carried out by a myriad of different entities, including corporations, non-governmental

9 International Assessment of Agricultural Knowledge Science and Technology for Development (IAASTD), *Agriculture at a Crossroads: Synthesis Report*, ed. Beverly D. McIntyre, Hans R. Herren, Judi Wakhungu and Robert T. Watson (Washington: UNEP, 2009), p. 3.

10 See Leslie Lipper, Philip Thornton, Bruce M. Campbell, Tobias Baedeker, Ademola Braimohm, Martin Bwalya, Patrick Caron, Andrea Cattaneol, Dennis Garrity, Kevin Henry, Ryan Hottle, Louise Jackson, Andrew Jarvis, Fred Kossam, Wendy Mann, Nancy McCarthy, Alexandre Meybeck, Henry Neufeldt, Tom Remington, Pham Thi Sen, Reuben Sessa, Reynolds Shula, Austin Tibu, and Emmanuel F. Torquebiau, "Climate-Smart Agriculture for Food Security," *Nature Climate Change* 4 (2014).

11 Food and Agriculture Organization of the United Nations (FAO), "Climate-Smart Agriculture Policies, Practices and Financing for Food Security, Adaptation and Mitigation: Background Paper for the Hague Conference on Agriculture, Food Security and Climate Change," (Rome: FAO, 2010).

12 World Bank, "Climate Smart Agriculture: A Triple Win," http://www.worldbank.org/en/news/video/2011/09/01/climate-smart-agriculture-a-triple-win.

13 "Policy Brief 73247: Opportunities and Challenges for Climate-Smart Agriculture in Africa," Agriculture, Forestry & Fisheries South Africa And African Union, https://ccafs.cgiar.org/sites/default/files/assets/docs/au_policybrief_opportunitieschallenges.pdf.

14 United Nations Framework Convention on Climate Change (UNFCCC), "Conference of the Parties: COP21 Paris Agreement," Article 4(5).

15 Ibid., Article 4(7).

organisations (NGOs) and community groups.[16] This chapter argues that non-state actors will play a large role in facilitating knowledge about CSA and uses a case study of an online platform called 'Mkulima Young' (means 'young farmer' in Swahili) to show how knowledge about CSA can be disseminated and acted upon. It is suggested that the UN Climate regime can enhance climate governance by seeking partnerships with non-state institutions, such as Mkulima Young, who disseminate CSA knowledge. Such partnerships will be crucial to see a wider take-up of CSA activity implementation.

Climate-smart agriculture: Farming practices and critiques

The UNFCCC has not developed terminology to describe agricultural practices, which deliver mitigation/adaptation benefits. As such, it is necessary to look at concepts developed by other international institutions with expertise in agriculture in order to gain an understanding of concepts that could be brought within the UNFCCC in respect of agriculture. This chapter focuses on the concept of CSA as opposed to other agriculture concepts such as food sovereignty and conservation agriculture, as CSA is the only concept which expressly connects production with climate change adaptation/mitigation.[17] CSA was first coined by the Food and Agricultural Organization of the United Nations ('FAO')[18] as an umbrella term for those on and off farm practices, policies, investments and technologies required to help farms mitigate and adapt to climate change while increasing productivity.[19] The FAO considers the dimensions of CSA to be sustainable farm and land management, including those practices that conserve ecosystem services, interventions in food processing and retail and consumption that promote CSA and services for farming communities that enable them to improve farm resilience while mitigating climate change.[20] In other words, CSA must be facilitated through enabling financial, policy and institutional arrangements, yet it is at the farm level that CSA will be realised.

Despite the importance of farm-level changes, some commentators argue that CSA does not encompass a specific set of practices that are considered

16 Chapter 2, this collection.

17 Hope Johnson, "From the Ground Up: An Analysis of the Public International Regulation of Agriculture Using a Rights-Based Approach to Food Security," (Queensland University of Technology, 2016).

18 The FAO used CSA in a document prepared as technical input for the Hague Conference on Food Security, Agriculture and Climate Change. "Questions and Answers: Knowledge of Climate-Smart Agriculture," Food and Agriculture Organization of the United Nations, https://ccafs.cgiar.org/climate-smart-agriculture-0#.V0e0oIR95hE.

19 See Food and Agriculture Organization of the United Nations (FAO), "Climate-Smart Agriculture Policies, Practices and Financing for Food Security, Adaptation and Mitigation: Background Paper for the Hague Conference on Agriculture, Food Security and Climate Change."

20 "Climate-Smart Agriculture," *Food and Agriculture Organization of the United Nations*, 1 September 2016.

'climate-smart'.[21] To be 'climate-smart', then, a farming practice must have the potential to contribute to the outcomes sought by CSA, that is, the practices must contribute to food security and climate change adaption and mitigation. In practice though, the term 'climate-smart' is applied to any farming practices that have the potential to do one of the following: increase yields, reduce GHG emissions from agriculture, increase carbon sequestered in soil and biomass or improve adaptation to climate change.[22]

Accordingly, most farming practices could be considered as contributing to food security where they form part of food production.[23] Likewise, the most efficient use of inputs and sustainable uses of natural resources could be considered 'climate-smart'. It seems then that most farming practices would be considered 'climate-smart' except for those practices that are clearly not 'climate-smart', such as intensive tillage or significant removal of vegetation or forests for agricultural purposes.

As this discussion illustrates, CSA is designed to be applicable to large and small farms and encompasses diverse farming methods, from, for instance, the use of genetically modified organisms to agroforestry.[24] By not being explicit about the kinds of farming models it promotes, CSA takes a middle-ground approach to the key debate in the area, which pits industrial agriculture against smaller-scale, ecologically based farms. In doing so, CSA is a concept that can be adopted by groups with different visions of how to farm in a changing climate. But the concept's flexibility also limits its ability to challenge 'business as usual' in order to adapt farming systems to changing climatic conditions.

Industrial or conventional agriculture is an approach to farm management that employs specialised and intensive practices based on lowering production costs by increasing output.[25] Beginning in the 1950s, industrial agriculture has been adopted worldwide and relies on synthetic external inputs such as fertilisers and by practices such as monoculture cropping or concentrated animal feeding operations.[26]

21 Henry Neufeldt, Molly Jahn, Bruce M. Campbell, John R. Beddington, Fabrice DeClerck, Alessandro De Pinto, Jay Gulledge, Jonathan Hellin, Mario Herrero, Andrew Jarvis, David LeZaks, Holger Meinke, Todd Rosenstock, Mary Scholes, Robert Scholes, Sonja Vermeulen, Eva Wollenberg, and Robert Zougmoré, "Beyond Climate-Smart Agriculture: Toward Safe Operating Spaces for Global Food Systems," *Agriculture & Food Security* 2, no. 12 (2013): p. 13; Doreen Stabinsky, Climate-Smart Agriculture: Myths and Problems, (Berlin: Heinrich-Böll-Stiffung, 2014), https://br.boell.org/sites/default/files/uploads/2014/09/epaper_climate_smart_eng_boll_brasil18.09.014.pdf.

22 Neufeldt et al., "Beyond Climate-Smart Agriculture: Toward Safe Operating Spaces for Global Food Systems," p. 13; Stabinsky, *Climate-Smart Agriculture: Myths and Problems.*

23 Note though that food security is a much broader concept than food production. See in particular, Amartya Sen, "Famines," *World Development* 8, no. 9 (1980), which led to a greater understanding of food insecurity, including a lack of economic resources to purchase food.

24 World Bank, "Climate-Smart Agriculture a Call to Action," (Washington: World Food Programme, 2012), pp. 4–5.

25 See e.g., Michael Duffy, "Economies of Size in Production Agriculture," *Journal of Hunger & Environmental Nutrition* 4, no. 3 (2009).

26 See Harriet Friedmann and Philip McMichael, "Agriculture and the State System: The Rise and Decline of National Agricultures, 1870 to the Present," *Sociologia Ruralis* 29, no. 2 (1989); Robert E. Evenson and D. Gollin, "Assessing the Impact of the Green Revolution, 1960 to 2000," *Science* 300, no. 5620 (2003).

Such farming methods significantly contribute to climate change by, for instance, involving deforestation for large plantations, degrading the carbon content in soils through the high use of annual (not perennial) crops and by emitting large amounts of methane from animal agriculture.[27] Industrial farming methods promoted by CSA include the use of genetically modified varieties (in particular, ones that are high yielding or drought tolerant), reproductive biotechnologies for livestock genetic improvement and the efficient use of pesticides and other synthetic inputs.[28]

On the other end of the spectrum is ecological-based agriculture, or 'agro-ecology'. Marsden et al. explained that '[a]gro-ecology emerged in the 1980s as an interesting attempt to establish a scientific basis for alternatives to industrialized agriculture'.[29] It is an approach to farm management that uses the interactions within an ecosystem to produce food and includes practices such as integrating crops and livestock, recycling biomass and enhancing soil biotic activity.[30] In fact, ago-ecological practices form the foundation of many alternative agricultural models, such as organic and conservation agriculture. Such farming practices have been shown to help farms mitigate and adapt to climate change by increasing the amount of carbon stored in soils, improving the resilience of farms to shocks and enhancing nearby natural resources and environmental services.[31]

CSA significantly promotes agro-ecology, as it emphasises the importance of adopting an ecosystems approach to farm management to achieve the multiple outcomes desired by CSA.[32] In particular, various FAO documents regarding

27 Henning Steinfeld, *Livestock's Long Shadow: Environmental Issues and Options* (Rome: FAO, 2006); Artemi Cerdà, Antonio Jordán, Lorena Zavala, María José Marqués, and Agata Novara, "Soil Erosion and Agriculture," *Soil and Tillage Research* 106 (2009); Brenda B. Lin and Michael Jahi Chappel, "Effects of Industrial Agriculture on Climate Change and the Mitigation Potential of Small-Scale Agro-Ecological Farms," *CAB Reviews: Perspectives in Agriculture, Veterinary Science, Nutrition, and Natural Resources* 6, no. 20 (2011).

28 See, for example, M.S. Swaminathan, "Role of Genetic Modification in Developing Climate Smart Agriculture to Ensure Sustained Food Security," *Agricultural Research* 1 (2012).

29 Terry Marsden, Jo Banks, Henk Renting, and Jan Douwe Van Der Ploeg, "The Road Towards Sustainable Rural Development: Issues of Theory, Policy and Research Practice," *Journal of Environmental Policy & Planning* 3, no. 2 (2001): p. 79.

30 See, for example, Steve Gliessman, "Agroecology and Food System Change," *Journal of Sustainable Agriculture* 35, no. 4 (2011); Miguel A. Altieri, "Linking Ecologists and Traditional Farmers in the Search for Sustainable Agriculture," *Frontiers in Ecology and the Environment* 2, no. 1 (2004).

31 J.N. Pretty, A.D. Noble, D. Bossio, J. Dixon, R.E. Hine, F.W. Penning De Vries, and J.I. Morison, "Resource-Conserving Agriculture Increases Yields in Developing Countries," *Environmental Science & Technology* 40, no. 4 (2006): pp. 1114–1119, in which 286 agro-ecological farming projects that took place in over 57 countries were analysed. The authors found that crop yield had increased by an average of 79 percent, water efficiency and carbon sequestration was significantly improved and the majority of projects saw a decline in pesticide use by 71 percent.

32 See, generally, Food and Agriculture Organization of the United Nations (FAO), *Climate-Smart Agriculture: Source Book* (Rome: FAO, 2013); Food and Agriculture Organization of the United Nations (FAO), "Climate-Smart Agriculture Policies, Practices and Financing for Food Security, Adaptation and Mitigation: Background Paper for the Hague Conference on Agriculture, Food Security and Climate Change."

CSA refer to 'ecological-based farm management' strategies.[33] Numerous agro-ecological farming practices are commonly associated with CSA, including crop rotation, inter-cropping, composting, integrated pest management and drip irrigation.[34] In essence, then, CSA promotes the uptake of agro-ecology without making it explicit and without significantly critiquing industrial agricultural systems.

The criticisms of CSA largely stem from civil society and academics, and are connected to the ways in which the concept is endorsed and governed. In particular, the Global Alliance for Climate-Smart (GACSA), which was launched by the FAO at the UN Climate Summit in 2014, is a multi-stakeholder initiative designed to promote CSA by creating a space for knowledge- and practice-sharing related to CSA.[35] The GASCA brings together various stakeholders, including governments, farmer organisations and corporations which are 'committed to incorporate climate-smart approaches encompassing all scales and types of agriculture systems'.[36]

It is the corporate membership and the structure of the GASCA that has raised the concerns of civil society organisations and led to claims of 'green washing'. Specifically, civil society has critiqued the governance structure of the GASCA for lacking monitoring and accountability mechanisms.[37] In a similar vein, various groups have raised concerns regarding the corporate influence on the activities carried out by GASCA.[38] For example, 17.2 percent of the GASCA's membership is from the private sector, with 60 percent of these members originating from the fertiliser industry.[39] Furthermore, some of the largest agricultural companies such

33 Food and Agriculture Organization of the United Nations (FAO), "Final Report for the International Symposium on Agroecology for Food Security and Nutrition," (Rome: FAO, 2015); Food and Agriculture Organization of the United Nations (FAO), "Climate-Smart Agriculture Policies, Practices and Financing for Food Security, Adaptation and Mitigation: Background Paper for the Hague Conference on Agriculture, Food Security and Climate Change"; Food and Agriculture Organization of the United Nations (FAO), *Climate-Smart Agriculture: Source Book.*

34 Giacomo Branca, Nancy McCarthy, Leslie Lipper, and Maria Christina Jolejole, "Climate-Smart Agriculture: A Synthesis of Empirical Evidence of Food Security and Mitigation Benefits from Improved Cropland Management," (Rome: Food and Agriculture Organization of the United Nations (FAO), 2011), pp. 22–23.

35 Global Alliance For Climate-Smart Agriculture (GACSA), "Framework Document: GACSA Series Document 1," (FAO, 1 September 2014).

36 "About GACSA," Global Alliance for Climate-Smart Agriculture, http://www.fao.org/gacsa/en/.

37 Friends of the Earth International, "Agroecology and Climate Justice: A People's Guide to Paris and Beyond," *Fact Sheet*, 13 November 2015.

38 Climate Technology Network (CTN), "Corporate-Smart Greenwash: Why We Reject the Global Alliance on Climate-Smart Agriculture," (Copenhagen: CTN, 1 January 2014), signed statement; Ben Lilliston, 4 December, 2015, http://www.iatp.org/blog/201512/the-clever-ambiguity-of-climate-smart-agriculture; Alana Mann, 20 January, 2015, http://sydney.edu.au/environment-institute/blog/whats-wrong-with-climate-smart-agriculture/.

39 CIDSE, "Climate-Smart Revolution . . . or a New Era of Green-Washing?" (Brussels: CIDSE, 2015).

as Monsanto, Kellogg Company and Yara International, which are dependent on industrial agricultural models to sell their products, have endorsed climate-smart agriculture.[40]

As a result, reputable NGOs operating in the area of agriculture and climate change, such as Oxfam and the International Planning Committee for Food Sovereignty, have not become members of the GASCA. In one joint statement, 350 civil society organisations argued:

> 'Climate-Smart Agriculture' may sound promising, but it is a politically-motivated term. The approach does not involve any criteria to define what can or cannot be called 'Climate Smart'. Agribusiness corporations that promote synthetic fertilisers, industrial meat production and large-scale industrial agriculture – all of which are widely recognised as contributing to climate change and undermining the resilience of farming systems – can and do call themselves 'Climate Smart'.[41]

Put differently, civil society organisations are focused on the organisational context of agricultural corporations, which tends to be one of market dominance dependent on industrial agriculture, distant dealings with farmers and economies of scale.[42] Thus, the concern is that corporate interests will prevent action and transitions towards genuinely CSA. Likewise, because CSA does not engage critically with power imbalances within the current food system and is applicable to both industrial and more ecologically managed farms, the concept risks being co-opted by groups that wish to enforce 'business as usual'.

Similar to 'sustainable development', then, CSA as a policy concept is useful because it is capable of having different meanings depending on the speaker and context, but these characteristics also limit its transformative value. Mooney and Hunt explain that some concepts become commonplace in policy because they do not 'immediately engender oppositional claims, making it difficult to mobilize opinion in favour of alternatives'.[43] Thus, Mooney and Hunt observe that being in support of, for example, unsustainable development 'is strategically dysfunctional under most conditions, even for those whose goals might effectively lead to objectively insecure or unsustainable outcomes'.[44] Based on this reasoning, some

40 "Agri-Business Leaders Get Climate Smart at COP21 and Aim to Make 50% More Food Available and Reduce Agricultural Emissions by 50% by 2030," World Business Council for Sustainable Development, 1 December 2015; Teresa Anderson, "Why 'Climate-Smart Agriculture' Isn't All It's Cracked up to Be," *The Guardian*, 17 October 2014.

41 "Don't Be Fooled! Civil Society Says No to 'Climate Smart Agriculture' and Urges Decision-Makers to Support Agroecology," http://www.climatesmartagconcerns.info/cop21-statement.html.

42 See e.g. Philip H. Howard, "Visualizing Consolidation in the Global Seed Industry: 1996–2008," *Sustainability* 1, no. 4 (2009).

43 Patrick H. Mooney and Scott A. Hunt, "Food Security: The Elaboration of Contested Claims to a Consensus Frame," *Rural Sociology* 74, no. 4 (2009): p. 470.

44 Ibid.

of the appeal of CSA as a policy concept for agriculture and climate change is its capacity to engender multiple meanings.

Regardless of whether CSA promotes industrial along with agro-ecological farming practices, the approach to farming that best improves long-term food security for the largest food insecure group, small-scale farmers, is agro-ecology.[45] The synthetic inputs and technologies relied upon for industrial farming practices are generally not affordable to poor, small-scale farmers in developing countries.[46] In addition, empirical evidence illustrates that smaller-scale farms employing largely ecological methods are better adapted to changing climatic conditions and may have yield advantages on industrial agriculture.[47]

Because CSA largely focuses on agro-ecological farming methods, it is a valuable policy concept for drawing attention to the types of farming practices that best enable small-scale farmers, and therefore food security. Furthermore, it is the only policy concept in food and agriculture that expressly interlinks, and considers the dynamics between, climate change, agriculture and food security. Overall then, CSA may not be a profoundly transformative concept, but it can play an important role in promoting and mobilising resources for ecologically based farm management and improving policies and regulatory responses to climate change adaptation, mitigation and food security.

The UN Climate regime has acknowledged that there is a gap in knowledge concerning CSA. The UNFCCC has encroached into the territory of agriculture by regulating land use, however, the regime has really not developed adequate policy in this area.[48] The IPCC held an expert meeting on CSA the first time

45 Food and Agriculture Organization of the United Nations (FAO), "Final Report for the International Symposium on Agroecology for Food Security and Nutrition," p. 8. It noted that '[a] transition to Agroecology is not optional. It is rather a context-dependent way forward with regards to sustainable agricultural production intensification and resilience to climate change'.

46 For instance, they may not have access to protective equipment or training on how to use the inputs. See Uma Khumairoh, Jeroen C.J. Groot, and Egbert A. Lantinga, "Complex Agro-Ecosystems for Food Security in a Changing Climate," *Ecology and Evolution* 2, no. 7 (2012).

47 Edwin A. Gyasi, Gordana Kranjac Berisavljevic, Essie T. Blay, and William Oduro, eds., *Managing Agrodiversity the Traditional Way Lessons from West Africa in Sustainable Use of Biodiversity and Related Natural Resources* (Tokyo: United Nations University Press, 2004); R.W. Willey, "Irrigation of Sugarcane and Associated Crops Resource Use in Intercropping Systems," *Agricultural Water Management* 17, no. 1–3 (1990); Miguel A. Altieri and Clara I. Nicholls, "Agroecology Scaling up for Food Sovereignty and Resiliency," in *Sustainable Agriculture Reviews*, ed. Eric Lichtfouse (Netherlands: Springer, 2012); Shibu Jose, "Agroforestry for Ecosystem Services and Environmental Benefits: An Overview," *Agroforestry Systems* 76, no. 1 (2009); Opender Koul, G.S. Dhaliwal, and Gerrit W. Cuperus, *Integrated Pest Management: Potential, Constraints and Challenges* (Cambridge: CABI, 2004); A. Nicholas E. Birch, Graham S. Begg, and Geoffrey R. Squire, "How Agro-Ecological Research Helps to Address Food Security Issues under New IPM and Pesticide Reduction Policies for Global Crop Production Systems," *Journal of Experimental Botany* 62, no. 10 (2011).

48 Rosemary Rayfuse and Shirley Scott, "Mapping the Impact of Climate Change on International Law," in *International Law in the Era of Climate Change*, ed. Roserymary Rafuse and Shirley Scott (London: Edward Elgar Publishers, 2012), pp. 3, 5.

last year.[49] It called for future research to close yield gaps in developing countries '[a]nd adaption gaps as well as knowledge for implementing climate-smart agriculture; and inclusion of socio-economic processes and linkages. . . .'[50] The UN Climate regime tends to follow the process of scientific and technical bodies defining a concept and gathering information, which is then used to inform wording with COP decisions. This chapter argues that the UN Climate regime must develop more concrete guidance concerning CSA methodologies and policies.

The co-production of knowledge for climate-smart agriculture governance

The UN Climate regime needs to identify and disseminate models of CSA using a bottom-up approach. The agro-ecological practices that inform part CSA are knowledge intensive. Instead of relying on synthetic inputs to kill pests or improve soil fertility, ecological methods of farm management work with and manipulate the on-farm conditions so that the various ecosystem dynamics and processes on the farm promote desirable outcomes.[51] Principles from biology and traditional or indigenous knowledge will provide the basic starting point for agro-ecological practices, while local farming knowledge, experimentations and practices adapt broad norms or strategies to the particular on-farm conditions.[52] Accordingly, appropriate agro-ecological farming practices generally emerge from the 'bottom-up' as a response to the specific biophysical environment on each farm, as well as local needs.[53]

Because of its knowledge-intensive nature, a recognised impediment to the adoption of CSA is a lack of local and ecological knowledge and skills.[54] Silici identified a key obstacle for the adoption of ecological farming methods as 'the ability of farmers to use and share their knowledge in innovative ways in order to adapt techniques to local conditions'.[55] Unfortunately, public and private investment into the ecological knowledge and practices required for CSA has been low.[56] This is partly due to declines in public funding for agriculture as well as

49 Inter Governmental Panel on Climate Change (IPCC), Meeting Report of the Intergovernmental Panel on Climate Change Expert Meeting on Climate Change, Food, and Agriculture (Switzerland: World Meteorological Organization, Geneva, 2015).

50 Ibid., p. 36.

51 Miguel A. Altieri and Clara I. Nicholls, *Agroecology and the Search for a Truly Sustainable Agriculture* (Mexico: United Nations Environment Programme, 2000), p. 31.

52 See, generally, V.G. Thomas and P.G. Kevan, "Basic Principles of Agroecology and Sustainable Agriculture," *Journal of Agricultural and Environmental Ethics* 6, no. 1 (1993).

53 Olivier De Schutter, "Report of the Special Rapporteur on the Right to Food: Agroecology and the Right to Food," (UN Doc A/HRC/16/49 20 December 2010), pp. 17–18.

54 See Food and Agriculture Organization of the United Nations (FAO), *Climate-Smart Agriculture: Source Book*, module 17.

55 Agroecology: What It Is and What It Has to Offer, (UK: IIED, 2014), http://pubs.iied.org/pdfs/14629IIED.pdf. p. 18.

56 Julian M. Alston, Philip G. Pardey and Johannes Roseboom, "Financing Agricultural Research: International Investment Patterns and Policy Perspectives," *World Development* 26,

the political support, for instance, through subsidies for industrial agricultural methods in large agricultural countries.[57]

Furthermore, the types of knowledge needed to develop CSA practices and to facilitate its adoption differ from the conventional way in which agricultural knowledge has been shared over the last decades. The dominant approach to agricultural research and development is linear; it tends to involve off-farm studies and technology development in laboratories and field trials that are eventually commercialised as inputs and sold to farmers who are positioned as end users.[58] In contrast, Warner explains that '[a]groecology cannot be transferred as a technological package. It can only be facilitated by social learning. It is inherently more knowledge-intensive than conventional approaches'.[59] The type of learning that Warner is referring to, social learning, involves interactive forms of learning that take place through people collectively building knowledge with others.[60] It results in the co-production of knowledge, that is, the sharing, merging and harmonisation of 'facts, interpretations, assumptions and causal relations from their different knowledge domains'.[61]

In the context of CSA, social learning involves farmers being developers and co-developers of farming practices and market access strategies, using local knowledge and experience in light of changing on-farm conditions.[62] Harvey and

---

no. 6 (1998); Philip G. Pardey, Julian M. Alston, and Roley Piggott, *Shifting Ground: Agricultural R&D Worldwide* (Washington: International Food Policy Research Institute, 2006), http://cdm15738.contentdm.oclc.org/utils/getfile/collection/p15738coll2/id/38632/filename/38627.pdf. Note, however, that low investment in agro-ecological knowledge and practices has not prevented the discipline from growing. See A. Wezel and V. Soldat, "A Quantitative and Qualitative Historical Analysis of the Scientific Discipline of Agroecology," *International Journal of Agricultural Sustainability* 7, no. 1 (2009).

57 UN agencies and the World Bank funded an international assessment of agricultural knowledge, science and technology. See Science and Technology for Development (IAASTD) International Assessment of Agricultural Knowledge, "Agriculture at the Crossroads: Global Report," ed. Beverly D. McIntyre, Hans R. Herren, Judi Wakhungu and Robert T. Watson (Washington: IAASTD, 2009). This report commented on page 72 that agricultural research funding 'tends to be concentrated in areas of commercial interest or in advanced sciences such as . . . nanotechnologies and genomics rather than in applications deeply informed by knowledge of farming practice and ecological contexts . . .'.

58 N.G. Roling and M.A.E. Wagemakers, "A New Practice: Facilitating Sustainable Agriculture," in *Facilitating Sustainable Agriculture: Participatory Learning and Adaptive Management in Times of Environmental Uncertainty*, ed. N.G. Roling and M.A.E. Wagemakers (Cambridge: Cambridge University Press, 2003), pp. 3, 12.

59 Keith Warner, *Agroecology in Action: Extending Alternative Agriculture through Social Networks* (MIT: MIT Press, 2007), p. 224.

60 Kevin Collins and Ray Ison, "Jumping Off Arnstein's Ladder: Social Learning as a New Policy Paradigm for Climate Change Adaptation," *Environmental Policy and Governance* 19, no. 6 (2009): p. 370.

61 Jurian Edelenbos, Arwin van Buuren and Nienke van Schie, "Co-Producing Knowledge: Joint Knowledge Production between Experts, Bureaucrats and Stakeholders in Dutch Water Management Projects," *Environmental Science & Policy* 14, no. 6 (2011): p. 627.

62 See, for example, Timothy Conley and Christopher Udry, "Social Learning through Networks: The Adoption of New Agricultural Technologies in Ghana," *American Journal of Agricultural Economics* 83, no. 3 (2001).

Ensor and explain that social learning comprises 'cycles of knowledge sharing and joint action to co-create knowledge, relationships and practices among diverse stakeholders. This results in learning and change that goes beyond the individual into communities, networks or systems'.[63] Put differently, social learning and the co-production of knowledge translate international climate change agreements into tailor-made, on-farm practices. These methods then can be incorporated into broader climate change strategies at the international level.

The International Assessment of Agricultural Knowledge, which was a project which brought together 400 international experts sponsored by UN agencies and the World Bank, concluded that social models of learning and the co-production of knowledge are essential to the uptake and promotion of sustainable or 'climate-smart' agriculture.[64] The initiative's final report explained:

> There is a growing consensus among researchers concerned with sustainable agriculture that no single group of actors should appropriate the right to define what type of combination should exist between scientific and 'local' forms of knowledge. As a consequence, participatory forms of coproduction of knowledge, based on social learning among actors involved, have become a key feature of sustainable agriculture and resource management.[65]

Such an approach is particularly appropriate in the context of adapting farms to changing climatic conditions. Adaptation to climate change on farms will necessarily entail trial and error.[66] This process can be made more effective and efficient by farmers and other stakeholders jointly problem solving and individually testing,

63 Blane Harvey, Jonathan Ensor, Ben Garside, John Woodend, Lars Otto Naess, and Liz Carlile, *Social Learning in Practice: A Review of Lessons, Impacts and Tools for Climate Change Adaptation: Working Paper No. 38* (Copenhagen: CGIAR Research Program on Climate Change, 2013), p. 509.

64 International Assessment of Agricultural Knowledge, "Agriculture at the Crossroads: Global Report."

65 Ibid., p. 17; Stephan Rist, Mani Chiddambaranathan, Cesar Escobar, and Urs Wiesmann, " 'It Was Hard to Come to Mutual Understanding . . .'—The Multidimensionality of Social Learning Processes Concerned with Sustainable Natural Resource Use in India, Africa and Latin America," Systemic Practice and Action Research 19 (2006); Iain J. Davidson-Hunt, "Adaptive Learning Networks: Developing Resource Management Knowledge through Social Learning Forums," *Human Ecology* 34, no. 4 (2006).

66 Harvey, Ensor, Garside, Woodend, Naess, and Carlile, *Social Learning in Practice: A Review of Lessons, Impacts and Tools for Climate Change Adaptation: Working Paper No. 38.* See also, Henk Van den Berg and Janice Jiggins, "Investing in Farmers – The Impacts of Farmer Field Schools in Relation to Integrated Pest Management," *World Development* 35, no. 4 (2007); Marco Barzman and Sylvie Desilles, "Diversifying Rice-Based Farming Systems and Empowering Farmers in Bangladesh Using the Farmer Field-School Approach," in *Agroecological Innovations: Increasing Food Production with Participatory Development*, ed. Norman Uphoff, 203–211. (London: Routledge, 2013); Elske van de Fliert, "Recognising a Climate for Sustainability: Extension Beyond Transfer of Technology," *Australian Journal of Experimental Agriculture* 43, no. 1 (2003).

refining, commenting on and sharing different strategies.[67] Reflective of the findings from the International Assessment on Agricultural Knowledge, a large body of empirical work positions the co-production of knowledge as the basis for effective environmental governance.[68] The next section will examine the governance arrangements that facilitate social learning and the co-production of knowledge required to translate international climate agreements into on-farm practices.

Networked, informational governance for climate-smart agriculture

Based on the knowledge-intensive nature of CSA and the ways in which appropriate farming practices are co-produced, the governance models that underpin CSA generally have to be decentralised, informational and networked to be effective at promoting CSA. As explained by Breakey, Cadman and Sampford, governance can take various forms, including non-hierarchical, cross-sectoral networks.[69] This type of governance is often described as 'nodal' or 'networked'. As these terms suggest, this model of governance occurs through networks between actors rather than 'a single centre of action', with the state being only one node or point in the network.[70]

In the context of CSA, networked governance will take different forms and exist at and between various scales, from local to international. For example, a group of farmers sharing techniques and experiences within the same ecological region will have outcomes on and off farm, which can be identified as 'climate-smart' or 'not climate-smart'.[71] Collectively, the uptake of climate-smart practices

67 For instance, in Kate G. Lonsdale, M.J. Gawith, K. Johnstone, R.B. Street, C.C. West, and A.D. Brown, "Attributes of Well-Adapting Organisations," 1–89 (London, 2010). Available at http://www.ukcip.org.uk/wp-content/PDFs/UKCIP_Well_adapting_organisations.pdf (accessed 30 August 2016). The authors observed that 'as social learning for change requires shifts in understanding either as individuals or as groups this type of learning seems to have great potential for exploring the process of adaptation to climate change. No one person has the whole answer, we all have a piece of the truth and there is a pressing need to come up with imaginative solutions'.

68 R. Hill, J. Davies, I.C. Bohnet, C.J. Robinson, K. Maclean and P.L. Pert, "Collaboration Mobilises Institutions with Scale-Dependent Comparative Advantage in Landscape-Scale Biodiversity Conservation," *Environmental Science & Policy* 51 (2015); Craig Mauelshagen, Pete Smith, Frank Schiller, David Denyer, Sophie Rocks, and Simon Pollard, "Effective Risk Governance for Environmental Policy Making: A Knowledge Management Perspective," *Environmental Science & Policy* 41 (2014); Carl Folke, Thomas Hahn, Per Olsson, and Jon Norberg, "Adaptive Governance of Social-Ecological Systems," *Annual Review of Environment and Resources* 30 (2005).

69 See Chapter 2 of this collection.

70 Clifford D. Shearing and Les Johnston, *Governing Security: Explorations of Policing and Justice* (London: Routledge, 2013).

71 A germinal work on nodal governance is Scott Burris, Peter Drahos, and Clifford Shearing, "Nodal Governance," *Australian Journal of Legal Philosophy* 30 (2005). Essentially, a person or an entity subjectively determines what will be considered as 'outcomes' because what the 'outcomes' are depends on what a person or an entity is considering the particular context (time and space) and the normative basis they are using to identify and categorise outcomes.

will depend on the changes individual farmers make to their farms or their products and how the knowledge gathered is shared and developed. More generally, Burris, Drahos and Shearing explain that the positive outcomes of a particular governance network depend upon

> [t]he degree to which people making up the [network] have made adaptations individually and in how they collectively manage the [network] that maximize their ability to tap and coordinate knowledge and capacity disseminated through the collectivity.[72]

For nodal governance scholars like Burris, Drahos and Shearing, a 'node' is a point within the network where 'knowledge, capacity and resources are mobilised to manage a course of events'.[73] Accordingly, in the context of CSA, 'nodes' may include the FAO, state governments, farming organisations, individual farmers, corporate actors and community groups. These actors can influence the outcomes and the behaviour and decisions of others in line with the objectives of CSA.

Along with networked governance, the knowledge-intensive, bottom-up nature of CSA practices also reflects the importance of informational governance for CSA. Informational governance coincides with a networked understanding of governance but is focused on the flow of information.[74] Mol explains that informational governance 'refers to the idea that information (and informational processes, technologies, institutions and resources linked to it) is fundamentally restructuring processes, institutions, and practices of environmental governance, in a way which is essentially different from conventional modes . . .'.[75] In other words, innovations in information communication technologies have fostered networked governance arrangements largely based on collecting, disseminating and using knowledge.

Such platforms are promoted as an avenue to CSA by the CGIAR, which is an agricultural research institution that receives funds from member governments and international institutions. The CGIAR explained:

> A crucial component of early action is enhancing the knowledge sharing and developing capacity. We need to create, designate or integrate national and

72 Burris, Drahos, and Shearing, "Nodal Governance," p. 36. Their approach draws on Hayek. See Friedrich Hayek, *Legislation and Liberty: A New Statement of the Liberal Principles of Justice and Political Economy* (Abingdon, UK: Routledge, 2012), pp. 176–78.

73 Burris, Drahos, and Shearing, "Nodal Governance." Note that Burris, Drahos and Shearing depicted 'nodes' as a site that could exist within or outside of a particular network or system, but for simplicity, we just refer to them as points within the network. See, for a critique of nodal governance, Ian Loader and Neil Walker, "State of Denial? Rethinking the Governance of Security," *Punishment & Society* 6, no. 2 (2004). They argue that some nodes, in particular state-based nodes, can have the ultimate authority depending on the situation.

74 See generally Manuel Castells, *The Rise of the Network Society: The Information Age: Economy, Society, and Culture*, 2 ed., vol. 1 (Chichester, UK: Wiley Blackwell, 2009).

75 Arthur P.J. Mol, "Environmental Governance in the Information Age: The Emergence of Informational Governance," *497* 24, no. 4 (2006): p. 501.

> regional knowledge networks or platforms for the identification and dissemination of climate-smart agricultural practices and technologies.[76]

The types of platforms referred to by the CGIAR enhance the flow of information and knowledge. Improved knowledge networks in turn promote 'thick' governance values discussed by Breakey, Cadman and Sampford, which includes participation (inclusiveness of all stakeholders and equality between them) and deliberation (including the mode of decision-making). Furthermore, the actors involved in such knowledge networks are producing, guiding and selecting information.[77] In the context of CSA, such platforms allow actors on the 'front line' of climate change and agriculture to develop and share knowledge and practices based on their needs and experiences, as well as the needs of their specific community.

Case study: Mkulima Young

Mkulima Young is an online platform that seeks to increase agriculture productivity and marketing linkages for smallholder farmers in East Africa, and more specifically, Kenya. It seeks to achieve this objective by providing a virtual space in which participants co-develop market access opportunities and agro-ecological farming practices. The platforms provided and managed by Mkulima Young are targeted at young, East African farmers (18–45) and aspiring farmers. Subsequently, the initiative is targeting one of the groups most vulnerable to the effects of climate change.

The initiative began as a response to the need for knowledge transfer, especially in Kenya, where the number of farmers has increased while the number of extension officers has decreased.[78] Hence, Mkulima Young came to address that gap, but in doing so, it has improved the transfer of knowledge by making the process more continuous, interactive and instantaneous. In other words, it has promoted social learning, while extension officers' visits are irregular and tend to be more instructive than collaborative. Unlike extension officers, online platforms can generate new CSA information that can be linked to new strategies that boost productivity while improving adaptation. Most of the time these innovations bypass smallholders who normally would have to travel to be exposed to new farming practices and ideas.

Furthermore, as information and communications technology (ICT) is increasingly used in the agricultural sector, more research and information is accessible

76 "Policy Brief 73247: Opportunities and Challenges for Climate-Smart Agriculture in Africa," p. 5.

77 Katrine Soma, M.C Onwezen, I.E Salverda, and R.I. van Dam, "Roles of Citizens in Environmental Governance in the Information Age – Four Theoretical Perspectives," *Current Opinion in Environmental Sustainability* 18 (2016).

78 Evelyne Kipot and Steven Franzel, "Voluntarism as an Investment in Human, Social and Financial Capital: Evidence from Farmer-to-Farmer Extension Program in Kenya," *Agriculture and Human Values* 31, no. 2 (2014).

for those with Internet access. Mkulima Young seeks to farmers who can then informally transfer the new research, technologies and knowledge to more marginalised farmers or farming communities. In order to do this, though, the initiative organisers select, package and deliver the information or technologies in ways that appeal to farmers. For instance, one of Mkulima Young's platforms is an online blog that uses narratives to teach farming and market access strategies by developing characters that the audience can relate to and shaping storylines around issues experienced by farmers.

By being based across multiple platforms and covering a range of farming and market-related topics, Mkulima Young is a one-stop shop for information whereby the users can read and contribute to the content in an interactive way. It comprises integration of various social media platforms, such as a website, mobile app and print media. This allows farmers to interact with each other and share agricultural knowledge, marketplace ideas and networking while modifying the information to suit the needs of the farmers. The Facebook page is the main online portal, with 90,000 followers. The Facebook links to other online services, such as the Mkulima Young app, which allows users to market their farm produce, and the option to join What'sApp groups, where farmers interact socially with an interest in agriculture as the common ground. The Facebook page was founded to provide an online space for young people to discuss issues concerning agriculture. The Mkulima Young Facebook page is currently used to share agriculture knowledge and tips and allows members to post questions and gain feedback from the community and links to the weekly 'Mkulima Moments' (weekly humour newspaper column on farming published by the *Daily Nation* newspaper).

Mkulima Young, being an efficient, cheap, online platform for the exchange of information, ideas and knowledge, is an enabling tool for wider socio-economic development. With its expansion and adoption, Mkulima Young has become an alternative beneficial tool to the farmers. The agricultural practices collaboratively developed and shared through the Mkulima Young platform are agro-ecologically based, and so are considered 'climate-smart'. However, the extent to which a practice is 'climate-smart' is not a specific concern for the farmers involved. Generally, 'climate-smart' agricultural practices are promoted by Mkulima Young because they tend to be innovative, practical and effective at increasing productivity. While concerned with the food security and other needs of smallholder farmers in East Africa, the promotion of modern innovative farming practices is helping farmers adapt to the impacts of climate change and take advantage of climate mitigation schemes. In this way Mkulima Young represents a subtle form of governance that enables progress towards the multiple objectives of CSA.

The Mkulima Young platform is farmer centric. In this platform a farmer is not a passive receiver of information, but rather an active co-creator: it is a way to build relationships, share information and connect marketing within a diverse community of practice in farming. The development and sharing of information on Mkulima Young's platforms is dynamic, iterative and demand driven. Subsequently, the information and knowledge exchanged is largely driven by the collective activity of the farmers involved. While a large group is generating information and co-producing knowledge, the organisers of Mkulima Young form a core group that develop more curated posts, verify information based on their agricultural science expertise and enforce implicit norms. The platform is self-regulated, with those engaging with the objective of information sharing and marketing.

One way in which knowledge is co-produced on Mkulima Young is by followers sharing their knowledge and experiences or asking a question to the online community of farmers. The farmers are then able to respond or ask further questions. If the discussion concerns a particular farming practice, Mkulima Young verifies the information and then conducts interviews with farmers using the strategy. These interviews involve photos and video footage and are shared across all platforms.

A strategy of Mkulima Young is to explain why such a practice has positive impacts on farm productivity, how to carry it out on farms and how it can benefit the farmer financially. Moreover, the posts regarding particular strategies are presented to followers at times that suit the current season of the region. For instance, a post on water harvesting would be posted just before the rainy season. The farmers involved then comment on the post and provide further advice or avenues to explore. Others may then take up the practice and share their experiences.

Overall, Mkulima Young is an example of the kinds of governance arrangements required to build capacity for and to promote CSA. It reveals the importance of understanding the targeted farming communities, their ecological conditions and their resource constraints in formulating such platforms. Governance processes can be seen by the role taken by Mkulima Young organisers and in the coproduction of knowledge by participants, while the platforms are the mechanisms that allow such processes to take place. Mkulima Young was invited to present at the Global Landscapes Forum at COP 19 and gave a presentation on Facebook Farming.[79] This shows that the regime is aware of the need to engage those working in the agriculture sector. This chapter argues that the agriculture knowledge being discussed on online platforms such as Mkulima Young should feed into developing CSA methods and policies within the UNFCCC in order to ensure that bottom-up knowledge informs international climate agricultural policy.

Conclusion

One of the key purposes of the UNFCCC is to prevent greenhouse gas emissions from reaching levels that adversely affect food production. CSA is a concept that links food production with adaption and mitigation and has been identified by the IPCC as a concept where more knowledge is needed. This chapter illustrates that the implementation of international climate law does not end with the

79 "The Facebook Farmer," http://www.landscapes.org/facebook-farmer/.

decisions and activities of the state, and instead, its goals can be translated and further developed at the community level. In fact, such translation and development is a critical part of mitigating and adapting to climate change and allows for bottom-up governance. Equally, this chapter illustrates the unique role of networked and informational governance models for contributing to the realisation of UNFCCC's objectives. In order to ground this analysis, the chapter explored the platforms created by Mkulima Young that have become places for the co-production of knowledge and social learning for young, small-scale farmers in Eastern Africa. Future research should empirically examine the impact of such governance models on the uptake of CSA.

References

"About GACSA." Global Alliance for Climate-Smart Agriculture, <http://www.fao.org/gacsa/en/> (accessed 14 June 2016).

"Agri-Business Leaders Get Climate Smart at COP21 and Aim to Make 50% More Food Available and Reduce Agricultural Emissions by 50% by 2030." *World Business Council for Sustainable Development*, 1 December 2015.

Alston, Julian M., Philip G. Pardey, and Johannes Roseboom. "Financing Agricultural Research: International Investment Patterns and Policy Perspectives." *World Development* 26, no. 6 (1998): 1057–71.

Altieri, Miguel A. "Linking Ecologists and Traditional Farmers in the Search for Sustainable Agriculture." *Frontiers in Ecology and the Environment* 2, no. 1 (2004): 35–42.

Altieri, Miguel A., and Clara I. Nicholls. "Agroecology Scaling up for Food Sovereignty and Resiliency." In *Sustainable Agriculture Reviews*, edited by Eric Lichtfouse, 1–29. Netherlands: Springer, 2012.

———. *Agroecology and the Search for a Truly Sustainable Agriculture*. Mexico: United Nations Environment Programme, 2000.

Anderson, Teresa. "Why Climate-Smart Agriculture Isn't All It's Cracked up to Be." *The Guardian*, 17 October 2014.

Asian Development Bank. "Making Grasslands Sustainable in Mongolia: Assessment of Key Elements in the Nationally Appropriate Mitigation Actions for Grassland and Livestock Management Development." In *Strengthening Carbon Financing for Grassland Management in the PRC and Mongolia*, 1–59. Manila: Asian Development Bank, 2014.

Bagnetto, Laura Angela. "Agriculture Neglected in COP21 Negotiation Says AU Expert: Interview with Esterine Fotabong." *Corporate Africa*, 15 April 2016.

Barzman, Marco, and Sylvie Desilles. "Diversifying Rice-Based Farming Systems and Empowering Farmers in Bangladesh Using the Farmer Field-School Approach." In *Agroecological Innovations: Increasing Food Production with Participatory Development*, edited by Norman Uphoff, 203–11. London: Routledge, 2013.

Birch, A. Nicholas E., Graham S. Begg, and Geoffrey R. Squire. "How Agro-Ecological Research Helps to Address Food Security Issues under New IPM and Pesticide Reduction Policies for Global Crop Production Systems." *Journal of Experimental Botany* 62, no. 10 (2011): 3251–61.

Branca, Giacomo, Nancy McCarthy, Leslie Lipper, and Maria Christina Jolejole. "Climate-Smart Agriculture: A Synthesis of Empirical Evidence of Food Security and Mitigation Benefits from Improved Cropland Management." 1–35. Rome: Food and Agriculture Organization of the United Nations (FAO), 2011.

Burris, Scott, Peter Drahos, and Clifford Shearing. "Nodal Governance." *Australian Journal of Legal Philosophy* 30 (2005): 30–58.

Calzadilla, Alvaro, Katrin Rehdanz, Richard Betts, Pete Falloon, Andy Wiltshire, and Richard S.J. Tol. "Climate Change Impacts on Global Agriculture." *Climatic Change* 120, no. 1–2 (2013): 357–74.

Castells, Manuel. *The Rise of the Network Society: The Information Age: Economy, Society, and Culture*. 2 ed. Vol. 1, Chicester, UK: Wiley Blackwell, 2009.

Cerdà, Artemi, Antonio Jordán, Lorena Zavala, María José Marqués, and Agata Novara. "Soil Erosion and Agriculture." *Soil and Tillage Research* 106 (2009): 107–08.

CIDSE. "Climate-Smart Revolution . . . or a New Era of Green-Washing?" 1–11. Brussels: CIDSE, 2015.

"Climate-Smart Agriculture". *Food and Agriculture Organization of the United Nations*, 1 September 2016.

Climate Technology Network (CTN). "Corporate-Smart Greenwash: Why We Reject the Global Alliance on Climate-Smart Agriculture." Copenhagen: CTN, 1 January 2014.

Collins, Kevin, and Ray Ison. "Jumping Off Arnstein's Ladder: Social Learning as a New Policy Paradigm for Climate Change Adaptation." *Environmental Policy and Governance* 19, no. 6 (2009): 357–73.

Conley, Timothy, and Christopher Udry. "Social Learning through Networks: The Adoption of New Agricultural Technologies in Ghana." *American Journal of Agricultural Economics* 83, no. 3 (2001): 668–73.

Davidson-Hunt, Iain J. "Adaptive Learning Networks: Developing Resource Management Knowledge through Social Learning Forums." *Human Ecology* 34, no. 4 (2006): 593–614.

De Schutter, Olivier. "Report of the Special Rapporteur on the Right to Food: Agroecology and the Right to Food." UN Doc A/HRC/16/49 20 December 2010.

"Don't Be Fooled! Civil Society Says No to 'Climate Smart Agriculture' and Urges Decision-Makers to Support Agroecology." <http://www.climatesmartagconcerns.info/cop21-statement.html> (accessed 14 June 2016).

Duffy, Michael. "Economies of Size in Production Agriculture." *Journal of Hunger & Environmental Nutrition* 4, no. 3 (2009): 375–92.

Edelenbos, Jurian, Arwin van Buuren, and Nienke van Schie. "Co-Producing Knowledge: Joint Knowledge Production between Experts, Bureaucrats and Stakeholders in Dutch Water Management Projects." *Environmental Science & Policy* 14, no. 6 (2011): 675–84.

Evenson, Robert E., and D. Gollin. "Assessing the Impact of the Green Revolution, 1960 to 2000." *Science* 300, no. 5620 (2003): 758–62.

"The Facebook Farmer." <http://www.landscapes.org/facebook-farmer/>.

Folke, Carl, Thomas Hahn, Per Olsson, and Jon Norberg. "Adaptive Governance of Social-Ecological Systems." *Annual Review of Environment and Resources* 30 (2005): 441–73.

Food and Agriculture Organization of the United Nations (FAO). "Climate Change and Food Security: A Framework Document." 1–93. Rome: FAO, 2008.

———. "Climate-Smart Agriculture Policies, Practices and Financing for Food Security, Adaptation and Mitigation: Background Paper for the Hague Conference on Agriculture, Food Security and Climate Change." 1–38. Rome: FAO, 2010.

———. "Climate-Smart Agriculture: A Call for Action." Paper Presented at the Asia-Pacific Regional Workshop, Bangkok, Thailand, 18 to 20 June 2015.

———. *Climate-Smart Agriculture: Source Book.* Rome: FAO, 2013.

———. "Final Report for the International Symposium on Agroecology for Food Security and Nutrition." 1–34. Rome: FAO, 2015.

Friedmann, Harriet, and Philip McMichael. "Agriculture and the State System: The Rise and Decline of National Agricultures, 1870 to the Present." *Sociologia Ruralis* 29, no. 2 (1989): 93–117.

Friends of the Earth International. "Agroecology and Climate Justice: A People's Guide to Paris and Beyond." *Fact Sheet*, 13 November 2015.

Gliessman, Steve. "Agroecology and Food System Change." *Journal of Sustainable Agriculture* 35, no. 4 (2011): 347–49.

Global Alliance For Climate-Smart Agriculture (GACSA). "Framework Document: GACSA Series Document 1." *FAO*, 1 September 2014.

Gyasi, Edwin A., Gordana Kranjac Berisavljevic, Essie T. Blay, and William Oduro, eds. *Managing Agrodiversity the Traditional Way Lessons from West Africa in Sustainable Use of Biodiversity and Related Natural Resources.* Tokyo: United Nations University Press, 2004.

Harvey, Blane, Jonathan Ensor, Ben Garside, John Woodend, Lars Otto Naess, and Liz Carlile. *Social Learning in Practice: A Review of Lessons, Impacts and Tools for Climate Change Adaptation: Working Paper No. 38.* Copenhagen CGIAR Research Program on Climate Change, 2013.

Hayek, Friedrich. *Legislation and Liberty: A New Statement of the Liberal Principles of Justice and Political Economy.* Abingdon, UK: Routledge, 2012.

Hill, R., J. Davies, I.C. Bohnet, C.J. Robinson, K. Maclean, and P.L. Pert. "Collaboration Mobilises Institutions with Scale-Dependent Comparative Advantage in Landscape-Scale Biodiversity Conservation." *Environmental Science & Policy* 51 (2015): 267–77.

Howard, Philip H. "Visualizing Consolidation in the Global Seed Industry: 1996–2008." *Sustainability* 1, no. 4 (2009): 1266–87.

Inter Governmental Panel on Climate Change (IPCC). *Meeting Report of the Intergovernmental Panel on Climate Change Expert Meeting on Climate Change, Food, and Agriculture.* Switzerland: World Meteorological Organization, Geneva, 2015.

International Assessment of Agricultural Knowledge, Science and Technology for Development (IAASTD). "Agriculture at the Crossroads: Global Report." edited by Beverly D. McIntyre, Hans R. Herren, Judi Wakhungu and Robert T. Watson. Washington: IAASTD, 2009.

———. "Agriculture at a Crossroads: Synthesis Report." edited by Beverly D. McIntyre, Hans R. Herren, Judi Wakhungu and Robert T. Watson, 1–91. Washington: IAASTD, 2009.

Johnson, Hope. "From the Ground Up: An Analysis of the Public International Regulation of Agriculture Using a Rights-Based Approach to Food Security." Queensland University of Technology, 2016.

Jose, Shibu. "Agroforestry for Ecosystem Services and Environmental Benefits: An Overview." *Agroforestry Systems* 76, no. 1 (2009): 1–10.

Khumairoh, Uma, Jeroen C.J. Groot, and Egbert A. Lantinga. "Complex Agro-Ecosystems for Food Security in a Changing Climate." *Ecology and Evolution* 2, no. 7 (2012): 1696–704.

Kipot, Evelyne, and Steven Franzel. "Voluntarism as an Investment in Human, Social and Financial Capital: Evidence from Farmer-to-Farmer Extension Program in Kenya." *Agriculture and Human Values* 31, no. 2 (2014): 231–41.

Koul, Opender, G.S. Dhaliwal, and Gerrit W. Cuperus. *Integrated Pest Management: Potential, Constraints and Challenges.* Cambridge: CABI, 2004.

Lilliston, Ben. "The Clever Ambiguity of Climate Smart Agriculture." Institute for Agriculture and Trade Policy, 2015.

Lin, Brenda B., and Michael Jahi Chappel. "Effects of Industrial Agriculture on Climate Change and the Mitigation Potential of Small-Scale Agro-Ecological Farms." *CAB Reviews: Perspectives in Agriculture, Veterinary Science, Nutrition, and Natural Resources* 6, no. 20 (2011): 1–15.

Lipper, Leslie, Philip Thornton, Bruce M. Campbell, Tobias Baedeker, Ademola Braimoh, Martin Bwalya, Patrick Caron, Andrea Cattaneo1, Dennis Garrity, Kevin Henry, Ryan Hottle, Louise Jackson, Andrew Jarvis, Fred Kossam, Wendy Mann, Nancy McCarthy, Alexandre Meybeck, Henry Neufeldt, Tom Remington, Pham Thi Sen, Reuben Sessa, Reynolds Shula, Austin Tibu, and Emmanuel F. Torquebiau. "Climate-Smart Agriculture for Food Security." *Nature Climate Change* 4 (2014): 1068–72.

Loader, Ian, and Neil Walker. "State of Denial? Rethinking the Governance of Security." *Punishment & Society* 6, no. 2 (2004): 221–28.

Lonsdale, Kate G., M.J. Gawith, K. Johnstone, R.B. Street, C.C. West, and A.D. Brown. "Attributes of Well-Adapting Organisations," 1–89. London: Committee on Climate Change, 2010. Available at http://www.ukcip.org.uk/wp-content/PDFs/UKCIP_Well_adapting_organisations.pdf (Accessed 30 August 2016).

Mann, Alana. "What's Wrong with Climate-Smart Agriculture?" Sydney Environment Institute, University of Sydney, 2015.

Marsden, Terry, Jo Banks, Henk Renting, and Jan Douwe Van Der Ploeg. "The Road Towards Sustainable Rural Development: Issues of Theory, Policy and Research Practice." *Journal of Environmental Policy & Planning* 3, no. 2 (2001): 75–82.

Mauelshagen, Craig, Pete Smith, Frank Schiller, David Denyer, Sophie Rocks, and Simon Pollard. "Effective Risk Governance for Environmental Policy Making: A Knowledge Management Perspective." *Environmental Science & Policy* 41 (2014): 23–32.

Mol, Arthur P.J. "Environmental Governance in the Information Age: The Emergence of Informational Governance." *497* 24, no. 4 (2006): 497–514.

Mooney, Patrick H., and Scott A. Hunt. "Food Security: The Elaboration of Contested Claims to a Consensus Frame." *Rural Sociology* 74, no. 4 (2009): 469–97.

Neufeldt, Henry, Molly Jahn, Bruce M. Campbell, John R. Beddington, Fabrice DeClerck, Alessandro De Pinto, Jay Gulledge, Jonathan Hellin, Mario Herrero, Andrew Jarvis, David LeZaks, Holger Meinke, Todd Rosenstock, Mary Scholes, Robert Scholes, Sonja Vermeulen, Eva Wollenberg, and Robert Zougmoré. "Beyond Climate-Smart Agriculture: Toward Safe Operating Spaces for Global Food Systems." *Agriculture & Food Security* 2, no. 12 (2013): 1–6.

Palma, Cheryl A., Sean M. Smuklera, Clare C. Sullivana, Patrick K. Mutuoa, Gerson I. Nyadzia, and Markus G. Walsha. "Identifying Potential Synergies and Trade-Offs for Meeting Food Security and Climate Change Objectives in Sub-Saharan Africa." *Proceedings of the National Academy of Sciences of the United States of America* 107, no. 46 (2010): 19661–66.

Pardey, Philip G., Julian M. Alston, and Roley Piggott. *Shifting Ground: Agricultural R&D Worldwide.* Washington: International Food Policy Research Institute, 2006.

<http://cdm15738.contentdm.oclc.org/utils/getfile/collection/p15738coll2/id/38632/filename/38627.pdf> (accessed 14 June 2016).

"Policy Brief 73247: Opportunities and Challenges for Climate-Smart Agriculture in Africa." Agriculture, Forestry & Fisheries South Africa and African Union, <https://ccafs.cgiar.org/sites/default/files/assets/docs/au_policybrief_opportunities challenges.pdf>.

Pretty, J.N., A.D. Noble, D. Bossio, J. Dixon, R.E. Hine, F.W. Penning De Vries, and J.I. Morison. "Resource-Conserving Agriculture Increases Yields in Developing Countries." *Environmental Science & Technology* 40, no. 4 (2006): 1114–19.

"Questions and Answers: Knowledge of Climate-Smart Agriculture." Food and Agriculture Organization of the United Nations, <https://ccafs.cgiar.org/climate-smart-agriculture-0#.V0e0oIR95hE> (accessed 15 June 2016).

Rayfuse, Rosemary, and Shirley Scott. "Mapping the Impact of Climate Change on International Law." In *International Law in the Era of Climate Change*, edited by Roserymary Rafuse and Shirley Scott, 84–117. London: Edward Elgar Publishers, 2012.

Rist, Stephan, Mani Chiddambaranathan, Cesar Escobar, and Urs Wiesmann. "'It was Hard to Come to Mutual Understanding . . .' – The Multidimensionality of Social Learning Processes Concerned with Sustainable Natural Resource Use in India, Africa and Latin America." *Systemic Practice and Action Research* 19 (2006): 219–37.

Roling, N.G., and M.A.E. Wagemakers. "A New Practice: Facilitating Sustainable Agriculture." In *Facilitating Sustainable Agriculture: Participatory Learning and Adaptive Management in Times of Environmental Uncertainty*, edited by N.G. Roling and M.A.E. Wagemakers, 3–22. Cambridge: Cambridge University Press, 2003.

Sanchez, Pedro, Philip Dobie, and Nolan Yuksel. *Halving Hunger: It Can Be Done.* London: United National Development Programme; UN Millennium Project, 2005.

Sen, Amartya. "Famines." *World Development* 8, no. 9 (1980): 612–21.

Shearing, Clifford D., and Les Johnston. *Governing Security: Explorations of Policing and Justice.* London: Routledge, 2013.

Silici, Laura. *Agroecology: What It Is and What It Has to Offer.* UK: IIED, 2014. <http://pubs.iied.org/pdfs/14629IIED.pdf> (accessed 14 June 2016).

Soma, Katrine, M.C Onwezen, I.E Salverda, and R.I. van Dam. "Roles of Citizens in Environmental Governance in the Information Age – Four Theoretical Perspectives." *Current Opinion in Environmental Sustainability* 18 (2016): 122–30.

Stabinsky, Doreen. *Climate-Smart Agriculture: Myths and Problems.* Berlin: Heinrich-Böll-Stiffung, 2014. <https://br.boell.org/sites/default/files/uploads/2014/09/epaper_climate_smart_eng_boll_brasil18.09.014.pdf> (accessed 14 June 2016).

Steinfeld, Henning. *Livestock's Long Shadow: Environmental Issues and Options.* Rome: FAO, 2006.

Sulda, Heidi, John Coveney, and Michael Bentley. "An Investigation of the Ways in Which Public Health Nutrition Policy and Practices Can Address Climate Change." *Public Health Nutrition* 13, no. 3 (2010): 304–13.

Swaminathan, M.S. "Role of Genetic Modification in Developing Climate Smart Agriculture to Ensure Sustained Food Security." *Agricultural Research* 1 (2012): 295–98.

Thomas, V.G., and P.G. Kevan. "Basic Principles of Agroecology and Sustainable Agriculture." *Journal of Agricultural and Environmental Ethics* 6, no. 1 (1993): 1–19.

United Nations Framework Convention on Climate Change (UNFCCC). "Conference of the Parties: COP21 Paris Agreement." FCCC/CP/2015/L.9: UNFCCC, 2015.

———. "The Kyoto Protocol." 37 ILM 22, 1998.

Urdal, Henrik. "Demographic Aspects of Climate Change, Environmental Degradation and Armed Conflict." 1–25. New York: United Nations, 2008.

van de Fliert, Elske. "Recognising a Climate for Sustainability: Extension Beyond Transfer of Technology." *Australian Journal of Experimental Agriculture* 43, no. 1 (2003): 29–36.

Van den Berg, Henk, and Janice Jiggins. "Investing in Farmers – The Impacts of Farmer Field Schools in Relation to Integrated Pest Management." *World Development* 35, no. 4 (2007): 663–86.

Vermeulen, Sonja J., Bruce M. Campbell, and John S.I. Ingram. "Climate Change and Food Systems." *Annual Review of Environment and Resources* 37, no. 1 (2012): 195–222.

Warner, Keith. *Agroecology in Action: Extending Alternative Agriculture through Social Networks.* MIT: MIT Press, 2007.

Wezel, A., and V. Soldat. "A Quantitative and Qualitative Historical Analysis of the Scientific Discipline of Agroecology." *International Journal of Agricultural Sustainability* 7, no. 1 (2009): 3–18.

Wheeler, Tim, and Joachim von Braun. "Climate Change Impacts on Global Food Security." *Science* 341, no. 6145 (2013): 508–13.

Willey, R.W. "Irrigation of Sugarcane and Associated Crops Resource Use in Intercropping Systems." *Agricultural Water Management* 17, no. 1–3 (1990): 215–31.

World Bank. "Climate-Smart Agriculture a Call to Action." 1–18. Washington: World Food Programme, 2012.

———. "Climate Smart Agriculture: A Triple Win." <http://www.worldbank.org/en/news/video/2011/09/01/climate-smart-agriculture-a-triple-win> (accessed 14 June 2016).

———. "World Development Report 2008: Agriculture for Development," 1–283. Washington: 2007. Available at http://go.worldbank.org/2DNNMCBGI0 (accessed 30 August 2016).

12 International climate change policy and the contribution of civil society organizations

Peter J. Glynn

Introduction

The negotiations for the successor treaty to the Kyoto Protocol have been undertaken at a time when the community, industry and governments largely accept the reality of climate change. Many stakeholders, including private sector entities, are also demonstrably contributing on their own initiative, as exemplified by the global retailer Ikea's pledge to clean energy and climate action,[1] and shareholder and investor groups' pledges to isolate fossil fuel–dependent activities.[2]

The Paris Conference of the Parties (COP) in December 2015 adopted the framework of the Lima Paris Action Agenda (LPAA) and its four pillars: the binding Paris Agreement, voluntary country contributions (INDCs), finance and an enhanced role for non-state actors. The Paris Agreement introduces a management model to ensure transparency in the reporting of voluntary undertakings, which will operate in concert with the binding Paris Agreement, and a modified system for monitoring and evaluation.[3] These arrangements will be integral to the effectiveness of the LPAA and are relevant to the governance and integrity framework presented in this volume.

The negotiations in the period leading up to the Paris COP were closely monitored and tense. The UNFCCC regular mid-term meetings of the Parties to the Convention in June 2015 in Bonn was the penultimate negotiating session, and while publicly the Parties expressed their commitment to a process, this did not translate to progress on streamlining the negotiating text. The meeting was also coming to terms with the need to share responsibility with non-Party actors, a reflection of acceptance that the targets for global warming and GHG emission containment could not be achieved by the states acting alone. This is a material shift whereby the concepts of shared responsibility and shared decision-making

1 P. Clarke, "Ikea Pledges 1bn Euro to Clean Energy and Climate Action," *Financial Times*, 4 June 2015.
2 Toby Heaps, "What Is Your Climate Change Business Plan?" *Financial Times FM Supplement*, 1 June 2015.
3 United Nations Framework Convention on Climate Change (UNFCCC), "Conference of the Parties: COP21 Paris Agreement," (FCCC/CP/2015/L.9: UNFCCC, 2015), Draft decision-/CP21, paragraphs 85–104.

were to be introduced into UNFCCC governance, a role beyond business as usual for civil society organizations (CSOs).

The international climate change agreements have a significant influence over domestic policy and initiatives of governments. This chapter discusses the process that takes those agreements into domestic policy and the effectiveness of civil society organizations as actors and contributors to that process.

Civil society

The UNFCCC rules permit non-Party organizations to participate in its proceedings, including the annual Conference of the Parties (COP) and meetings of Parties (governments). The non-Party organizations permitted to participate include intergovernmental organizations such as the International Federation of the Red Cross and Red Crescent Societies, the International Labour Organization, the World Bank and civil observers. Civil observers are defined as non-governmental, civil society organizations that meet the requirements of the UNFCCC for formality of their establishment and that are representative.[4] Registered civil observers are then nominated to one of the nine major groups and interface with the UNFCCC Secretariat through the group's nominated focal point. For example, the International Chamber of Commerce is the focal point for the Business and Industry Major Group (BINGO).

For the purposes of this discussion, 'civil society' is the collective term for civil society organizations (CSOs). The World Bank provides a representative view, referring to civil society as '. . . the wide array of non-governmental and not-for-profit organizations that have a presence in public life, expressing the interests and values of their members or others based on ethical, cultural, political, scientific, religious or philanthropic considerations'.[5] In climate change forums the perception of civil society is as environmental and rights activists,[6] whereas the UNFCCC register of accredited CSOs lists a full range of interests including economic, commercial, labour, youth and academic stakeholders, demonstrating the recognition that climate is impacting all aspects of our lives.

Environmental CSOs have a record of being prominent in shaping public policy on the environment and climate change at domestic and international levels. They provide informed advocacy, bring expertise related to their specific interests and provide information that is often otherwise not commonly available.[7]

4 United Nations Framework Convention on Climate Change (UNFCCC), "Parties and Observers in United Nations Framework Convention on Climate Change [Database Online]," (2011), http://unfccc.int/parties_and_observers/items/2704.php.

5 World Bank, "Defining Civil Society," *World Bank [database online]* (2013), p. 1 http://web.worldbank.org/WBSITE/EXTERNAL/TOPICS/CSO/0.

6 Peter J. Glynn, "The Role of Employers' Organisations and Trade Unions in the Development of Climate Change Policy; a European Perspective," (Gold Coast: Bond University, 2014).

7 Ibid.

CSOs' contribution to public debate and public service may be sought by government and regulators because of their specialist knowledge and their relationship with a stakeholder group. In the UK, for example, some of the major environmental initiatives by the government have been attributed to the activities of CSOs. The World Wildlife Fund (WWF) campaigned for UK energy companies to commit to low-emission technology,[8] and the Friends of the Earth 'Big Ask' campaign convinced a cross-party group of Members of Parliament to table its drafted Climate Change Bill. The Bill provided a legal framework to manage future emissions and formed the basis of legislation that was enacted on 28 October 2008.[9]

In international climate negotiations, CSOs have traditionally been very active, with the UNFCCC giving formal recognition to over 1,300 civil society organizations as observers.[10] These organizations represent a broad spectrum of interests, embracing representatives from business and industry, environmental groups, farming and agriculture, indigenous populations, local governments and municipal authorities, research and academic institutes, labour unions, women, gender and youth groups.[11]

Despite their commitment to the UNFCCC process and to addressing climate change, CSO participation since the 2009 Copenhagen COP had been declining.[12] There were a number of reasons attributed to this; some CSOs felt disenfranchised from the process, with fewer opportunities for effective advocacy.[13] Possibly more telling was the view expressed by BusinessEurope after the 2012 Doha COP that progress in the negotiations was slow and inefficient,[14] a view also held by the Climate Group that wrote of the weak outcome, and that some tough issues were simply ignored.[15] This came on the back of the

8 World Wildlife Fund, *Power Switch: Impacts of Climate Policy on the Global Power Sector* (London: World Wildlife Fund, 2003).

9 Friends of the Earth, "History of the Big Ask," *Friends of the Earth [database online]* (2008), www.foe.co.uk/news/big_ask_history_15798.html; Nina L. Hall and Ros Taplin, "Solar Festivals and Climate Bills: Comparing NGO Climate Change Campaigns in the UK and Australia," *Voluntas: International Journal of Voluntary and Nonprofit Organizations* 18, no. 4 (2007); G. Lock, "The Big Ask Gets Top-Quality Response," *Third Sector* 420 (2006): 14.

10 United Nations Framework Convention on Climate Change (UNFCCC), "Participation Breakdown," (2015), http://unfccc.int/files/parties_and_observers/ngo/application/pdf/participation_breakdown_at_cops.pdf.

11 United Nations Framework Convention on Climate Change (UNFCCC), "Parties and Observers in United Nations Framework Convention on Climate Change [Database Online]."

12 United Nations Framework Convention on Climate Change (UNFCCC), "Participation Breakdown".

13 Dana R. Fisher, "COP-15 in Copenhagen: How the Merging of Movements Left Civil Society out in the Cold," *Global Environmental Politics* 10, no. 2 (2010).

14 BusinessEurope, *Doha Climate Conference: Absence of Level Playing Field Requires EU Climate and Energy Policy Rethink*, Press Release (Brussels: BusinessEurope, 10 December 2012).

15 Damian Ryan, "COP 18 Wrap-Up: Weak Doha Outcome Underlines Importance of Clean Revolution Leadership," (10 December 2012), http://www.theclimategroup.org/what-we-do/news-and-blogs/cop-18-wrap-up-weak-doha-outcome-underlines-importance-of-clean-revolution-leadership/.

2011 Durban COP, which was also perceived as having done little to address the ambition gap.[16]

There was similarly evidence that governments were less committed to their undertakings, such as honouring financial commitments.[17] Agreements to address climate-related problems were being negotiated and struck outside the UNFCCC process as well. There was an initiative led by the then US Secretary of State Hilary Clinton to form the Climate and Clean Air Coalition, comprising twenty-six governments, with the objective of reducing pollutants that were not covered under UNFCCC accounting.[18] Compounding the apparent lack of cohesion in the negotiations was the shifting membership of the traditional political blocks, redefining the geographic and political spheres of interest and the strategic balances of power.[19]

Eastwood,[20] in research on UN-based policy negotiations, observed that government participants and civil society actors alike were starting to admit that as international negotiations stalled on the global level, local actions and smaller-scale projects would be the locus of climate change mitigation and adaptation. The Paris COP was effectively the ultimate deadline; there could be no more extensions. The Parties responded, and as physical evidence of the impacts of climate change mounted and realities of the consequences were documented, the Parties were able to find the common ground necessary to deliver a framework that is suitable for the move to the post-Kyoto environment.

Meaningful opportunities for civil society to participate effectively have been a challenge, a situation acknowledged by the Parties who at the 2012 Rio + 20 Summit and UNFCCC Doha COP[21] committed to stronger and more productive engagement.[22] On taking up the Presidency of the 2015 Paris COP in December 2014, Laurent Fabius declared his intention to also honour this commitment and to extend the engagement to non-state actors (NSAs), including major corporations. He announced plans for forums at which the corporation's CEOs would be the invited participants.[23] These are significant shifts in the strategic approach by

16 Climate Action Network, "ECO 7, Rio+20, English Version," *CAN Newsletter Rio + 20* 7 (2012); Lauren Eastwood, "Climate Change Negotiations and Civil Society Participation: Shifting and Contested Terrain," *Theory in Action* 4, no. 1 January (2011); Fisher, "COP-15 in Copenhagen: How the Merging of Movements Left Civil Society out in the Cold."

17 Eastwood, "Climate Change Negotiations and Civil Society Participation: Shifting and Contested Terrain."

18 Climate and Clean Air Coalition (CCAC), "The History of the CCAC," http://www.ccacoalition.org/en/content/about-us.

19 Climate Action Network, "ECO 7, Rio+20, English Version," p. 7.

20 Eastwood, "Climate Change Negotiations and Civil Society Participation: Shifting and Contested Terrain."

21 Ryan, "COP 18 Wrap-Up: Weak Doha Outcome Underlines Importance of Clean Revolution Leadership."

22 United Nations, "Outcome Document: 'The Future We Want'," in *Report of the United Nations Conference on Sustainable Development* (Rio de Janeiro, Brazil: 20–22 June 2012).

23 Laurent Fabius, "Discours De M. Laurent Fabius, Ministre Des Affaires Étrangères Et Du Développement International. Seul Le Prononcé Fait Foi," *UNFCCC SBSTA42 Opening Plenary* (2015).

the Parties to the Convention on two levels, firstly by sharing the policy development and implementation responsibility, and then by moving to engage directly with the stakeholders rather than only through their representative CSOs. They may still have no decision-making powers within the negotiations themselves, but their capacity to influence has been increased.

The term non-state actor (NSA) refers to non-governmental organizations, which may be civil society organizations (CSOs), corporations and intergovernmental organizations, and is a designation with broader application than CSOs.[24] A non-state actor has been defined as an individual or organization that has significant political influence but is not allied to any particular country or state.[25] Another more descriptive definition refers to non-governmental organizations and other civil society groups, and also other non-state entities having some impact upon international relations, including multinational corporations, terrorist groups, and organized criminal gangs.[26] In the context of this chapter, the main considerations are CSOs as representative membership-based organizations; and NSAs as any of CSOs, individual persons, corporations and non-governmental organizations. In essence, while CSOs are also non-state actors, not all non-state actors are CSOs.

In the lead-up to the 2014 Lima COP, there were concerns that the negotiators might not resolve an agreement in time for presentation to the government ministers at the COP in 2015. The 2014 COP, however, produced a glimmer of hope, declaring

> the elements of the new agreement . . . [and] the ground rules on how all countries can submit contributions to the new agreement during the first quarter of next year. These Intended Nationally Determined Contributions (INDCs) will form the foundation for climate action post 2020 when the new agreement is set to come into effect.[27]

This was a positive indicator, and if the voluntary contributions through the INDCs proved to be insufficient to deliver the 2°C commitment, the engagement with the broader community of NSAs could be called upon to bridge the gap. For many CSOs this was an unsatisfactory model: the environmental activists

24 A. Joey, "Introduction," *The Role of Non-state Actors in International Relations* (academic.edu, 2015), https://www.academia.edu/5124220/The_Role_of_Non-state_Actors_in_International_Relations; Thomas G. Weiss, D. Conor Seyle, and Kelsey Coolidge, *The Rise of Non-State Actors in Global Governance: Opportunities and Limitations*, One Earth Future Discussion Paper (Broomfield, USA: One Earth Future Foundation, 2013).

25 Oxford Dictionary. Non-state actors. 2015 Available from http://www.oxforddictionaries.com/definition/english/non-state-actor.

26 Australian Legal Dictionary. Definition: Non-state actors. 2015 Available from http://bond.libguides.com/dictionaries-encyclopaedia-commentaries/dictionaries.

27 United Nations Framework Convention on Climage Change (UNFCCC), "Lima Call for Climate Action Puts World on Track to Paris 2015," (2015), http://newsroom.unfccc.int/lima/lima-call-for-climate-action-puts-world-on-track-to-paris-2015/.

could not accept that the business community, which in their view had flaunted the gift of nature, was to be brought back into the fold and that they could be trusted to contribute to the solution.[28] Business, on the other hand, was providing material demonstrations of its preparedness to conform to the new regime of climate governance and to contribute by reducing emissions for its operations.[29]

In summary, the traditional role of CSOs in the process has been as activists, lobbyists and a source of expert information whose interest in the process declined after the Copenhagen COP, when progress in the negotiations was slow and the CSOs felt disenfranchised. The 2012 Rio + 20 and UNFCCC Doha COP revived the engagement with promises of more recognition and opportunity to input. The initiative by the Presidency of the 2015 COP to extend this engagement to include the broader community of non-state actors and the decisions from the Paris COP,[30] which reference the role of business and civil society, change the dynamic and the nature of the respective influences, a move supported by some CSOs but which crosses the ideological boundaries of others. The incoming President for the 2016 COP 22 has already declared his intention to further advance the engagement, and the May 2016 meeting of the Parties has set in place the 'SBI contact group on Arrangement for Intergovernmental Meetings (AIM)' to explore how this can be done. Resolving the questions of process, mandates, policy integrity and many other issues in order to take account of the respective ideologies of the CSOs could make negotiating tensions between the Parties in the years leading to the Paris COP seem like a walk in park.

International agreements and the UNFCCC process

The objective of the United Nations Framework Convention on Climate Change (UNFCCC) is to stabilize greenhouse gas concentrations 'at a level that would prevent dangerous anthropogenic (human induced) interference in the climate system'.[31] The 2010 COP resolved to commit to action that would contain warming to a maximum temperature rise of 2°C above pre-industrial levels, and to consider lowering that maximum to 1.5°C in the near future,[32] a commitment

28 P. Clarke, "Green Campaigners Angered as Shell Joins Climate Change Advisory Body," *Financial Times*, 25 September 2015.

29 World Economic Forum, "Sustainable Development Goals and Paris Climate Agreement: Now It's Time to Act," (2016), https://www.weforum.org/press/2016/01/sustainable-development-goals-and-paris-climate-agreement-now-it-s-time-to-act/.

30 United Nations Framework Convention on Climate Change (UNFCCC), "Conference of the Parties: COP21 Paris Agreement," paras. 117–121.

31 United Nations Framework Convention on Climate Change (UNFCCC), "United Nations Framework Convention on Climate Change," (Rio: United Nations, 1992).

32 United Nations Framework Convention on Climate Change – Secretariat, "Climate Change Conference in Cancun Delivers Balanced Package of Decisions, Restores Faith in Multilateral Process," news release, 2010, https://unfccc.int/files/press/news_room/press_releases_and_advisories/application/pdf/pr_20101211_cop16_closing.pdf.

reiterated at the 2015 Paris COP.[33] The IPCC Fifth Assessment Report,[34] however, provides a sobering review, reporting that without additional efforts to reduce GHG emissions beyond those in place today, global emissions growth is expected to persist,[35] putting the 2°C target out of reach.

The negotiations for the agreement to succeed the Kyoto Protocol began in 2007 with the Bali Action Plan,[36] intended to be finalized for resolution at the 2009 UNFCCC COP in Copenhagen. The Parties at Copenhagen did not adopt the negotiated text, and subsequent COPs also failed to agree to terms for the next generation treaty. The Parties, however, rose to the occasion of the ultimate deadline, and the 2015 Paris COP delivered a strong outcome. Time will tell whether the outcome, and the enthusiasm from the COP, is sufficient to deliver the 2°C global warming objective and to bridge the gap between the voluntary commitments of the INDCs and the emission reductions required.

Prior to the 2015 Paris COP, there had been concerns that the multilateral management arrangements of the COP were no longer capable of putting in place the measures necessary to deliver the objectives of the Convention or targeted outcomes. As stated above, civil society organizations observed that progress has been slow and inefficient,[37] the outcomes weak and that some tough issues were simply ignored.[38] The outcome agreements from the 2012 Rio + 20 Summit and the UNFCCC COPs prior to the Paris COP were either lauded as positive demonstrations of multilateralism or criticized as inadequate to address the environmental concerns.[39] Agreements from the COPs had become more difficult to finalize, and then only after the formal end date of the Conference and late night/early morning negotiating sessions. Increases to existing commitments, such as emission reduction targets, and new commitments had been modest and obtained in only a few instances.

33 United Nations Framework Convention on Climate Change (UNFCCC), "Conference of the Parties (COP): COP21 Paris Decisions," (Paris: UNFCCC, 2015).

34 Intergovernmental Panel on Climate Change, "Climate Change 2014: Impacts, Adaptation, and Vulnerability," *IPCC WGII AR5 Phase I Report*, ed. Christopher B. Field, Vicente R. Barros, David Jon Dokken, Katharine J. Mach, Michael D. Mastrandrea, T. Eren Bilir, Monalisa Chatterjee, Kristie L. Ebi, Yuka Otsuki Estrada, Robert C. Genova, Betelhem Girma, Eric S. Kissel, Andrew N. Levy, Sandy MacCracken, Patricia R. Mastrandrea, Leslie L. White, 1–1101 (Tokyo: Intergovernmental Panel on Climate Change, 2014).

35 Renate Christ, "Statement of Renate Christ, Secretary of the IPCC at the Opening of Sbsta-40 Bonn," (Geneva, Switzerland: Intergovernmental Panel on Climate Change Secretariat, 4 June 2014).

36 United Nations Framework Convention on Climage Change (UNFCCC), "Bali Action Plan Decision 1/CP.13," *Report of the Conference of the Parties on Its Thirteenth Session, Held in Bali from 3 to 15 December 2007 Addendum Part Two: Action Taken by the Conference of the Parties at Its Thirteenth Session* (FCCC/CP/2007/6/Add.1: 14 March 2008).

37 BusinessEurope, *Doha Climate Conference: Absence of Level Playing Field Requires EU Climate and Energy Policy Rethink.*

38 Ryan, "COP 18 Wrap-Up: Weak Doha Outcome Underlines Importance of Clean Revolution Leadership."

39 Glynn, "The Role of Employers Organisations and Trade Unions in the Development of Climate Change Policy; a European Perspective."

Time-consuming and unresolved issues from the UNFCCC COPs at Durban and Doha included intellectual property rights, finance and claims by developing nations for disaster compensation, issues which can only be described as incidental to the environmental objective. As well, the Cancun Agreement[40] and Rio + 20[41] added commitments to labour reform. The merits or otherwise of these additions are not at issue; at issue is whether the international environmental forum is the appropriate place for them to be negotiated and whether they encumber the ability to negotiate an agreement that could address the climate change ambition gap. The 2012 Doha COP heard arguments by developed countries that other forums such as the World Trade Organization (WTO) and World Intellectual Property Organization (WIPO) would be better placed to resolve the intellectual property issues that are being brought to the COP.[42] The same could be said of the International Labour Organization in respect of the emerging labour issues. It is noted that these issues have not been carried forward into the decisions from the Paris COP, which may bode well for efficiency of future COPs and the implementation strategies. A legally binding core agreement stripped of the 'clutter' of the Cancun and later COPs may improve the efficiency of decision-making.

The intended structure of the outcomes from the Paris COP (known as the Lima Paris Action Agenda – LPAA) was foreshadowed by the incoming COP 21/CMP 11 by President Laurent Fabius, Foreign Minister, France in his address to the Plenary of the UNFCCC meetings of June 2015. He spoke of a model in four pillars: a universal legally binding agreement; INDCs; finance, technology and capacity building; and the role and contribution of non-state actors.[43] The stated vision of the French government for Paris 'focuses on actions pre-2020, as well as afterward'. Its Action Agenda sought to trigger a growing engagement of non-state actors and proposed dialogue during the COP with national governments, cities and businesses,[44] effectively harnessing the willingness of the major corporations to act[45] and tap the private-sector finance opportunities.[46] The dialogue was conducted as proposed, and

40 United Nations Framework Convention on Climate Change – Secretariat, "Climate Change Conference in Cancun Delivers Balanced Package of Decisions, Restores Faith in Multilateral Process".

41 United Nations, "Outcome Document: 'The Future We Want'."

42 Climate Action Network, "ECO 7, Rio+20, English Version," p. 7.

43 Fabius, "Discours De M. Laurent Fabius, Ministre Des Affaires Étrangères Et Du Développement International. Seul Le Prononcé Fait Foi," p. 1.

44 COP20/CMP10 Peruvian Presidency, COP21/CMP11 French Presidency, Office of the Secretary-General of the United Nations, and UNFCCC Secretariat, "The Lima-Paris Action Agenda. Vision and Approach," (Paris: UN Climate Change Conference COP 21-CMP 11, 2015).

45 Edward Cameron, Cammie Erickson, Emile Prattico, and Ryan Schuchard, *Business in a Climate-Constrained World (Second Edition): Creating an Action Agenda for Private-Sector Leadership on Climate Change*, BSR Working Paper (San Francisco: BSR, 2015).

46 Business and Climate Summit, *Business & Climate Summit Conclusions: Towards a Low-Carbon Society* (Paris: Business and Climate Summit, 2015).

the declaration post-COP by these non-party stakeholders is testament to the vision and effectiveness of the strategy.[47]

Reaching out to the business community and others who, by their own actions, can reduce GHG emissions and impact the effectiveness of adaptation and mitigation strategies is a logical and mature rationale. It does, however, introduce other considerations, importantly the monitoring and evaluation process. It also has the effect of subordinating the pre-eminence of the agreement that will be binding on governments, being only one of the four pillars of action, and then with the key role as a technical-standard-setting authority for measurement and reporting. It can be said that the four-pillar model is a reflection of society's concerns for the issue and its resolve to act (and the inability of the state alone to deliver the targeted outcome), but this comes at the expense of the capacity to enforce binding limits and timeframes.

In sum, the 2015 Paris COP agreed to a framework for the management of climate change that is less centralized, relying more on the voluntary initiatives of the Parties individually: the private sector through the greening of their own activities and those in their supply chains, the financing of adaptation and mitigation and the broader network of non-state actors. This is a bottom-up model, rather than the top-down model of Kyoto, and civil society resources need to be redirected accordingly, a change that may benefit many CSOs and allow many to play to their strengths in domestic policy. The converse may also be the case: CSOs may become less relevant as the UNFCCC creates an interface for direct dialogue with their constituents under the banner of non-state actors (NSA) engagement.

Theoretical framework: Governance and integrity

This chapter has so far considered the dynamics of the negotiations leading to the Paris Agreement, notably the involvement of non-state actors in achieving the objectives of the Convention. However, efforts to date to deliver the outcomes of the targeted climate change agreements have proved problematic. It now turns to governance and integrity considerations, and whether the theoretical framework presented in this volume could facilitate the work of the UNFCCC and the domestic policy development process. Theoretical frameworks serve as guides for practitioners and researchers in that they organize existing knowledge and aid in making new discoveries to aid practice.[48] They are formulated to explain the problem under study, to predict and understand phenomena and to challenge

47 Food and Beverage Leaders, "Dear US and Global Leaders: This Could Be the Turning Point," *Financial Times*, 1 October 2015, p. 5; Justin Gillis and Nicholas St. Fleur, "Major Companies Join Climate Change Efforts," *International New York Times*, 24 September 2015, p. 15; Gaia Pianigiani, "World Mayors Join Pope in Push for Climate Pact," ibid., 22 July.

48 Bonnie L. Metzger, "Theoretical Framework," in *Encyclopedia of Nursing Research* (3 ed.), ed. Joyce J. Fitzpatrick, 508–13 (New York: Springer, 2012).

and extend existing knowledge within the limits of the critical bounding assumptions.[49] For the purposes of the discussion about governance and integrity, the institutional 'regime' is the Convention, the committees formed under the Convention and the institutions that give effect to its objectives and which are nested and sit inside a regime.[50] In Chapter 2, Breakey, Cadman and Sampford refer to governance as the structures and processes used to steer and coordinate interaction within an institution or institutional complex. They present the concepts of 'thin', 'thickish' and 'thick' governance values, where thin values reflect qualities that ensure reliability and basic oversight, and thick values capture the full gamut of social and moral qualities that can be demanded of institutional mechanisms of steering and coordination. Institutional integrity refers to actions that show the public that the institution deserves their support. Presenting the idea of comprehensive integrity, they explore the extent to which an institution's activities are both internally coherent and consistent with its enabling environment, thereby rendering it capable of fulfilling public institutional justification (PIJ). The PIJ is the measure of what the institution is 'for', and is what an institution's members and representatives use to justify the institution's existence to the public at large.[51]

It is contended here that the governance values of the UNFCCC regime have to date been thin and its institutional integrity weak. The thinness and weakness are reflected in the inability of the Parties to find sufficient common ground to conclude the negotiations initiated through the 2007 Bali Action Plan[52] until the 2015 Paris COP; the science that foreshadows global warming of greater than 3°C based on the current trajectory;[53] reports from observers that progress has been slow and inefficient[54] and tough issues that are simply ignored.[55] While appearing to be a modest evaluation, it is also a reflection of the complexity of the UNFCCC mandate and the enormity of the task. The objectives of the Convention require changes to patterns of consumption and production, the basis of social and corporate behaviour. In addition are the compliance exemptions afforded non-Annex 1 countries under the Kyoto Protocol principle of common but differentiated responsibilities, an exemption that has extended to the largest

49 USC. University of Southern California definition: Theoretical framework. 2014 Available from http://libguides.usc.edu/content.php?pid.

50 Timothy Cadman, "Stakeholder Perspectives on the Integrity of the Climate Regime," in *Ethical Values and the Integrity of the Climate Change Regime*, ed. Hugh Breakey, Vesselin Popovski, and Rowena Maguire (Surrey: Ashgate, 2015).

51 Hugh Breakey and Timothy Cadman, "A Comprehensive Framework for Evaluating the Integrity of the Climate Regime Complex," ibid.

52 United Nations Framework Convention on Climage Change (UNFCCC), "Bali Action Plan Decision 1/CP.13."

53 Intergovernmental Panel on Climate Change, "Climate Change 2014: Impacts, Adaptation, and Vulnerability."

54 BusinessEurope, *Doha Climate Conference: Absence of Level Playing Field Requires EU Climate and Energy Policy Rethink.*

55 Ryan, "COP 18 Wrap-Up: Weak Doha Outcome Underlines Importance of Clean Revolution Leadership."

GHG emitters; the regulatory responsibility remains vested in the state, and is therefore subject to the intervention of domestic political tensions and differences, and domestic climate-change policies are often linked to energy security, strategies which are not always complementary. That said, the problem being addressed deserves and requires the unqualified commitment of all the Parties.

The measure of institutional integrity and the Public Institutional Justification suffers on the same grounds as governance values. The gap between the commitments to the decisions of the COPs and implementation through domestic policy is quite significant in too many instances, which is also a reflection of a lack of good faith in the negotiations. Further, the failure to honour funding commitments to support developing countries, the tensions between the parties over intellectual property rights in technology transfer arrangements and the requirement to implement labour conventions provide challenges to delivering the outcome and therefore the awarding of a satisfactory measure of integrity.

The thinness of the regime's governance value and its suboptimal institutional integrity assessment provide some further explanation for the view in the year leading up to the Paris COP that the international negotiations could stall out on the global level, and the consequence would be that local actions and smaller-scale projects would be the locus of climate change mitigation and adaptation.[56] In the end, the Parties and other stakeholders at the Paris COP revised the objective and introduced a work plan that reflected an increase in their commitment to the regime that on paper should see a shift in governance value beyond thin to thick, as well as greater comprehensive institutional integrity to support the PIJ.

The platform established by the outcomes of the 2015 Paris COP and the commitments to the Paris Agreement[57] has the potential to deliver thickish or even possibly thick governance values. Institutions with thickish governance are effective at getting things done, and thick governance organizations produce effective results.[58] There are indications that the Parties responded to address the governance weaknesses and at the 2015 COP produced ambitious commitments and a detailed work plan, as well as improvements to transparency, capacity building and compliance/implementation consistent with a model of thick governance. The Lima Paris Action Agenda, the Paris Agreement and the Decisions from the 2015 Paris COP provide a framework that, if implemented as intended by the Parties, should provide measures of thick governance and institutional integrity, almost as if the templates developed in the governance and integrity literature were used as guides by the Parties to the COP.

In essence, the theoretical framework provides a validation and logical sequence of important conditions to contextualize the work. To that end, delegates of the Parties, CSOs and other stakeholders would be better positioned to contribute

56 Eastwood, "Climate Change Negotiations and Civil Society Participation: Shifting and Contested Terrain."

57 United Nations Framework Convention on Climage Change (UNFCCC), "Lima Call for Climate Action Puts World on Track to Paris 2015."

58 Breakey and Cadman, "Integrity of the Climate Regime Complex."

constructively to the structure and processes of the climate regime by applying thick governance values and thereby producing effective results. Society and all stakeholders should expect that the implementation strategies to give effect to the Lima Paris Action Agenda would be models of thick governance and would demonstrate institutional integrity. At this stage, it is still too early to tell. The potential is there, but there is a long way to go before the post-Paris action agenda is properly implemented in time for emissions to come down and global economies to be decarbonized by 2020.

Findings and recommendations

This research project was motivated by the concern that the international climate agreements may not deliver the objectives of the Convention and the targets for containing the growth of global warming. The assessment methodology applied a theoretical framework that considered the governance and integrity issues confronting the climate regime. UNFCCC agreements are required by the Convention to combat climate change, and in doing so complement poverty eradication, sustainable development and equity. At present, the regime's policy objective may not yet fully align with thick governance values, nor demonstrate comprehensive integrity.

Nevertheless, the Convention and agreements under the Lima Paris Action Agenda do demonstrate features of reliability and oversight, as well as social and moral qualities that could produce effective results – a measure of thick governance. As such, there is some progress towards a greater degree of institutional integrity. However, while this preliminary evaluation supports the governance and integrity framework as a guide for assessing current policy development processes, the evaluation also establishes that emerging policies may be required to achieve multiple objectives and/or to recognize other factors that may be encumbrances on the pathway to the desired outcome. In the present instances, those factors are (i) the requirement in the agreements that climate policies facilitate economic growth, which conflicts with the parallel need to conserve the finite available resources and to change methods of production and consumption, (ii) the principle of common but differentiated responsibilities and (iii) the limited decision-making powers of non-state actors. It could be countered that while honourable and possibly even appropriate in 1992 when the Convention was adopted, it has provided an exemption from the Agreements for countries that are now major economies as well as the largest GHG emitters. Further, the market mechanisms have not been effectively implemented, and measures have been insufficient to bridge the ambition gap.

The 2015 Paris COP has formally adopted an agreement in two parts: the core agreement and decisions. The core agreement, the Paris Agreement, will be legally binding on all Parties and will address the ambition, requirements for adaptation and mitigation, financing and implementation. The Decisions of the COP contain the detail that may be varied as required and appropriate. The further voluntary commitments by each Party to emission reduction will be through the INDC. The four pillars of the Action Agenda as described earlier in this chapter are the legal agreement, ambitious INDCs, long-term finance and a commitment from non-state actors. This is a model that could deliver strong and appropriate governance and integrity. The success of the model and its governance and integrity value is, however, contingent on bridging the outstanding gap between the aggregate of the INDCs and the emission reductions required for containing global warming to less than the 2°C objective. It is contended that this is not a weakness in the theoretical model, but rather a factor of the implementation decision.

Civil society has been a constant and respected contributor to the work of the UNFCCC. Civil society is often seen as environmental and rights activists, a concept that has evolved and matured to now embrace all stakeholders. It is yet to be seen what the impact will be on the contribution by civil society with the emergence of the concept of non-state actors. A factor that may impact the role and effectiveness of civil society in international activities is that the core agreement is only one of the four pillars; the other three are voluntary and are decisions made in domestic and corporate headquarters, away from the UNFCCC. They are also the action items on which civil society would traditionally be most active at the UNFCCC.

In sum, this chapter finds that an appropriate theoretical framework can ground policymaking and also provide an empirical method of analysis. It has established that civil society is an important and respected contributor to the policy development in international and domestic forums, and while it has made a significant commitment to the international process, this could and should change if the major decisions made in domestic and corporate headquarters are to be productively incorporated into the climate regime. It is recommended that further research be considered in the monitoring and evaluation requirements for the proposed climate change management regime beyond 2015, when it is expected that the commitments to achieving the ambition will become voluntary and decentralized rather than mandatory and centralized.

References

Breakey, Hugh, and Timothy Cadman. "A Comprehensive Framework for Evaluating the Integrity of the Climate Regime Complex." Chap. 2 In *Ethical Values and the Integrity of the Climate Change Regime*, edited by Hugh Breakey, Vesselin Popovski and Rowena Maguire, 17–27. Surrey: Ashgate, 2015.

Business and Climate Summit. *Business & Climate Summit Conclusions: Towards a Low-Carbon Society*. Paris: Business and Climate Summit, 2015.

BusinessEurope. Doha Climate Conference: Absence of Level Playing Field Requires EU Climate and Energy Policy Rethink. Press Release. Brussels: BusinessEurope, 10 December 2012.

Cadman, Timothy. "Stakeholder Perspectives on the Integrity of the Climate Regime." Chap. 4 In *Ethical Values and the Integrity of the Climate Change Regime*, edited by Hugh Breakey, Vesselin Popovski and Rowena Maguire, 43–59. Surrey: Ashgate, 2015.

Cameron, Edward, Cammie Erickson, Emile Prattico, and Ryan Schuchard. *Business in a Climate-Constrained World (Second Edition): Creating an Action Agenda for Private-Sector Leadership on Climate Change*, BSR Working Paper. San Francisco: BSR, 2015.

Christ, Renate. "Statement of Renate Christ, Secretary of the IPCC at the Opening of Sbsta-40 Bonn." Geneva, Switzerland: Intergovernmental Panel on Climate Change Secretariat, 4 June 2014.

Clarke, P. "Green Campaigners Angered as Shell Joins Climate Change Advisory Body." *Financial Times*, 25 September 2015, 15.

———. "Ikea Pledges 1bn Euro to Clean Energy and Climate Action." *Financial Times*, 4 June 2015.

Climate Action Network. "ECO 7, Rio+20, English Version." *CAN Newsletter Rio + 20 7* (2012): 1–12.

Climate and Clean Air Coalition (CCAC). "The History of the CCAC." <http://www.ccacoalition.org/en/content/about-us> (accessed 15 June 2016).

COP20/CMP10 Peruvian Presidency, COP21/CMP11 French Presidency, Office of the Secretary-General of the United Nations, and UNFCCC Secretariat. "The Lima-Paris Action Agenda: Vision and Approach." Paris: UN Climate Change Conference COP 21-CMP 11, 2015.

Eastwood, Lauren. "Climate Change Negotiations and Civil Society Participation: Shifting and Contested Terrain." *Theory in Action* 4, no. 1 January (2011): 8–38.

Fabius, Laurent. "Discours De M. Laurent Fabius, Ministre Des Affaires Étrangères Et Du Développement International: Seul Le Prononcé Fait Foi." *UNFCCC SBSTA42 Opening Plenary*, 2015.

Fisher, Dana R. "COP-15 in Copenhagen: How the Merging of Movements Left Civil Society out in the Cold." *Global Environmental Politics* 10, no. 2 (2010): 11–17.

Food and Beverage Leaders. "Dear US and Global Leaders: This Could Be the Turning Point." *Financial Times*, 1 October 2015, 5.

Friends of the Earth. "History of the Big Ask." *Friends of the Earth [database online]* (2008). <http://www.foe.co.uk/news/big_ask_history_15798.html> (accessed 15 June 2016).

Gillis, Justin, and Nicholas St. Fleur. "Major Companies Join Climate Change Efforts." *International New York Times*, 24 September 2015, 15.

Glynn, Peter J. "The Role of Employers' Organisations and Trade Unions in the Development of Climate Change Policy; a European Perspective." Gold Coast: Bond University, 2014.

Hall, Nina L., and Ros Taplin. "Solar Festivals and Climate Bills: Comparing NGO Climate Change Campaigns in the UK and Australia." *Voluntas: International Journal of Voluntary and Nonprofit Organizations* 18, no. 4 (2007): 317–38.

Heaps, Toby. "What Is Your Climate Change Business Plan?" *Financial Times FM Supplement*, 1 June 2015.

Intergovernmental Panel on Climate Change. "Climate Change 2014: Impacts, Adaptation, and Vulnerability." *IPCC WGII AR5 Phase I Report*. Tokyo: Intergovernmental Panel on Climate Change, 2014.

Joey, Aw. "Introduction." In *The Role of Non-state Actors in International Relations* academic.edu, 2015. <https://www.academia.edu/5124220/The_Role_of_Non-state_Actors_in_International_Relations> (accessed 15 June 2016).

Lock, G. "The Big Ask Gets Top-Quality Response." *Third Sector* 420 (2006): 14.

Metzger, Bonnie L. "Theoretical Framework." In *Encyclopedia of Nursing Research* (3 ed.), edited by Joyce J. Fitzpatrick, 508–13. New York: Springer, 2012.

Pianigiani, Gaia. "World Mayors Join Pope in Push for Climate Pact." *International New York Times*, 22 July 2015.

Ryan, Damian. "COP 18 Wrap-Up: Weak Doha Outcome Underlines Importance of Clean Revolution Leadership." (10 December 2012). <http://www.theclimategroup.org/what-we-do/news-and-blogs/cop-18-wrap-up-weak-doha-outcome-underlines-importance-of-clean-revolution-leadership/> (accessed 15 June 2016).

United Nations. "Outcome Document: 'The Future We Want'." Chap. 1 In *Report of the United Nations Conference on Sustainable Development*, 1–3. Rio de Janeiro, Brazil, 20–22 June 2012.

United Nations Framework Convention on Climate Change (UNFCCC). "Bali Action Plan Decision 1/CP.13." In *Report of the Conference of the Parties on Its Thirteenth Session, Held in Bali from 3 to 15 December 2007 Addendum Part Two: Action Taken by the Conference of the Parties at Its Thirteenth Session*, 3–6. FCCC/CP/2007/6/Add.1, 14 March 2008.

———. "Conference of the Parties (CO. P): COP21 Paris Decisions." Paris: UNFCCC, 2015.

———. "Conference of the Parties: COP21 Paris Agreement." FCCC/CP/2015/L.9: UNFCCC, 2015.

———. "Lima Call for Climate Action Puts World on Track to Paris 2015." (2015). <http://newsroom.unfccc.int/lima/lima-call-for-climate-action-puts-world-on-track-to-paris-2015/> (accessed 15 June 2016).

———. "Participation Breakdown." (2015). <http://unfccc.int/files/parties_and_observers/ngo/application/pdf/participation_breakdown_at_cops.pdf> (accessed 15 June 2016).

———. "Parties and Observers in United Nations Framework Convention on Climate Change [Database Online]." (2011). <http://unfccc.int/parties_and_observers/items/2704.php> (accessed 15 June 2016).

———. "United Nations Framework Convention on Climate Change." Rio: United Nations, 1992.

United Nations Framework Convention on Climate Change – Secretariat. "Climate Change Conference in Cancun Delivers Balanced Package of Decisions, Restores Faith in Multilateral Process." news release, 2010, <https://unfccc.int/files/press/news_room/press_releases_and_advisories/application/pdf/pr_20101211_cop16_closing.pdf> (accessed 15 June 2016).

Weiss, Thomas G., D. Conor Seyle, and Kelsey Coolidge. "The Rise of Non-State Actors in Global Governance: Opportunities and Limitations." *One Earth Future Discussion Paper*. Broomfield, USA: One Earth Future Foundation, 2013.

World Bank. "Defining Civil Society." *World Bank [database online]* (2013). <http://web.worldbank.org/WBSITE/EXTERNAL/TOPICS/CSO/0> (accessed 15 June 2016).

World Economic Forum. "Sustainable Development Goals and Paris Climate Agreement: Now It's Time to Act." (2016). <https://www.weforum.org/press/2016/01/sustainable-development-goals-and-paris-climate-agreement-now-it-s-time-to-act/> (accessed 15 June 2016).

World Wildlife Fund. *Power Switch: Impacts of Climate Policy on the Global Power Sector*. London: World Wildlife Fund, 2003.

Afterword

The long road to Paris: Insider and outsider perspectives

Klaus Radunsky and Tim Cadman

Introduction

The Paris Agreement was adopted on 12 December 2015, the goal of which is to put the world on track towards a global effort to reduce greenhouse gas GHG emissions and to slow global warming. In this final chapter, the authors provide some personal perspectives on the outcomes of Paris with respect to earlier agreements under the UNFCCC and the Kyoto Protocol. They look particularly at a number of historical issues relating to ambition (for example, mitigation of GHG emissions); adaptation (including loss and damage); finance; transparency; differentiation; and implementation and compliance. The intention is to help better understand – and qualify – the Paris Agreement. Other important issues deserving a similar analysis include market mechanisms and sinks (including emissions trading and REDD+) and finance, and in these instances readers are directed to the other relevant chapters in this volume. Rather than loading the text with analysis in addition to commentary, a number of footnotes have been inserted throughout on issues of integrity and governance.

The way to Paris

A key factor that triggered the development of a convention on climate change was the Climate Conference of 1988 in Toronto, Canada – also the year of the creation of the Intergovernmental Panel on Climate Change (IPCC) by the United National Environment Programme (UNEP) and the World Meteorological Association (WMO). IPCC's first assessment report (FAR) on the state of the climate was produced in 1990. Toronto was one of several conferences held over the period prior to the 1992 UN Conference on Environment and Development (UNCED). Scientists suggested curbing CO_2 emissions from the use of fossil fuels by 20 percent by 2005 as a first step to controlling climate change. However, only a few countries subscribed to such a voluntary goal, and even fewer lived up to any genuine commitments. The first genuinely intergovernmental attempt to combat climate change was achieved in 1992 with the adoption of the UN Framework Convention on Climate Change (UNFCCC) at the Rio 'Earth' Summit (UNCED). The Convention established a long-term objective, as well as general principles, common but differentiated commitments for developed and developing countries, and a basic governance arrangement, including an annual Conference of Parties (COP).[1]

COP 1 in Berlin, Germany, in 1995 was largely preoccupied with the concerns expressed by all Parties over the extent to which they could meet commitments to reduce emissions – discussions which largely occurred within the two permanent subsidiary bodies under the Convention: one which covered scientific and technical advice (SBSTA), and the other more focussed on implementation (SBI).[2] COP 2 in 1996 in Geneva is to be remembered for its acceptance of the findings of the second IPCC report of 1995, the call for 'legally binding' targets for emissions reductions over the 'mid term', and the scrapping of 'harmonised' arrangements for reducing emissions in favour of 'flexible' approaches.[3] This paved the way for a successful COP 3 in Kyoto, Japan, in 1997. Five years after Rio, countries finally agreed to legally binding commitments to reduce GHG emissions, but only for 'Annex I' (developed country) Parties. As a result, the structure of the Kyoto Protocol (KP) was hamstrung over the longer term (see Copenhagen), and it took another eighteen years more to agree a more appropriate structure (that is, not until Paris). However, elements of KP, such as the 'flexible' market mechanisms, in particular the Clean Development Mechanism (CDM), proved to be a success and may survive longer (note the emphasis on *may*; the fate of the CDM remains unclear). Kyoto took a more top-down and highly differentiated approach between developed and developing countries, but it did establish negotiated, binding emissions targets for developed countries, although there were no new commitments for developing countries. Because the United States did not join, some countries would not set targets beyond 2012, the consequence being that the Protocol now covers less than 15 percent of global emissions.[4] COP 4 in 1998 in Buenos Aires, Argentina, and COP 5 in 1999 in Bonn, Germany, were largely caught up with technicalities associated with completing some of the unexpectedly complex outcomes of COP 3, much of which was hived off to SBSTA and SBI during their inter-COP meetings. They concluded their work by adopting a number of draft conclusions, containing text for decisions by COP 6. However, much of the text transmitted by the subsidiary bodies

1 Here we see the emergence of a 'proto' integrity system, including a participatory structure.

2 At least procedurally we can see an effort being put into processes leading to the desired 'thick' governance value of implementation (or how else could commitments be put into practice?).

3 Here we see the conflict between the Convention trying to meet its Public Institutional Justification (PIJ) to reduce emissions and disagreements over governing values.

4 This demonstrates the adoption of a 'restricting' governing value (in this case, common but differentiated responsibilities – CBDR – around climate mitigation), resulting in actual conflict with 'thick' governance values around implementation – an example of the climate regime demonstrating a lack of consistency-integrity. Here, issues of contingency in the KP (a nested institution within the regime) and CDM (a sub-institution of the KP) impact the Convention fulfilling its PIJ. But we note partial progress nonetheless, which reflects the 'realpolitik' of the regime.

to the COP lacked complete agreement by delegates, and differences on many crucial issues remained.

COP 6 represented one of the first truly 'mega' climate conferences, with over 7,000 participants from 182 governments, 323 intergovernmental and non-governmental organizations, and 443 media outlets in attendance. The conference was notable for the large wooden ship constructed outside by Dutch environmental NGOs, highlighting the fate for The Hague if Parties did not reach an agreement over binding targets for emissions reductions. The meeting aimed to set the operational details for commitments on reducing emissions of greenhouse gases under the 1997 Kyoto Protocol, and to reach an agreement on actions to strengthen implementation of the UNFCCC itself. In attempting to achieve these goals, the meeting was intended to bring to a close more than two years of preparations and negotiations set out in the UNFCCC's 1998 Buenos Aires Plan of Action. During the first week, delegates met in informal contact groups and other negotiating groups mandated by the UNFCCC subsidiary bodies. These meetings aimed to reduce differences on text for decisions on a range of issues related to the Protocol and the UNFCCC, including the transfer of technology and capacity building to assist developing countries and countries with economies in transition;[5] the adverse effects of climate change and the impact of implementation of response measures; best practices in domestic policies and measures to address greenhouse gas emissions; the mechanisms outlined under the Kyoto Protocol; a compliance system for the Kyoto Protocol;[6] and issues relating to the land use, land-use change, and forestry (LULUCF) sector. During the second week, President Jan Pronk (the Netherlands) attempted to facilitate progress on the many disputed political and technical issues by convening high-level informal plenary sessions to address the key political issues, which he grouped into four 'clusters', or 'boxes', as follows: (a) capacity building, technology transfer, adverse effects, and guidance to the Global Environment Facility (GEF); (b) mechanisms; (c) LULUCF; and (d) compliance, policies and measures, and accounting, reporting, and review under Articles 5 (methodological issues), 7 (communication of information), and 8 (review of information). Ministers and other senior negotiators convened in four groups to negotiate on these clusters in an attempt to reach consensus. However, by Thursday, 23 November, negotiations appeared stalled, and President Pronk distributed a note containing his proposals on key issues in an attempt to force a breakthrough that would lead to consensus. Ministerial negotiations took place throughout Friday and into Saturday. However, after almost thirty-six hours of intense talks on the President's proposals, negotiators did not achieve a breakthrough, with LULUCF (notably around

5 Here we note an effort by the regime to come to grips with the 'thin' value of transparency, and the 'thicker' value of resources for developing countries to participate more meaningfully in the regime.

6 Here we also note an effort by the regime to come to grips with the 'thin' value of transparency and the 'thicker' value of resources for developing countries to participate more meaningfully in the regime.

the inclusion of forests as carbon 'sinks' eligible for inclusion in the CDM) proving to be a particular sticking point. On Saturday afternoon, 25 November, President Pronk convened a final high-level informal plenary in which he announced that delegates had failed to reach agreement.[7]

Delegates agreed to suspend COP 6 and expressed a willingness to resume their work at COP 6.5 (or 'bis') in 2001 in Bonn, Germany, where some broad principles, known as the 'Bonn Agreements', were drafted.[8] Much more political momentum was evident at COP 7, held later in the same year in Marrakech, Morocco – despite the fact that the US confirmed that it did not intend to ratify the Protocol, an issue that was already evident in The Hague. Negotiators reached agreement on a complex set of decisions spelling out rules for implementing KP and providing detailed and suitably 'legal' text, elaborating on the broad principles of the Bonn Agreement. The Marrakech Accords effectively completed the work under the Buenos Aires Plan of Action, adopted at COP 4, and set the stage for countries to ratify KP and to bring it into force. KP took effect only when ratified by at least fifty-five countries, accounting for at least 55 percent of developed country emissions of carbon dioxide in 1990.[9]

Major areas covered in the accords included:

- Operating rules for international emissions trading and the Protocol's two other flexibility mechanisms (the Clean Development Mechanism and Joint Implementation), and rules defining a party's eligibility to participate in the mechanisms.
- A compliance regime that set consequences for failing to meet an emissions target, but in which the question of whether the consequences were legally binding is deferred until a later Conference.[10]
- Accounting procedures, meaning that emissions units under all three mechanisms were fungible and could be transferred several times as equal units.
- Creation of a new type of emissions unit for 'sinks' credits that could not be banked for future commitment periods.
- In addition, the Conference appointed 10 members and 10 alternates to the CDM Executive Board, nearly doubled Russia's allocation for forest management sinks credit, and approved a declaration to the World Summit on Sustainable Development in September 2002 in Johannesburg, South Africa ('Rio+10').

7 Here we see the consequences of the absence of 'thick' procedural governance values around methods for reaching agreements and settling disputes that negatively impact negotiations.

8 Significantly, we note a concerted attempt to introduce a 'thick' governance value (agreement) into the regime's operating discourse.

9 In both the Bonn Agreement and the Marrakech Accords there is evidence of an ongoing and serious effort to insert the 'thick' value of implementation into the KP – contingent on the absence of developing countries.

10 This demonstrates the difficulty of achieving implementation in the absence of a clearly defined governing value (legally binding vs. non-legally binding – which was it to be?).

COP 8 in 2002 in New Delhi, India, and COP 9 in 2003 in Milan, Italy, were largely follow-up events relating to decisions also made at COP 7 and focussing on procedures concerning the review of commitments made by developed country Parties that would frame discussions for future efforts.[11]

COP 10 held in 2004 in Buenos Aires was a reflective event, looking back at ten years of achievements and looking forward to what to do after the expiry of commitments made under the KP in 2012. As well as the conventional discussions around mitigating climate change, adaptation to its impacts – an increasing reality for many low-lying countries vulnerable to sea-level rise – also came to the fore. This included a proposal for a five-year programme of work on impacts, vulnerability, and adaptation.[12] The proposal was taken to and agreed at COP 11 in 2005 in Montreal, Canada, which was notable for the first meeting of the Parties to KP (CMP 1). The conference also succeeded in reaching an agreement via the Montreal Action Plan to extend the life of KP beyond 2012, as well as securing deeper cuts to GHG emissions.[13]

The adaptation programme was taken to COP 12 in Nairobi in 2006, where it was re-named the Nairobi Work Programme, but the conference itself was characterised by a lack of political momentum as government negotiators merely continued negotiating the processes launched in Montreal, agreeing only to 'consider' the next steps in the international climate effort and opening another track to 'review' the Kyoto Protocol in the final hours.[14] Nevertheless, over the two weeks of talks, in addition to the modest steps on adaptation, Parties debated approaches to reducing deforestation and accelerating technology transfer, and heard proposals from South Africa and Brazil on ways to promote stronger action by developing countries.[15]

The foundations for the outcomes of Copenhagen (see later) were laid at COP 13 in Bali in 2007. Here, the Parties remained far apart on fundamental issues, even after tense and chaotic talks that ran a full day longer than planned. They could agree only to launch a loosely framed negotiating process, albeit with the ambitious goal of achieving a new global climate agreement in 2009. Bali highlighted both the urgency and extraordinary difficulty of forging a comprehensive new agreement to reduce global greenhouse gas emissions. On one hand, the scientific case for action laid out unambiguously in the latest assessment report of the IPCC (AR4) was underscored time and again by a myriad of high-level

11 Further evidence of an ongoing and serious effort to insert the 'thick' value of implementation into the KP – contingent on the absence of developing countries.

12 Here we see an expansion of governing values, moving beyond the need to mitigate (stop) climate change to adapting to it (coping mechanisms).

13 This is also further evidence of an ongoing and serious effort to insert the 'thick' value of implementation into the KP (contingent on the absence of developing countries).

14 This looks like a breakdown of context integrity. By taking such a weak position, and essentially backtracking on previous commitments, the surrounding negotiating environment to the regime appears to be hampering its potential to fulfil its PIJ.

15 By contrast, this appears to be an attempt to 'thicken' the regime by further addressing the governance value of resources/capacity and by expanding the governing value of CBDR by means of the governance value of inclusiveness.

speakers, including UN Secretary General Ban Ki-moon and former Vice President Al Gore. On the other hand, in adopting the so-called Bali Roadmap, governments could agree only on the most general parameters for the process going forward, leaving virtually all the key issues to future talks.[16]

Nevertheless, the Bali Road Map included the Bali Action Plan, which launched a 'comprehensive process to enable the full, effective and sustained implementation of the Convention through long-term cooperative action, now, up to and beyond 2012',[17] with the aim of reaching an agreed outcome and adopting a decision at COP 15 in Copenhagen.[18] Governments divided the plan into five main categories: shared vision, mitigation, adaptation, technology, and financing.

Other elements in the Bali Road Map included:

- A decision on deforestation and forest management (reducing emissions from deforestation and forest degradation – or REDD);
- A decision on technology for developing countries;[19]
- The establishment of the Adaptation Fund Board;
- A review of the Financial Mechanism and extending it beyond the GEF (a child of Rio).[20]

This consensus became possible because the US also finally accepted the Bali mandate.

COP 19, in 2014 in Warsaw, took a more positive step forward, calling on parties to submit intended, nationally determined contributions (INDCs) to reducing national GHG emissions well before the Paris conference (which was by now being seen as a make-or-break event), signalling an important development in the 'bottom-up' features of the emerging agreement.[21] But the conference was marked by a strong presence from the fossil fuel industries and a 'coal summit' held at the same time, which was condemned by many NGOs.[22] Nevertheless,

16 Here we also see the consequences of the absence of 'thick' procedural governance values around methods for reaching agreements and settling disputes that are negatively impacting the negotiations.

17 United Nations Framework Convention on Climate Change (UNFCCC), "Report of the Conference of the Parties on Its Thirteenth Session, Held in Bali from 3 to 15 December 2007," *United Nations Framework Convention on Climate Change* (UNFCCC), 2008.

18 However, here we note the effort to reach agreement, albeit over a longer-term timeframe than desirable. This is an example of the regime's surrounding environment (national interests) positively impacting its consistency-integrity and thereby improving its coherence-integrity vis-à-vis its capacity to fulfil its PIJ.

19 This indicates a thickening along the governance values continuum for both resources and inclusiveness, in the sense that the provision of technological resources allows developing countries to participate more meaningfully in reducing emissions and thereby help fulfil the Convention's PIJ.

20 This also concerns resources beyond technology.

21 Demonstrating a 'thickening' of inclusiveness.

22 A possible example of 'unscrupulous actors' attempting to undermine the coherence and context integrity of the regime (Timothy Cadman, "Stakeholder Perspectives on the Integrity of the Climate Regime," in *Ethical Values and the Integrity of the Climate Change Regime*, ed. Hugh Breakey, Vesselin Popovski, and Rowena Maguire (Surrey: Ashgate, 2015), p. 53.

more than 180 countries producing more than 90 percent of global emissions had submitted INDCs, a much broader response than many had anticipated. This made Warsaw an important milestone because without the progress achieved on INDCs, as well as positive discussions on loss and damage, the Paris outcome might have been significantly weaker.

COP 15 in 2009 in Copenhagen did not deliver as anticipated by the Danish COP Presidency. An overly ambitious agenda, the arrival of President Obama to 'fix' things at the last hour, and his perceived cavalier treatment of China, as well as crowded venues and freezing weather, all conspired to make the event one of the most chaotic in COP history.[23]

But it is nevertheless impossible to fully understand and value the outcomes of Paris without taking the joint outcomes of both COP 15 and COP 16 in Cancun in 2010 into consideration. The Copenhagen Accord and the Cancun Agreements established a parallel 'bottom-up' framework to the existing architecture, with countries undertaking national 'pledges' for emissions reductions for 2020 that represented political rather than legal commitments. This approach attracted much wider participation than KP, including specific mitigation pledges by developing countries for the first time.[24] However, countries' pledges fell far short of the reductions needed to meet the goal set in Copenhagen and Cancun of keeping average warming below 2°C above pre-industrial levels.

With respect to mitigation/effort to reduce emissions, the Copenhagen Accord Parties:

- Endorsed the continuation of the Kyoto Protocol;
- Recognised 'the scientific view that the increase in global temperature should be below 2 degrees Celsius'[25] to combat climate change in the context of sustainable development;
- Agreed that developed countries (Annex I Parties) would 'commit to economy-wide emissions targets for 2020',[26] to be submitted by 31 January 2010 – and agreed that these Parties to the Kyoto Protocol would strengthen their existing targets. Delivery of reductions and finance by developed countries were also to be measured, reported, and verified (MRV) in accordance with COP guidelines;[27]

23 Copenhagen represents a good example of contingency issues negatively impacting the regime's consistency-integrity overall.

24 Note the progress in increasing meaningful participation, but also in making the regime more inclusive. Note also that the Cancun Agreements also referred to 'transparent and effective national governance structures' for REDD+ – a clear reference to 'thick' governance values for this mechanism (UNFCCC, "Report of the Conference of the Parties on Its Sixteenth Session, Held in Cancun from 29 November to 10 December 2010," (FCCC/CP/2010/7/Add. 1, 15 March 2011). Available from: http://unfccc. int/files/na/application/pdf/07a01–1. pdf (accessed 1 January 2012), 2011). 26.

25 United Nations Framework Convention on Climate Change (UNFCCC), "Conference of the Parties Fifteenth Session (Copenhagen Accord)," (2009). 1.

26 Ibid. 2.

27 This would appear to recognise the governance value of transparency.

- Agreed that developing nations ('non-Annex I Parties' in UN speak) would 'implement mitigation actions'[28] (Nationally Appropriate Mitigation Actions, or NAMAs) to slow growth in their carbon emissions, and also submitting these by 31 January 2010.[29] Least Developed Countries (LDCs) and (Small Island Developing States) SIDS should be allowed to undertake actions voluntarily and on the basis of (international) support ('may' in the text);[30]
- Agreed that developing countries would report those actions once every two years via the UN Climate Change Secretariat (CCS), subject to their domestic MRV. NAMAs seeking international support were to be subject to international MRV ('will');[31]
- Recognised 'the crucial role of reducing emission from deforestation and forest degradation and the need to enhance removals of greenhouse gas emission by forests'[32] and the need to establish a mechanism to enable the mobilisation of financial resources from developed countries to help achieve this (including reducing emissions from deforestation and forest degradation, now enhanced and referred to as 'REDD+'); and
- Decided to pursue opportunities to use markets to enhance the cost-effectiveness of and to promote mitigation actions.

With respect to adaptation, the Copenhagen Accord stated that:

- 'Enhanced action and international cooperation on adaptation is urgently required to . . . reduc[e] vulnerability and build . . . resilience in developing countries, especially in those that are particularly vulnerable, especially least developed countries, small island developing states and Africa',[33] and
- 'Developed countries shall provide adequate, predictable and sustainable financial resources, technology and capacity-building to support the implementation of adaptation action in developing countries'.[34]

With respect to finance, Parties agreed in the Copenhagen Accord:

- That 'scaled up, new and additional, predictable and adequate funding as well as improved access shall be provided to developing countries . . . to enable and support enhanced action';[35]
- That developed countries would raise funds of USD 30 billion of new and additional resources from 2010 to 2012('will');

28 UNFCCC, "Conference of the Parties Fifteenth Session (Copenhagen Accord)." 2.
29 This would appear to recognise the governance value of implementation.
30 However, this governing value (again reflecting the issue of CBDR) impacts the regime's role fulfilment.
31 But here, in the light of the previous comments, we see an attempt to insert some form of 'legal' transparency into these voluntary actions.
32 UNFCCC, "Conference of the Parties Fifteenth Session (Copenhagen Accord)."
33 Ibid.
34 Ibid. Good examples of the interrelationship between the governance values of resources/capacity and the ability for countries to undertake implementation.
35 UNFCCC, "Conference of the Parties Fifteenth Session (Copenhagen Accord)." 3.

- To set a 'goal' for the world to raise USD 100 billion per year by 2020 from 'a wide variety of sources'[36] to help developing countries cut carbon emissions (mitigation);
- Establish a new multilateral funding arrangement for adaptation with a governance structure; and
- To establish a Green Climate Fund (GCF), as an operating entity of the financial mechanism, 'to support projects, programme, policies and other activities in developing countries related to mitigation',[37] and to this end create a High Level Panel.[38]

Despite these positive developments, Copenhagen COP ultimately failed to ensure the longer-term viability of a comprehensive KP, the longer-term consequence of which was to deliver six years of uncertainty that was only partially resolved at Paris, and which partially contributed to the demise of the CDM. COP 16 helped to rescue some of these issues, however, at least in terms of adaptation (where there was more consensus than mitigation), resulting in the Cancun Adaptation Framework.

Progress towards resolving many of the contentious issues from Copenhagen, and which ultimately led to the Paris Agreement, was given a boost at COP 17 in 2011 in Durban when Parties agreed to initiate the Durban Platform for Enhanced Action (ADP). ADP called for 'a protocol, another legal instrument or an agreed outcome with legal force under the Convention applicable to all Parties'[39] to apply from 2020, but provided no further substantive guidance.

COP 18 in Doha, Qatar, was remarkable only for the enormity and isolation of the venue, and the hours-long bus rides along congested streets to get there. Very little of a positive nature was achieved, with developed (donor) countries pushing for MRV requirements for REDD+, and developing (recipient) countries resisting. The package of outcomes referred to as the Doha Gateway nevertheless contained reference to loss and damage for states affected by climate change, but there was little agreement over climate finance, and notably the GCF. The Doha Amendment to KP helped keep the protocol on life support, but with very few first-commitment-period countries signing up to the second commitment period (2012–2020), such as Japan; and with no room for developing countries previously excluded from KP-1 (such as Brazil, India, and China), it was limited to a fraction of global GHG emissions.

Given the buildup to Paris, it was not surprising that COP 20 in Lima, Peru, in 2014 was a relatively small event. The fact that the various marquees and pavilions were constructed inside a fenced compound within a military base, with

36 Ibid.
37 Ibid.
38 These are further examples of the governance value of resources/capacity.
39 UNFCCC, "Report of the Conference of the Parties on Its Seventeenth Session, Held in Durban from 28 November to 11 December 2011," (United Nations Framework Convention on Climate Change: Bonn, 2012), 2.

restricted access to country delegations for non-governmental observers, did not help.[40] The government-organised civil society event was also on the same base, but at the opposite gate several kilometres away, making exchange between the two events almost impossible. However, the ceviche was excellent, which helped delegates digest some of the less-than-overwhelming discussions around REDD+ and loss and damage. But the conference helped set the scene for COP 21, which was well advertised with a central booth in the main plaza. Public transport to and from the venue once again got stuck in bad traffic, and tended to drop passengers at some distance from their hotels. This meant that taxi drivers fared well, quite the opposite of what the authorities intended.

COP 21 was held in 2015 in Paris, France, the main outcome of which was the Paris Agreement. The goal of this UN agreement is to put the world on track towards a global effort to reduce GHG emissions and to slow global warming.

The Paris agreement 2015

Ambition

With respect to mitigation/level of ambition the Paris Agreement:

- Reaffirms the goal of limiting global temperature increase well below 2°C, while urging efforts to limit the increase to 1.5°C;[41]
- Establishes binding commitments by all parties to make nationally determined contributions (NDCs) and to pursue domestic measures aimed at achieving them;[42]
- Commits all countries to report regularly on their emissions and 'progress made in implementing and achieving'[43] their NDCs and to undergo international review;[44]
- Commits all countries to submit new NDCs every five years, with the clear expectation that they will represent a 'progression' beyond previous ones;[45]

40 This is possibly an example of the absence or 'thickish' (rather than 'thick') application of the governance value of inclusiveness for non-state actors.

41 A clear governing value, contributing to the Convention's PIJ fulfilment.

42 Contributes significantly – if commitments are put into practice – to the governance value of implementation and of the coherence-integrity of this 'nested' aspect of the regime to the Convention fulfilling its PIJ. There is also a linkage to the 'thin' value of accountability – as it has the potential to render countries accountable to each other under the Convention – but what are the sanctions for non-compliance?

43 UNFCCC, "Paris Agreement as Contained in the Report of the Conference of the Parties on Its Twenty-First Session" (FCCC/CP/2015/L.9 2015) 28.

44 Contributes significantly – if commitments are put into practice – to the governance value of transparency and of the coherence-integrity of this 'nested' aspect of the regime to the Convention fulfilling its PIJ.

45 Also contributes significantly – if commitments are put into practice – to the governance value of transparency and of the coherence-integrity of this 'nested' aspect of the regime to the Convention fulfilling its PIJ.

- Requires parties engaging in international emissions trading to avoid 'double counting';[46] and
- Calls for a new mechanism, similar to the Clean Development Mechanism under the Kyoto Protocol, enabling emission reductions in one country to be counted towards another country's NDC.

Thus, in broad structure, the Paris Agreement reflects a 'hybrid' approach blending bottom-up flexibility to achieve broad participation, with top-down rules to promote accountability and ambition (and hopefully transparency and capacity-building, as well as technology and so forth). It is a treaty under international law, but only certain provisions contain language to create legal obligations. The issue of which provisions to make binding (expressed as 'shall', as opposed to 'should') was a central concern for many countries, in particular, the United States, which wanted an agreement the President could accept without seeking congressional approval. Meeting that test precluded binding emissions targets and new binding financial commitments. A final step in Paris was negotiating a 'technical correction' substituting 'should' for 'shall' in Article 4.4, calling on developed countries to undertake 'absolute economy-wide emissions targets', which was a form of recognising differential commitments.[47]

Adaptation

With respect to adaptation, a major priority for many developing countries was strengthening adaptation efforts under the UNFCCC. The agreement does that by:

- Establishing a global goal of 'enhancing adaptive capacity, strengthening resilience and reducing vulnerability to climate change';[48]
- Requiring all parties, 'as appropriate' to plan and implement adaptation efforts;[49]
- Encouraging all parties to report on their adaptation efforts and/or needs;[50]
- Committing enhanced adaptation support for developing countries;[51] and

46 Also contributes significantly to the governance value of transparency and potentially of accountability, as well as the coherence-integrity of this 'nested' aspect of the regime to the Convention fulfilling its PIJ (if commitments are put into practice).

47 UNFCCC, "Paris Agreement as Contained in the Report of the Conference of the Parties on Its Twenty-First Session." 21.

48 UNFCCC, "Paris Agreement as Contained in the Report of the Conference of the Parties on Its Twenty-First Session." 24. Possible new emerging governing values for adaptation, an important 'nested' element within the regime; also further recognition of the governance value of resources ('enhancing capacity').

49 Dependent on how 'appropriate' is understood, which could positively or negatively impact on governance value.

50 Also dependent on what 'encourage' means in the context of transparency.

51 This represents some recognition of the values of resources and inclusiveness.

- Including a review of the adaptation progress and of the adequacy and effectiveness of adaptation support in the 'global stocktake'[52] to be undertaken every five years.

The agreement also extends the Warsaw international mechanism to address loss and damage resulting from climate change, which explicitly will not 'involve or provide a basis for any liability or compensation'.[53]

Finance

With respect to finance, the Paris agreement:

- Reaffirms the binding obligations of developed countries under the UNFCCC to support the efforts of developing countries, while for the first time encouraging voluntary contributions by developing countries too;[54]
- Extends the current goal of mobilising $100 billion a year in support by 2020 through 2025, with a new, higher goal to be set for the period after 2025.[55]

In addition to reporting on finance already provided and received, developed countries commit to submit every two years 'indicative quantitative and qualitative information'[56] on future support, including, 'as available', projected levels of public finance;[57] and other countries are encouraged to do so voluntarily. Finance will also be considered in the global stocktake.

Transparency

The Paris Agreement rests heavily on transparency as a means of holding countries accountable. In another move beyond bifurcation, it establishes a new transparency system with common binding commitments for all parties and 'built-in flexibility' to accommodate varying national capacities. All countries are required to submit emissions inventories with all the required information to track progress in reaching their NDCs. The COP decision says that, with the exception of least developed and small island countries, these reports are to be submitted at least every two years. In addition, developed countries 'shall' report on

52 This is an important transparency development.

53 UNFCCC, "Paris Agreement as Contained in the Report of the Conference of the Parties on Its Twenty-First Session." 8.

54 This is an important resourcing development (and compliance of resourcing), and recognition of the importance of this governance value in contributing to the regime's role fulfilment.

55 This is equally important – so long as it is complied with.

56 UNFCCC, "Paris Agreement as Contained in the Report of the Conference of the Parties on Its Twenty-First Session." 26.

57 This represents an important transparency initiative.

support provided; developing countries 'should' report on support received; and all 'should' report on their adaptation efforts. Information reported by countries on mitigation and support will undergo 'expert technical review', and each party must participate in 'a facilitative, multilateral consideration of progress'[58] in implementing and achieving its NDC (a form of peer review).

Developing countries are promised capacity-building support to help them meet the new transparency requirements. The COP decision says they will be given flexibility in the scope, frequency, and detail of their reporting, and in the scope of review. Details of the new transparency system are to be negotiated by 2018 and formally adopted once the agreement enters into force.[59]

Differentiation

A crosscutting issue was how to reflect the UNFCCC's principle of 'common but differentiated responsibilities and respective capabilities'.[60] On the whole, the Paris Agreement represents a fundamental shift away from the categorical binary approach of the Kyoto Protocol towards more nuanced forms of differentiation, reflected differently in different provisions.

The agreement includes references to developed and developing countries, stating in several places that the former should take the lead. But it notably makes no mention of the Annex I (developed) and non-Annex I (developing) categories contained in the UNFCCC.

Many provisions establish common commitments while allowing flexibility to accommodate different national capacities and circumstances – either through self-differentiation, as implicit in the concept of nationally determined contributions, or through more detailed operational rules still to be developed.[61]

Implementation/Compliance

The agreement establishes a new mechanism to facilitate implementation and to promote compliance. The mechanism – a committee of experts – is to be 'facilitative, non-intrusive, non-punitive manner, respectful of national sovereignty, and avoid placing undue burden on Parties.'[62] It will report annually to the

58 UNFCCC, "Paris Agreement as Contained in the Report of the Conference of the Parties on Its Twenty-First Session." 28.

59 Here we can see that transparency has perhaps become a governing value of the regime – a very important development, but it should be remembered that transparency is a 'thin' governance value, i.e., one we would *expect* to see in the regime.

60 UNFCCC, "Paris Agreement as Contained in the Report of the Conference of the Parties on Its Twenty-First Session." 28.

61 This is a significant development in that it seeks to break the impasse of developing country contributions under KP and may represent a substantive 'reshaping' of this governing value.

62 UNFCCC, "Paris Agreement as Contained in the Report of the Conference of the Parties on Its Twenty-First Session." 27.

COP. Details are to be decided at the first meeting of parties after the agreement takes force.[63]

Next steps

The Paris Agreement was open for signature on 22 April 2016, and will remain so for a year. In order to become a party to the agreement, a country must then express its consent to be bound through a formal process of ratification, acceptance, approval or accession (different terms for essentially the same thing). Each country has its own domestic procedures for deciding whether to join an international agreement. The agreement establishes a 'double trigger' for entry-into-force: it requires approval by at least fifty-five countries accounting for at least 55 percent of global greenhouse gas emissions (as with KP, but all countries). If states ratify quickly, these conditions could be satisfied pre-2020, allowing the COP to begin meeting as the Meeting of the Parties to the Paris Agreement, to be known by the acronym CMA.

In the meantime a new Ad Hoc Working Group on the Paris Agreement (APA) has begun meeting to consider issues requiring further rules or guidance. This new ad hoc working group met for the first time when the UNFCCC subsidiary bodies convened in Bonn, Germany, on 16–26 May 2016.

Conclusion

Putting the Paris Agreement into the historic perspective, it becomes evident that it has been a great success that has been made possible because of the political will of all major groups that had to cross red lines to reach the final consensus. In the Paris Agreement, countries were able to agree to the basic structure of the new climate change regime – the cycle of NDCs, reporting, review, stocktaking, and updating.

However, the Paris agreement should be seen as a starting point, as it cannot guarantee that the risks of climate change will be managed adequately, as it remains to be seen whether countries are willing to reduce emissions coherent with the intention to limit global warming well below 2°C but to target 1.5°C. As in the early days of the climate process, it seems necessary that the IPCC Special Report requested by Parties in the Paris Agreement (paragraph 21) provides relevant information for policy makers/decision makers to further drive the speed and willingness to act.

It remains to be seen whether there is the willingness to pay for the implementation of the quite expensive technologies that allow even negative global emissions, which will have to be achieved later in the twenty-first century in order to

63 The extent to which the qualifications put in place in the Agreement will positively or negatively impact the governing value of compliance, and the governance value of implementation, remains to be seen – previous practices under the CDM were not especially inspiring.

meet the temperature goals of the Paris Agreement. The alternative would be a growing amount of people suffering from climate change impacts, which might imply significant consequences – a scenario without the prospect of sustainable development.

Over the years, it is possible to see a degree of 'thickening' in the governance values of UNFCCC. This is evident in the gradual increase in the recognition attributed to such important aspects of governance as transparency. There is also acknowledgement of the need for capacity-building (resources) to enable transparency of reporting under the new agreement. On this view, it would be expected that compliance around implementation should be stronger post-Paris than under the KP, whose compliance mechanism was a toothless tiger. But whether the new mechanism will have any teeth is a matter of speculation. The implication of the whole agreement structure is that Parties' public reporting against their NDCs will encourage a mutual reinforcement and increase of commitments over time. But there is still no recognition of the need for Parties to be held to account for their failure to comply. Here it might be argued that national sovereignty (and the manner in which national citizens can or cannot hold their governments to account) has the upper hand. By contrast, however, there is also some increased recognition of the role of non-state actors in the agreement (section V, 'non-Party stakeholders'), but they still have no formal deliberative role in the negotiations, and all decision-making powers reside with the nation states. In this sense, the 'procedural equity' that one would hope to see embodied in the climate regime remains absent. This is perhaps unavoidable given the structures of the UN, but it leaves Paris at best as a document embodying only 'thickish' governance values. Despite promoting sustainable development and poverty alleviation whilst keeping global temperatures under control, the legitimacy of the regime as a whole remains in question. To ultimately determine this, the Paris Agreement will need to reach enough signatories/emissions reductions targets to make a demonstrable contribution to combatting anthropogenic climate change effectively.

References

Cadman, Timothy. "Stakeholder Perspectives on the Integrity of the Climate Regime." Chap. 4 In *Ethical Values and the Integrity of the Climate Change Regime*, edited by Hugh Breakey, Vesselin Popovski and Rowena Maguire, 43–59. Surrey: Ashgate, 2015.

United Nations Framework Convention on Climate Change (UNFCCC). "Conference of the Parties Fifteenth Session (Copenhagen Accord)." (FCCC/CP/2009/11/Add.1 30 March 2010) 2009.

———. "Paris Agreement as Contained in the Report of the Conference of the Parties on Its Twenty-First Session." (FCCC/CP/2015/L.9) 2015.

———. "Report of the Conference of the Parties on Its Seventeenth Session, Held in Durban from 28 November to 11 December 2011." United Nations Framework Convention on Climate Change: Bonn, 2012.

———. "Report of the Conference of the Parties on Its Sixteenth Session, Held in Cancun from 29 November to 10 December 2010." FCCC/CP/2010/7/Add. 1, 15 March 2011. Available from: <http://unfccc.int/files/na/application/pdf/07a01-1>pdf> (accessed 14 June 2016).

———. "Report of the Conference of the Parties on Its Thirteenth Session, Held in Bali from 3 to 15 December 2007." *United Nations Framework Convention on Climate Change* (UNFCCC), 2008.

Index

Adaptation Fund 13, 203–5, 207, 255
Ad Hoc Working Group on Long-Term Cooperative Action (AWG–LCA) 125
Ad Hoc Working Group on the Durban Platform for Enhanced Action (ADP) 8, 10, 258
Ad Hoc Working Group on the Paris Agreement (APA) 10, 263
agriculture *see* Climate Smart Agriculture (CSA)
Alliance of Small Islands States (AOSIS) 124–5

Bali Action Plan 100, 241, 244, 254–5
Bali Road Map 255
Ban Ki-moon 255
behavioural change 30, 39, 181
bilateral financial institutions (BFIs) 103
Bonn Agreement 253
Buenos Aires Plan of Action 252–3
Business and Industry Major Group (BINGO) 236

Cadman Framework 28, 30
Cambrian explosion 49
Cancun Adaptation Framework: climate displacement/migration 119, 123–4, 126, 132
Cancun Agreement (2010) 140, 189–90, 242, 256
Cancun Safeguards 85–6
cap and trade mechanism: Kyoto Protocol 156–7
carbon credits 9, 10, 84, 156, 160
carbon offsets 9, 10
carbon taxation 167
Carbon Trust Fund 205
CDM *see* Clean Development Mechanism (CDM)
certified emissions reductions (CERs) 10
CGIAR 225–6
civil society: international climate change and 236–40
Civil Society Organisation (CSO) 14, 236–40
Clean Development Mechanism (CDM) 9–10, 16, 39, 76, 159, 161, 251, 253, 258; Kyoto Protocol 156–7
Clean Technology Trust Fund 205
climate agreements: governance comparisons of 166–70
Climate and Clean Air Coalition 238
climate change: applying human rights and equity in practice 147–51; human rights and 139–41, 177–82; principle of equity 136–8
Climate Change Bill 237
climate displaced person (CDP) 120–1
climate displacement/migration: adaptation funding for people at risk 130–2; Cancun Adaptation Framework 119, 123–4, 126, 132; compensation for 128–30; governance of persons in 121–8; Paris Agreement 119, 127–8, 132; understanding 119–21; UNFCCC 122–3; Warsaw Loss and Damage Mechanism 124–7, 128, 133
climate finance: assessing global governance system (GGS) 100–107; distributive fairness and equity 102–3; evaluating global governance system for 107–9; global governance of 93–5; Paris Agreement missing 112–14; restricted geographical deployment of 105–7; thin ↔ thicker governance of multilateral 202–9
climate governance regime complex: actors and agency in 53, 55–6, 59;

analysing governance systems 48–50; bipartite networks 55–6; centrality of institutions in 58; governance decagon 52; governance triangle 51; institutions 50–3, 54, 55, 57, 58; measurements of institutional fragmentation/integration 67; meta-discourses and narratives in 59–66; network analysis of 66–7; one-mode network 57; *see also* meta-discourses of climate governance
Climate Investment Fund 13, 204–5, 209
Climate Policy Initiative (CPI) 105–7
climate regime: equity and differential obligations in 144–6; governing 3–4; human rights and climate change 139–41; human rights and good governance in 189–92; human rights enhancing good governance in 192–3
Climate Smart Agriculture (CSA) 13–14, 214–15; case study for 226–8; co-production of knowledge for, governance 221–4; definition of 214; farming practices and critiques 215–21; networked governance for 224–6
Clinton, Hilary 238
coherence-integrity 33, 181
Combatting Desertification (CCD) 19
Common But Differentiated Responsibilities (CBDR) principle 12–13, 137, 141–3, 144–7, 149, 151
comprehensive-integrity 34, 38, 93
Comprehensive Integrity Framework 180
Conference of the Parties (COP) 76, 85, 119, 213, 221, 236, 242, 245, 252–4, 261; COP1 (Berlin) 251; COP2 (Geneva) 251; COP3 (Kyoto) 251; COP7 (Marrakesh) 131; COP11 (Montreal) 76, 77; COP13 (Bali) 125; COP14 (Copenhagen) 71, 83, 105, 140, 237, 240, 241; COP 16 (Cancun) 86, 105, 112, 131, 189; COP17 (Durban) 83, 84, 238, 242; COP18 (Doha) 126, 237, 238, 242; COP19 (Warsaw) 83, 128, 228; COP20 (Lima) 39, 239; COP21 (Paris) 3, 8, 10, 11, 25, 31, 70, 95, 105, 109, 112, 170, 235, 238, 239–44, 246, 263
consistency-integrity 6, 34, 37, 88, 93, 188–9
context-integrity 7, 13, 33–7, 39, 87, 93–5, 108, 111, 143
Conventions on Biological Diversity (CBD) 19
Copenhagen climate summit (2009) 14, 71, 83, 100, 140, 199; CSO participation since 237, 240–1; outcomes 251, 254–8
core-periphery 59, 67
CSA *see* Climate Smart Agriculture (CSA)

dangerous interference: concept of 191
Development Bank of South Africa (DBSA) 209
displaced persons *see* climate displacement/migration
distributive fairness: climate change policies 102–3, 110–11
Doha Amendment (2012) 10, 126, 199, 237–8, 240, 242, 258
Durban Platform for Enhanced Action (ADP) 8, 10, 258

Ecological Modernization (EM): meta-discourse 59, 60–5, 67
Emission Reduction Units (ERUs) 156
emissions reduction: market mechanisms under Paris Agreement 158–62
emission trading 155–7, 167
Environmental Integrity Group 84
equity, principle of: application in practice 148–51; climate change 102–3, 136–8; Fair Shares approach 149, 150; in international climate regime 143–6; international environmental law 141–3, 151–2; Paris Agreement 150–1
European Emission Trading Scheme (EU ETS) 156, 161, 165, 170

Fabius, Laurent 242
fairness: climate change policies 102–3, 110–11; *see also* equity
FAO *see* Food and Agriculture Organisation of the United Nations (FAO)
farming practices: climate-smart agriculture (CSA) 215–21
FCPF *see* Forest Carbon Partnership Facility (FCPF)
Fifth Assessment Report (FAR) 16, 241
finance 93; multilateral climate funds 202–9; Paris Agreement 149–51; *see also* climate finance
Food and Agriculture Organisation of the United Nations (FAO) 103, 212, 215, 217–18

Forest Carbon Partnership Facility (FCPF) 13, 77, 85, 204–5
free, prior and informed consent (FPIC) 24
functional realism 27–8

GIS (Geographic Information System) 80
Global Alliance for Climate-Smart Agriculture (GACSA) 218
Global Carbon Integrity System/Regime 3
global climate change regime: Reducing Emissions from Deforestation and Forest Degradation (REDD+) in 76–87; thin ↔ thicker governance of processes 198–202
Global Energy Efficiency and Renewable Energy Fund 103
Global Environment Facility (GEF) 10, 130, 203, 205–7, 209, 252, 255
global governance system: architecture of 96, 98, 103–5; assessment for climate finance 100–107; climate financing 93–5; evaluation for climate finance 107–9; framework for analysis 95–9; geographical scope 98–9; international relations theory (IR) 97; key policy recommendations 109–11; restricted geographical deployment of climate finance 105–7; theoretical model 5; variations in norms 97–8
good governance 19, 21; in climate regime 4, 189–92; fundamental principles of human rights for 187–8; human rights and 179–80; human rights principles achieving 182–8; human rights values enhancing 192–3; procedural rights supporting 183–5; substantive rights supporting 185–7
Gore, Al 255
governance: applications 20–2; climate regime 3–4; climate-smart agriculture 224–6; comparing old and new climate agreements 166–70; institutional integrity and 4–8; integrity and 243–6; laying foundations for 164–5; for legitimacy 28–9, 31; overview of arrangements of UNFCCC 11–14; Paris Agreement 162–6; thick governance values 28–9, 31; thickish governance values 27–8; thin governance values 26–7; triangle 50, 51; as value-neutral term 18–20; *see also* good governance
governance values 19; institutional integrity and 31–40; *vs* 'governing values' 22–5
governing values 7; governance values *vs* 22–5
Green Climate Fund (GCF) 10, 103, 157, 200, 203–5, 207–9, 258
Green Governmentality (GG): meta-discourse 59, 60–5, 67
greenhouse gas (GHG): emissions 3, 16–17, 198, 216; emission trading 155–6; goal of curbing 72, 76, 78; historical responsibility 144–5; Kyoto Protocol 76, 252, 253, 254

Huberts, Leo 23
Human Rights Council 140, 147, 148, 150
human rights principles 181–2; achieving good governance 182–8; application in practice 148–51; climate change and 138–41, 177–82; in climate regime 189–92; contribution to achieving Public Institutional Justification (PIJ) 188–9; differential obligations 145–8; enhancing good governance in climate regime 192–3; fundamental 187–8; Paris Agreement 150–1; procedural rights 183–5; right to access to justice 185, 192; right to equality before law and fair trial 186–7, 192; right to freedom of opinion and expression 185–6; right to information 184; right to participate 184–5, 192; right to self-determination 187, 192; substantive rights 185–7
Humphreys, Stephen 150

Independent Review Mechanism (IRM) 207
individual ethics 5
institutional ethics 5
institutional integrity 93; governance values and 31–40; relationship to governance 4–8
institutional legitimacy 4, 7, 13, 17, 24
institutions: centrality of global climate change 58; climate governance regime complex 50–3; definition of 71–2; distribution across zones 55; global climate change per zone and per role 54; governance 20–2; governance decagon 52; governance triangle 51; one-mode network of 57; power

struggle of 74–5; REDD+ practices and 78–80
integrity: governance value 7
Intended Nationally Determined Contributions (INDCs) 112, 235, 239, 241–2, 247, 255–6
Intergovernmental Panel on Climate Change (IPCC) 16, 191, 220, 228, 241, 250–1, 254, 263
Internal Organizational Arrangements (IOA) 167–8
International Assessment of Agricultural Knowledge 223–4
international climate change: civil society and 236–40
International Covenant on Civil and Political Rights 139
International Covenant on Economic, Social and Cultural Rights (ICESCR) 139, 148
international emissions trading (IET) 9, 16, 83, 253, 260
International Federation of the Red Cross and Red Crescent Societies 236
International Institute for Environment and Development (IIED) 107
International Labour Organization 236, 242
International Migration Organization (IOM) 119–20
International Monetary Fund (IMF) 163
international organizations (IOs) 70, 79–80
International Planning Committee for Food Sovereignty 219
international relations (IR): institutions and climate change regime 73–6
International Tropical Timber Organization 79

joint implementation (JI) 16, 83, 156, 253; Kyoto Protocol 157–8
justice: climate change governance 75, 85–6, 88; *see also* equity

Knox, John 147
Kyoto Protocol 8–10, 13, 50; Clean Development Mechanism (CDM) 156–7; emission trading option 155–7, 252, 254; flexible mechanisms of 16; history of tradable permits 155; human rights of 190; Joint Implementation (JI) 157–8; market practices of 75–6; outcomes of 199–200

land use, land-use change and forestry (LULUCF) 252
Least Developed Countries (LDCs): climate finance 100–103, 204–6; priority areas 131–2
Least Developed Country Fund 13, 131
legitimacy: climate regime governance 4, 7, 17; governance for 28–9, 31
Like-Minded Developing Countries (LMDCs) 8–9
Lima Paris Action Agenda (LPAA) 235, 242, 245–6
loss and damage mechanism 12, 124, 150–1; *see also* Warsaw International Mechanism for Loss and Damage (WIM)

Marrakech Accords 253
meta-discourses of climate governance: Ecological Modernization (EM) 59, 60–5, 67; Green Governmentality 59, 60–5, 67; Radical Civic Environmentalism 59, 61–6; Reformist Civic Environmentalism 59, 61–6
migration *see* climate displacement/migration
Mkulima Young (online platform) 215, 226–8
monitoring, reporting and verifying (MRV) system 71, 78–80, 88, 113; Copenhagen Accord 256–7; towards international standard 168–70
Montreal Protocol 76–7, 163, 254
MRV *see* monitoring, reporting and verifying (MRV) system
multilateral climate funds: thin ↔ thicker governance of 202–9
Multilateral Environmental Agreements 55

Nairobi Work Programme 254
National Adaptation Programs of Action (NAPAs) 131
National Development Banks (NDBs) 103
Nationally Appropriate Mitigation Actions (NAMAs) 257
Nationally Determined Contributions (NDCs) 9, 159–61, 259–64; Paris Agreement 159–61
New Institutionalism 73
non-state actor (NSA) 239, 243

Obama, Barack 113, 256
official development assistance (ODA) 102

Old Institutionalism 73, 74
Ostrom, Elinor 104
Oxfam 129, 219

Paris Agreement (2015) 259–63; accounting system in 84; adaptation of 260–1; ambition of 259–60; background to 8–11; climate displacement/migration 119, 127–8, 132; collective actions of states 197, 199–200, 202; differentiation of 262; finance of 150–1, 261; food production 213–14; governance arrangements under 162–6; governance model in 87–8; human rights 190; implementation/compliance of 262–3; integrity and fragmentation 70; involvement of non-state actors 243–6; laying foundations for effective governance 164–5; loss and damage 150–1; market mechanisms for emissions reduction 158–62; missing climate finance 112–14; mitigation and long-term goal of 149–50; monitoring, review and verification (MRV) system 168–70; Public Institutional Justification (PIJ) 167–8; recommendations of 246–7; regulation and monitoring 165–6; steps on the way to 250–9; technology transfer 150; transparency of 261–2; transparency of finance flows 203, 235
persons displaced by climate *see* climate displacement/migration
Pillay, Navi 146
Polluter Pays principle 167
polycentrism 59, 104
precautionary principle 75, 202
procedural equity 25, 264
procedural rights: right to access to justice 185, 192; right to information 184; right to participate 184–5, 192; supporting good governance 183–5
Pronk, Jan 252
Public Institutional Justification (PIJ) 6, 17, 32–8, 244; contribution of human rights principles to 180, 188–9; global climate regime 212–15; governance goal 50, 244–5; Kyoto Protocol 166–7; Paris Agreement 167–8; UNFCCC 32–40, 136–7, 212–15

Qualitative Data Analysis (QDA) 59

Radical Civic Environmentalism: meta-discourse 59, 61–6
Readiness Trust Fund 205
Reducing Emissions from Deforestation and Forest Degradation (REDD+) 9–12, 71; approaches by country 81; double accounting 83–4; evolution of practices 80–8; funding mechanisms 82–4; in global climate change regime 76–87; international institutions and scope of 76–8; measuring, reporting and verification (MRV) 71, 78–80, 88; mechanism of 27, 39; practices and international institutions 78–80; safeguards 81, 86–7
Reducing Emissions from Forest Deforestation (RED) 80
Reformist Civic Environmentalism: meta-discourse 59, 61–6
regime complex 19; conceptualization of 49; *see also* climate governance regime complex
Rio Declaration 141–5
Rio 'Earth Summit' (1992) 141, 250

Sands, Philippe 142
Schmidt, Vivien 27–8
Shelton, Dinah 143, 144–5, 147–8
Shue, Henry 145
Small Island States (SIS): climate finance 100–103
Special Climate Change Fund 13, 204–6
Standing Committee on Finance (SCF) 10, 106
Subsidiary Body for Scientific and Technological Advice (SBSTA) 12, 77–8, 80–8, 251
substantive rights: right to equality before the law 186–7, 192; right to fair trial 186–7; right to freedom of opinion and expression 185–6; right to self-determination 187, 192; supporting good governance 185–7
sustainable development 7, 19, 38; accords 253, 256; climate change 38–40; climate smart agriculture 219; concept of 141–3; Paris Agreement 24, 31, 149, 159, 167, 197, 246; UNFCCC 38–40, 76, 83, 136

technology transfer: Paris Agreement 150
thick governance values 7, 30, 72, 94, 136; governance for legitimacy 28–9, 31, 36–8, 40

thickish governance values 27–8, 29, 30
thin governance values 26–7, 29, 30, 36, 40, 72, 94
throughput legitimacy 27–8
tradable permits: Kyoto Protocol 155
Transparency Accountability Initiative 201
Transparency International 6, 13, 204

UN Conference on Environment and Development (UNCED) 75, 141–3, 250
UN Declaration of Human Rights (UNDHR) 139
UN Framework Conference on Climate Change (UNFCCC) 4, 118–19; climate displacement/migration 122–3; climate governance 70–3, 88; evolution of REDD practices in 80–7; governance and integrity 243–6; governing persons displaced by climate change 121–8; human rights 110; integrity framework of 143–9; international agreements and 240–3; market practices 75–6, 88; precautionary principle 75, 202; Public Institutional Justification (PIJ) 32–40, 136–7, 212–15; sovereignty 75, 88; stabilising greenhouse gas concentrations 191; sustainable development 38–40, 76, 83, 136; thin ↔ thicker governance of processes 198–202; understanding climate displacement 119–21
United Nations Development Programme (UNDP) 103
United Nations Environment Programme (UNEP) 103, 250
United Nations High Commissioner for Human Rights (OHCHR) 130, 140, 143

Verified Carbon Standard (VCS) 165

Warsaw International Mechanism for Loss and Damage (WIM) 8, 10, 261; climate displacement/migration 124–7, 128, 133
World Bank 80, 85, 103, 130, 204–7, 223, 236
World Intellectual Property Organization (WIPO) 242
World Meteorological Association (WMO) 250
World Trade Organization (WTO) 163, 242
World Wide Fund for Nature 128
World Wildlife Fund (WWF) 237